UNITEXT

La Matematica per il 3+2

Volume 136

The **UNITEXT - La Matematica per il 3+2** series is designed for undergraduate and graduate academic courses, and also includes advanced textbooks at a research level.

Originally released in Italian, the series now publishes textbooks in English addressed to students in mathematics worldwide.

Some of the most successful books in the series have evolved through several editions, adapting to the evolution of teaching curricula.

Submissions must include at least 3 sample chapters, a table of contents, and a preface outlining the aims and scope of the book, how the book fits in with the current literature, and which courses the book is suitable for.

For any further information, please contact the Editor at Springer: francesca.bonadei@springer.com

THE SERIES IS INDEXED IN SCOPUS

More information about this subseries at https://link.springer.com/bookseries/5418

Farook Rahaman

The Special Theory of Relativity

A Mathematical Approach

Second Edition

 Springer

Farook Rahaman
Department of Mathematics
Jadavpur University
Kolkata, West Bengal, India

ISSN 2038-5714   ISSN 2532-3318   (electronic)
UNITEXT
ISSN 2038-5722   ISSN 2038-5757   (electronic)
La Matematica per il 3+2
ISBN 978-981-19-0496-7   ISBN 978-981-19-0497-4   (eBook)
https://doi.org/10.1007/978-981-19-0497-4

This Springer imprint is published by the registered company Springer Nature Singapore Pte Ltd.
The registered company address is: 152 Beach Road, #21-01/04 Gateway East, Singapore 189721, Singapore

To my parents
Majeda Rahaman and
Late Obaidur Rahaman
and my wife Pakizah Yasmin &
my son Md Rahil Miraj

Preface to the Second Edition

The purpose of this revised second edition is to bring out some new materials, worked examples that create the book more comprehensive and useful. However, the book remains as sketched in the preface to first edition. A new chapter on electromagnetic waves is added for the reason that most of the universities have this as a part of the course. In the new chapter we have discussed electromagnetic waves. Maxwell equations indicate that the electric and magnetic fields can be traveled in the space in the form of waves. These waves are known as electromagnetic waves. We have discussed wave equations for magnetic intensity and electric field strength and have shown that the wave equations for both the cases are of the same form. Electromagnetic waves in a non-conducting dielectric medium and in a conducting medium are discussed. We have also discussed Poynting's theorem. Usually, electric and magnetic field vectors experience a modification while passing from one medium to other. We have explored the boundary conditions of the electromagnetic fields at the boundary surface between different media. We have explained skin depth and wave guides. It is shown that Maxwell's equations are equivalent to a single wave equation in terms of Hertz vector. Finally, we have given a brief introduction of relativistic wave equation namely Dirac equation. We have added 54 new problems with hints in different chapters mostly taken from the question papers from various universities. It is hoped that the second edition of the book will be more beneficial to the students in physics and mathematics at the undergraduate level. It is pleasure to acknowledge my students for helpful comments, correcting typos and questions they have raised. Particular thanks are due to Lipi Baskey, Nayan Sarkar, Md. Rahil Miraj, Sabiruddin Molla, Susmita Sarkar, Dr. Tuhina Manna, Dr. Sayeedul Islam, Ksh. Newton Singh, Somi Aktar, Bidisha Samanta, Dr. Anil Kumar Yadav, Antara Mapdar, Dr. Banashree Sen, Moumita Sarkar, Abhisek Dutta.

Finally, I am thankful to the engineer Md. Mehedi Hasan for preparing the cover page.

Kolkata, India

Farook Rahaman

Preface to the First Edition

In 1905, Albert Einstein wrote three papers which started three new branches in physics; one of them was the special theory of relativity. The subject was soon included as a compulsory subject in graduate and postgraduate courses of physics and applied mathematics across the world. And then various eminent scientists wrote several books on the topic.

This book is an outcome of a series of lectures delivered by the author to graduate students in mathematics at Jadavpur University. During his lectures, many students asked several questions which helped him to know the needs of students. It is, therefore, a well-planned textbook, whose contents are organized in a logical order where every topic has been dealt with in a simple and lucid manner. Suitable problems with hints are included in each chapter mostly taken from question papers of several universities.

In Chap. 1, the author has discussed the limitations of Newtonian mechanics and Galilean transformations. Some examples related to Galilean transformations have been provided. It is shown that the laws of mechanics are invariant but laws of electrodynamics are not invariant under Galilean transformation. Now, assuming both Galilean transformation and electrodynamics governed by Maxwell's equations are true, it is obvious to believe that there exists a unique privileged frame of reference in which Maxwell's equations are valid and in which light propagates with constant velocity in all directions. Scientists tried to determine relative velocity of light with respect to earth. For this purpose, the velocity of earth relative to some privileged frame, known as ether frame, was essential. Therefore, scientists performed various experiments involving electromagnetic waves. In this regard, the best and most important experiment was performed by A. A. Michelson in 1881.

In Chap. 2, the author has described the Michelson–Morley experiment. In 1817, J. A. Fresnel proposed that light is partially dragged along by a moving medium. Using the ether hypothesis, Fresnel provided the formula for the effect of moving medium on the velocity of light. In 1951, Fizeau experimentally confirmed this effect. The author has discussed Fizeau's experiment. Later, the author provided the relativistic concept of space and time. To develop the special theory of relativity, Einstein used two fundamental postulates: the principle of relativity or equivalence and the principle of constancy of the speed of light. The principle of relativity states that the laws of nature are invariant under a particular group of space-time coordinate transformations. Newton's laws of motion are invariant under Galilean transformation, but Maxwell's equations are not, and Einstein resolved this conflict by replacing Galilean transformation with Lorentz transformation.

In Chap. 3, the author has provided three different methods for constructing Lorentz transformation between two inertial frames of reference. The Lorentz transformations have some interesting mathematical properties, particularly in measurements of length and time. These new properties are not realized before. Mathematical properties of Lorentz transformations—length contraction, time dilation, relativity of simultaneity and their consequences—are discussed in Chaps. 4 and 5. Also, some paradoxes like twin paradox, car–garage paradox have been discussed in these chapters.

Various types of intervals as well as trajectories of particles in Minkowski's four-dimensional world can be described by diagrams known as space–time diagrams. To draw the diagram, one has to use a specific inertial frame in which one axis indicates the space coordinate and the other effectively the time axis. Chapter 6 includes the geometrical representation of space–time.

Chapter 7 discusses the relativistic velocity transformations, relativistic acceleration transformations and relativistic transformations of the direction cosines. The author has also discussed some applications of relativistic velocity and velocity addition law: (i) explanation of index of refraction of moving bodies (Fizeau effect), (ii) aberration of light and (iii) relativistic Doppler effect.

In Newtonian mechanics, physical parameters like position, velocity, momentum, acceleration and force have three-vector forms. It is natural to extend the three-vector form to a four-vector form in the relativistic mechanics in order to satisfy the principle of relativity. Therefore, a fourth component of all physical parameters should be introduced. In Euclidean geometry, a vector has three components. In relativity, a vector has four components.

The world velocity or four velocity, Minkowski force or four force, four momentum and relativistic kinetic energy have been discussed in Chaps. 8 and 10. It is known that the fundamental quantities, length and time are dependent on the observer. Therefore, it is expected that mass would be an observer-dependent quantity. The author provides three methods for defining relativistic mass formula in Chap. 9. The relativistic mass formula has been verified by many scientists; however, he has discussed the experiment done by Guye and Lavanchy.

A short discussion on photon and Compton effect is provided in Chap. 11. A force, in general, is not proportional to acceleration in case of the special theory of relativity. The author has discussed force in the special theory of relativity and covariant formula of the Newton's law. Relativistic Lagrangian and Hamiltonian are discussed in Chap. 12. He has also explained Lorentz transformation of force and relativistic harmonic oscillation. When charges are in motion, the electric and magnetic fields are associated with this motion, which will have space and time variation. This phenomenon is called *electromagnetism*. The study involves time-dependent electromagnetic fields and the behaviour of which is described by a set of equations called *Maxwell's equations*. In Chap. 13, the author has discussed relativistic electrodynamics of continuous medium. In Chap. 14, the author has provided a short note on relativistic mechanics of continuous medium (continua). Finally, in appendixes, the author has provided some preliminary concepts of tensor algebra and action principle.

The author expresses his sincere gratitude to his wife Mrs. Pakizah Yasmin, mother-in-law (Begum Nurjahan), younger brother (Mafrook Rahaman), sister in law (Asmina Rahaman) and niece (Ayat Nazifa) for their patience and support during the gestation period of the manuscript. It is a pleasure to thank Dr. M. Kalam, Indrani Karar, Iftikar Hossain Sardar and Mosiur Rahaman for their technical assistance in preparation of the book. Finally, he is thankful to the painter Ibrahim Sardar for drawing the plots.

Farook Rahaman

Contents

About the Author

Farook Rahaman (FRAS) is Professor at the Department of Mathematics, Jadavpur University, Kolkata, India. With over 21 years of experience in teaching and working in various fields of application of relativity, his area of research interests are topological defects, compact stars, wormholes, gravastars, galactic dark matter and cosmology. He has supervised more than 28 PhD scholars and has more than 300 research papers published in renowned international journals. Professor Rahaman has been elected as Fellow of the Royal Astronomical Society (FRAS), London, UK, in 2019, as well as he has been selected for the TWAS UNESCO Associateship, Trieste, Italy, from 2011–2014. Further, he has also been selected for Associate of the Inter-University Centre for Astronomy and Astrophysics (IUCAA), Pune, India, and Associate of the Institute of Mathematical Sciences (IMSc), Chennai, India, in 2009. He has been offered the UGC Post-Doctoral Research award by the Government of India in 2010. He has been included to the top 2-percent scientists, 2020 and 2021, in the research field of nuclear and particle physics prepared by Stanford University, USA. He is the author/coauthor of three books with several publishers: *Finsler Geometry of Hadrons and Lyra Geometry: Cosmological Aspects (Lambert Academic Publishing), General Theory of Relativity: A Mathematical Approach (Cambridge University Press)*, and *Special Theory of Relativity: A Mathematical Approach* (Springer).

Chapter 1
Pre-relativity and Galilean Transformation

1.1 Failure of Newtonian Mechanics

Isaac Newton proposed three laws of motion in the year 1687. Newton's first law states: Every body continues in its state of rest or of uniform motion in a straight line unless it is compelled by external forces to change that state. The law requires no definition of the internal structure of the body and force, but simply refers to the behaviour of the body. I can remember, my teacher's example. Suppose, one puts a time bomb in a place. It is at rest and according to Newton's first law, it remains at rest. But after some time, it explodes and the splinters run here and there. Now, we cannot say the body (i.e. the bomb) is at rest as it was some time ago and no external force was applied. This is not a breakdown of Newton's first law, rather we should define the internal structure of the body. Also, we note that uniform motion is not an absolute property of the body but a joint property of the body and reference system used to observe it. Therefore, one should specify the frame of references relative to which the position of the body is determined. Moreover, Newton's first law is true only for some observers and their associated reference frame and not for others. Assume we have an observer A_1 sitting in a reference frame S_1 for which Newton's first law is true, i.e. he determines that the motion of a forceless body is uniform and rectilinear. Now, we consider an observer A_2 sitting in another reference frame S_2 moving with an acceleration with respect to the former. For him, the same body (not acted upon by external forces) will appear to have an accelerated motion. He concludes that Newton's first law is not true. Therefore, one should mention the frame of reference in developing Mechanics based on Newton's laws of motion. In the macroscopic world, the velocity of particles (v) with respect to any observer is always less than the velocity of light c ($=3 \times 10^8$ m/s). In fact, $\frac{v}{c} \ll 1$ in our ordinary experiences. However, in the microscopic world, particles can be found frequently whose velocities are very close to the velocity of light. Experiments show that these high-speed particles do not follow Newtonian mechanics. In Newtonian mechanics, there is no upper bound of the particle's speed. As a result, speed of light plays no special role at all in Newtonian mechanics.

© The Author(s), under exclusive license to Springer Nature Singapore Pte Ltd. 2022
F. Rahaman, *The Special Theory of Relativity*, La Matematica per il 3+2 136,
https://doi.org/10.1007/978-981-19-0497-4_1

For example: An electron accelerated through a 10 million-volt (MeV) potential difference, then its speed 'v' equal to $0.9988c$. If this electron is accelerated through a 40 MeV potential difference, experiment shows that its speed equals .9999c. The speed is increased from $0.9988c$ to $0.9999c$, a change of 11 % and remains below c. However, according to Newtonian mechanics ($K = \frac{1}{2}\mathrm{mv}^2$), we expect its velocity should be doubled to $1.9976c$. Hence, Newtonian mechanics may work at low speeds, it fails badly as $\frac{v}{c} \to 1$. Einstein extended and generalized Newtonian mechanics. He correctly predicted the results of mechanical experiments over the complete range of speeds from $\frac{v}{c} \to 0$ to $\frac{v}{c} \to 1$.

Inertial Frame: The frames relative to which an unaccelerated body appears to be unaccelerated are called Inertial Frames. In other words, the frame in which the Newton's laws of motion are valid.

Non-inertial Frame: The frames relative to which an unaccelerated body appears to be accelerated are called non-inertial frames. In other words, the frame in which the Newton's laws of motion are not valid.

Loosely speaking, Newtonian frame of reference fixed in the stars is an inertial frame. The coordinate system fixed in earth is not an inertial frame since earth rotates about its axis and also about the sun. In fact, rotational frame is not an inertial frame.

The laws of Newtonian mechanics are applicable only for the inertial frames. Galileo pointed out that it is impossible to distinguish between rest and motion in the case of bodies moving with uniform velocities along any straight line relative to each other. In other words, one inertial frame is as good as any other.

1.2 Galilean Transformations

We try to find the relationship of the motions as observed by two observers in two different inertial frames. Let us consider two different Cartesian frame of references S and S_1 comprising the coordinate axes (x, y, z) and (x_1, y_1, z_1) which are inertial. Here, the axes (x_1, y_1, z_1) of S_1 are parallel to the axes (x, y, z) of S. We further assume that at time $t = 0$, the origin O_1 of S_1 system coincides with the origin O of S system and all axes overlapped. The frame S_1 is supposed to move at a uniform velocity v along the x-axis. Let an event occurs at a point P whose coordinates are (x, y, z) and (x_1, y_1, z_1) in S and S_1 frames, respectively. Let t and t_1 be the time of occurrence of P that observer S and S_1 record their clocks. Then, the coordinates of two systems can be combined to each other as (from Fig. 1.1)

$$x_1 = x - vt \tag{1.1}$$

$$y_1 = y \tag{1.2}$$

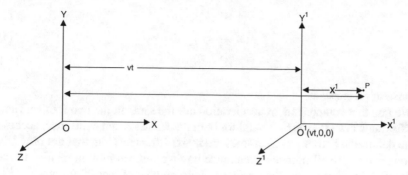

Fig. 1.1 S_1 frame is moving with velocity v along x-direction

$$z_1 = z \qquad (1.3)$$

It is assumed that 'time' can be defined independently of any particular frame of reference. As a result, time of the event recorded by the clocks of both systems will be the same. Hence, one can write the equation of transformation of time:

$$t_1 = t \qquad (1.4)$$

Actually, our own common senses confirm this conclusion as we do not detect any influence of the motion on clock rate in daily life. The above equation indicates that the time interval between occurrence of two events, say A and B, remains the same for each observer, i.e.

$$t_1^A - t_1^B = t^A - t^B$$

The transformation Eqs. (1.1)–(1.4) are valid only for the relative position of the reference frame S and S_1 and known as Galilean Transformations. Note that these are the transformations of the event coordinates from reference S to the reference S_1. To describe an event, one has to say where and when and this occurs. As a result, time turns out to be the fourth coordinate. Therefore, the coordinates of an event are defined by four numbers (x, y, z, t).

Differentiation Eqs. (1.1)–(1.3) with respect to 'time', we get

$$u_x^1 = u_x - v \qquad (1.5)$$
$$u_y^1 = u_y \qquad (1.6)$$
$$u_z^1 = u_z \qquad (1.7)$$

A second differentiation gives

$$a_x^1 = a_x \tag{1.8}$$

$$a_y^1 = a_y \tag{1.9}$$

$$a_z^1 = a_z \tag{1.10}$$

because v is a constant.

We find that components of acceleration are the same in the two frames. Hence, the Newton's law ($P = mf$) is valid for both the frames S and S_1. Hence, according to the definition, both S and S_1 frames are inertial frames. Note that not only S and S_1 frames but also all frames of reference having uniform motion relative to them are inertial. In other words, there exists infinite number of inertial frames.

Inverse of these transformations will be

$$x = x_1 + vt \tag{1.11}$$

$$y = y_1 \tag{1.12}$$

$$z = z_1 \tag{1.13}$$

$$t = t_1 \tag{1.14}$$

1.3   Galilean Transformations in Vector Form

Consider two systems S and S_1, where S_1 moving with velocity v relative to S. Initially, origins of two systems coincide. Let $\vec{r}$ and $\vec{r_1}$ be the position vectors of any particle P relative to origins O and O_1 of systems S and S_1, respectively, after time t. Then, $\overrightarrow{OO_1} = \vec{v}\,t$. So, from Fig. 1.2, using the law of triangle of vector addition

$$\vec{r_1} = \vec{r} - \vec{v}\,t \tag{1.15}$$

Also

$$t_1 = t \tag{1.16}$$

The inverse of Galilean Transformation in vectorial form is

$$\vec{r} = \vec{r_1} + \vec{v}\,t \tag{1.17}$$

Also

$$t = t_1 \tag{1.18}$$

Since Newtonian laws of motion remain the same in any two inertial frames S and S_1 (where one frame is moving with uniform velocity with respect to other), therefore,

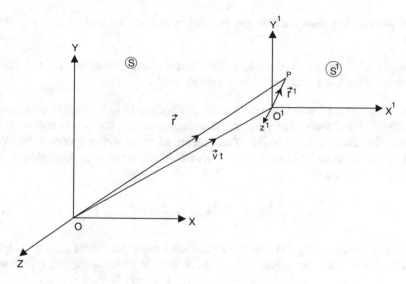

Fig. 1.2 S_1 frame is moving with velocity v in arbitrary direction

one could not able to distinguish mechanical experiments following Newtonian laws of motion.

Since at the time of Galileo, the laws of mechanics were the only known physical laws, therefore, the following principle is known as *Principle of Galilean Relativity*:

If an inertial system is given, then any other system that moves uniformly and rectilinear with respect to it is itself inertial.

Example 1.1 Using Galilean Transformation, show that the length of a rod remains the same as measured by the observers in frames S and S^1.

Hint: Let coordinates of the end points P and F of a rod in S frame are (x_1, y_1, z_1), (x_2, y_2, z_2) and coordinates of the end points P and F of the same rod in S^1 frame are (x_1^1, y_1^1, z_1^1), (x_2^1, y_2^1, z_2^1).

Then according to G.T.,

$$x_2^1 - x_1^1 = (x_2 - vt) - (x_1 - vt) = x_2 - x_1;$$
$$y_2^1 - y_1^1 = y_2 - y_1;$$
$$z_2^1 - z_1^1 = z_2 - z_1;$$

Therefore,

$$l^{1^2} = [(x_2^1 - x_1^1)^2 + (y_2^1 - y_1^1)^2 + (z_2^1 - z_1^1)^2]$$
$$= [(x_2 - x_1)^2 + (y_2 - y_1)^2 + (z_2 - z_1)^2] = l^2$$

In general, $(\vec{r}_2^1 - \vec{r}_1^1)^2 = (\vec{r}_2 - \vec{r}_1)^2$.

Example 1.2 The distance between two points is invariant under Galilean Transformation.

Example 1.3 Law of conservation of momentum is invariant under Galilean Transformation if the law of conservation of energy holds.

Hint: We prove it by means of collision of two bodies. Conservation of momentum states that in the absence of external forces, sum of momenta before and after collision remains the same. Let two bodies of masses m, M move with respective velocities u, U before impact and let u_1, U_1 be their velocities after impact. According to the law of conservation of energy,

$$\frac{1}{2}mu^2 + \frac{1}{2}MU^2 = \frac{1}{2}mu_1^2 + \frac{1}{2}MU_1^2 + \triangle W, \tag{1.1}$$

where $\triangle W$ is the change of internal excitation energy due to the collision. Now, one can observe the same collision from the S_1 system which is moving with v with respect to S system. Therefore, according to the Galilean Transformation,

$$\begin{aligned} u^1 &= u - v \\ U^1 &= U - v \\ u_1^1 &= u_1 - v \\ U_1^1 &= U_1 - v. \end{aligned} \tag{1.2}$$

Since law of conservation of energy holds in both system, then we have

$$\frac{1}{2}m(u^1)^2 + \frac{1}{2}M(U^1)^2 = \frac{1}{2}m(u_1^1)^2 + \frac{1}{2}M(U_1^1)^2 + \triangle W, \tag{1.3}$$

where u^1, U^1, u_1^1, U_1^1 represent the corresponding velocities in S^1 system. Here, internal excitation energy $\triangle W$ remains the same in both systems. Using the values of the velocities in S_1 system in (1.3), we get

$$\frac{1}{2}m(u - v)^2 + \frac{1}{2}M(U - v)^2 = \frac{1}{2}m(u_1 - v)^2 + \frac{1}{2}M(U_1 - v)^2 + \triangle W, \tag{1.4}$$

Using Eqs. (1.1) and (1.4) yields

$$mu + MU = mu_1 + MU_1. \tag{1.5}$$

Hence, the proof.

It is important to note that in an elastic collision, $\triangle W = 0$. Therefore, if a collision is either elastic or inelastic, the conservation law of momentum holds good.

Example 1.4 Derive Galilean Transformation for total momentum.

Hint: Galilean Transformation of momentum $\vec{p}$ is

$$\vec{p}^{\,1} = m\vec{u}^{\,1} = m(\vec{u} - \vec{v}) = \vec{p} - m\vec{v}$$

Let us consider a system of particles. We define the total momentum and total mass as

$$P = \sum p_i, \qquad M = \sum m_i.$$

Galilean Transformation of total momentum $\vec{P}$ is

$$\vec{P}^{\,1} = \sum \vec{p}_i^{\,1} = \sum \vec{p}_i - \vec{v} \sum m_i = \vec{P} - M\vec{v}.$$

Example 1.5 Let a reference frame S^1 is moving with velocity v relative to S frame which is at rest. A rod is at rest relative to S, at an angle θ to the x-axis. Find the angle θ^1 that makes with the x^1-axis of S^1.

Hint: Let at any time the rod lies in xy-plane with its ends (x_1, y_1), (x_2, y_2). Then the angle θ is given by

$$\theta = \tan^{-1} \left[\frac{y_2 - y_1}{x_2 - x_1} \right].$$

Similarly, the angle θ^1 is defined by

$$\theta^1 = \tan^{-1} \left[\frac{y_2^1 - y_1^1}{x_2^1 - x_1^1} \right].$$

From Galilean Transformation, we have

$$x_2^1 - x_1^1 = x_2 - x_1, \qquad y_2^1 - y_1^1 = y_2 - y_1$$

Therefore, we get

$$\theta^1 = \tan^{-1} \left[\frac{y_2^1 - y_1^1}{x_2^1 - x_1^1} \right] = \tan^{-1} \left[\frac{y_2 - y_1}{x_2 - x_1} \right] = \theta.$$

Example 1.6 Show that the electromagnetic wave equation

$$\frac{\partial^2 \phi}{\partial x^2} + \frac{\partial^2 \phi}{\partial y^2} + \frac{\partial^2 \phi}{\partial z^2} - \frac{1}{c^2} \frac{\partial^2 \phi}{\partial t^2} = 0$$

is not invariant under Galilean Transformation.

Hint: Applying chain rule, we get

$$\frac{\partial \phi}{\partial x} = \frac{\partial \phi}{\partial x^1}\frac{\partial x^1}{\partial x} + \frac{\partial \phi}{\partial y^1}\frac{\partial y^1}{\partial x} + \frac{\partial \phi}{\partial z^1}\frac{\partial z^1}{\partial x} + \frac{\partial \phi}{\partial t^1}\frac{\partial t^1}{\partial x}.$$

Using G.T., one can get

$$\frac{\partial \phi}{\partial x} = \frac{\partial \phi}{\partial x^1}$$

In a similar manner, one can find

$$\frac{\partial \phi}{\partial y} = \frac{\partial \phi}{\partial y^1}, \qquad \frac{\partial \phi}{\partial z} = \frac{\partial \phi}{\partial z^1}, \qquad \frac{\partial \phi}{\partial t} = \frac{\partial \phi}{\partial t^1} - v\frac{\partial \phi}{\partial x^1}.$$

The second derivative yields

$$\frac{\partial^2 \phi}{\partial x^2} = \frac{\partial^2 \phi}{\partial x^{12}}, \qquad \frac{\partial^2 \phi}{\partial y^2} = \frac{\partial^2 \phi}{\partial y^{12}}, \qquad \frac{\partial^2 \phi}{\partial z^2} = \frac{\partial^2 \phi}{\partial z^{12}},$$

$$\frac{\partial^2 \phi}{\partial t^2} = \frac{\partial^2 \phi}{\partial t^{12}} - 2v\frac{\partial^2 \phi}{\partial x^1 \partial t^1} + v^2\frac{\partial^2 \phi}{\partial x^{12}},$$

etc.

Example 1.7 A particle moves in xy plane with speed v plane at an angle $\theta = 30°$ to the x-axis in a reference frame S. What θ' does it velocity v' makes with x'-axis of S'-frame if $v = u$.

Hint: According to the Galilean Transformation, $v'_x = v_x - u$, $v'_y = v_y$.
   So,

$$\theta' = \tan^{-1}\left(\frac{v'_y}{v'_x}\right),$$

or,

$$\theta' = \tan^{-1}\left(\frac{v_y}{v_x - v}\right).$$

Here, $u = v$ and $\theta = 30°$.

1.4   Non-inertial Frames

Consider two systems S and S^1, the latter moving with an acceleration f relative to the former. If the first frame to be inertial, then the second frame will be non-inertial and vice versa. Let us take a Cartesian frame of reference S, which is inertial. Let

S^1 be another Cartesian frame whose axes x^1, y^1, z^1 are parallel to axes x, y, z of S and is moving with an acceleration f along the x-axis. We further assume that at the origin at time $t = 0$, the point O coincides with O^1 and all axes overlapped. Then, the x^1 coordinate of a particle P will be related to the x-coordinate by the relation:

$$x^1 = x - \frac{1}{2}ft^2.$$

The y, y^1 and z, z^1 coordinates will be identical, i.e.

$$y^1 = y, \quad z^1 = z.$$

It is assumed that 'time' can be defined independently of any particular frame of reference. That is, time of the event recorded by the clocks of both systems will be the same. Hence, one can write the equation of transformation of time:

$$t^1 = t.$$

Differentiating twice the above equations with respect to time, we get

$$a_x^1 = a_x - f \; ; \quad a_y^1 = a_y \; ; \quad a_z^1 = a_z.$$

Thus, total force acting on the particle as observed in S^1 is given by

$$F^1 = m(a_x - f) = ma_x - mf.$$

Clearly, additional force on particle due to acceleration of frame S_1 is given by

$$F_p = -mf = -\text{mass} \times \text{acceleration of non-inertial frame}.$$

The force experienced by the particle due to acceleration of non-inertial frame is called *fictitious force*.

1.5 Galilean Transformation and Laws of Electrodynamics

The only physical laws known at the time of Galileo were the laws of mechanics. And the laws of physics (laws of mechanics) are invariant in Galilean Transformation (inertial frame). Lately, the laws of electrodynamics were discovered and are not invariant under Galilean Transformation. Electrodynamics is governed by Maxwell's equations. According to Maxwell's theory, light is identified as electromagnetic wave (ψ) and electromagnetic waves propagate in empty space with constant velocity $c = (3 \times 10^8 \, \text{m/s})$. In S frame,

$$\psi = f[\,\overrightarrow{\mathbf{k}}\cdot\overrightarrow{\mathbf{r}} - ct\,],$$

where $\overrightarrow{\mathbf{k}} = l\hat{i} + m\hat{j} + n\hat{k}$ is unit vector along the wave normal. Now, we consider another inertial frame S^1 which is moving with velocity $\overrightarrow{v}$ relative to S. Therefore, in S^1 frame,

$$\psi^1 = f(\overrightarrow{\mathbf{k}}\cdot(\overrightarrow{\mathbf{r}}^1 + \overrightarrow{v}\,t) - ct) = f[\,\overrightarrow{\mathbf{k}}\cdot\overrightarrow{\mathbf{r}}^1 - (c - \overrightarrow{\mathbf{k}}\cdot\overrightarrow{v})t\,].$$

Hence, in S^1 frame, the speed of the wave will be $(c - \overrightarrow{\mathbf{k}}\cdot\overrightarrow{v})$. Therefore, the velocity will be different in different directions. Therefore, Maxwell's equations are not preserved in form by the Galilean Transformation, i.e. Maxwell's equations are not invariant under Galilean Transformation.

1.6 Attempts to Locate the Absolute Frame

It is shown that the laws of mechanics are invariant but the laws of electrodynamics are not invariant under Galilean Transformation. Now, if we assume, both the Galilean Transformation and Electrodynamics governed by Maxwell's equations are true, then it is obvious to believe there should exist a unique privileged frame of reference in which Maxwell's equations are valid and in which light propagates with constant velocity c in all directions.

The theory of Ether: After the advancement of modern techniques at the beginning of the twentieth century, scientists were trying to measure the speed of light in refined manner. Then, it was questioned what was the reference system or coordinate frame with respect to which the speed of light was measured? Also it was once believed that wave motion required a medium through which it propagates as sound in air. Since light evidently travels through a vacuum between the stars, it was believed that the vacuum must be filled by a medium for wave propagation. The imaginary medium filled the entire Universe is called ETHER (weightless). The obvious hypothesis that arises from this supposition is that the 'privileged' observers are those at rest relative to the ether, i.e. scientists thought that a frame of reference fixed relative to this ether medium was the *special privileged frame of reference*. Thus in their opinion, ether was the frame absolutely at rest. The preferred medium for light was ether, which filled all space uniformly. It is assumed that the ether is perfectly transparent to light and it exerts no resistance on material bodies passing through it. The rotational velocity of earth is about 3×10^4 m/s; therefore, it was supposed that according to Galilean Transformation, the speed of light should depend upon the direction of motion of light with respect to earth's motion through the ether. Since ether remains fixed in space, therefore it was considered to be possible to find the absolute velocity of any body moving through it. Several Physicists such as Fizeau, Michelson and Morley, Trouton and Noble, and many others conducted a number of experiments to find the absolute velocity of earth through ether. However, their attempts were failed. Later, Hertz proposed that ether was not at rest rather it is moving with the same velocity of body moving through it. This assumption was inconsistent with the phenomenon of aberration and hence the assumption was rejected.

Aberration: The phenomenon of apparent displacement in the sky is due to finite speed of light and the motion of earth in its orbit about the sun is called aberration.

After this, Fizeau and Fresnel predicted that light would be partially dragged along a moving medium, i.e. ether was partially dragged by the body moving through it.

Chapter 2
Michelson–Morley Experiment and Velocity of Light

2.1 Attempts to Locate Special Privileged Frame

Scientists tried to determine relative velocity of light with respect to earth. For this purpose, the velocity of earth relative to ether frame of special privileged frame was essential. Therefore, scientists performed various experiments involving electromagnetic waves. In this regard, the best and most important experiment was first performed by Michelson in 1881 and repeated the experiment more carefully in collaboration with Morley in 1887 and also reiterated many times thereafter. Actually, Michelson and Morley experiments laid the experimental foundations of special relativity. Michelson was awarded the Nobel prize for his experiment in 1907.

2.2 The Michelson–Morley Experiment (M–M)

The experiment of M–M was an attempt to measure the velocity of earth through the ether. The experiment failed in the sense that it showed conclusively that ether does not exist. More generally, it demonstrated that there are no privileged observer that the velocity of light is c irrespectively of the state of motion of the observer who measures it.

The apparatus of the experiment are Monochromatic source S of light, Interferometer A, one semi-silvered plate B, two mirrors, M_1, M_2 (see Fig. 2.1).

The light from a source at S falls on a semi-silvered plate B inclined at 45° to the direction of propagation. The plate B, due to semi-silvered, divides the beam of light into two parts, namely, reflected and transmitted beams. The two beams, after reflecting at mirrors M_1 and M_2 are brought together at A, where interference fringes are formed due to the small difference between the path lengths of two beams. Let us assume that earth (carrying the apparatus) is moving with velocity v along $BM_1 = l_1$ relative to the privileged frame. The time (t_1) required for the transmitted light ray to traverse the length l_1 both ways, i.e. along BM_1B is

© The Author(s), under exclusive license to Springer Nature Singapore Pte Ltd. 2022
F. Rahaman, *The Special Theory of Relativity*, La Matematica per il 3+2 136,
https://doi.org/10.1007/978-981-19-0497-4_2

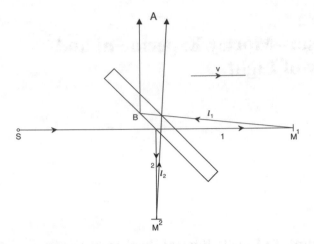

Fig. 2.1 Experimental set-up of Michelson–Morley Experiment

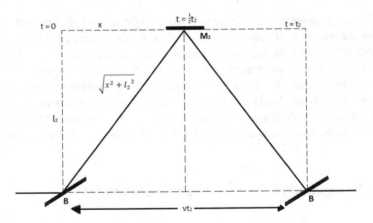

Fig. 2.2 Mirror moves some distance in time t

$$t_1 = \frac{l_1}{(c - v)} + \frac{l_1}{(c + v)} = \frac{\left(\frac{2l_1}{c}\right)}{\left(1 - \frac{v^2}{c^2}\right)} \tag{2.1}$$

[c is the velocity of light in the ether. So, downstream speed is $(c + v)$ and upstream speed is $(c - v)$].

Now, consider the path of the reflected light ray, i.e. the path BM_2B: As the light travels from B to M_2 ($BM_2 = l_2$), the whole apparatus has moved a distance x along BM_1. The light has therefore travelled a distance $\sqrt{(l_2^2 + x^2)}$. We have seen that earth moves some distance during the same time. According to Fig. 2.2, the mirror moves the distance $x = vt$ in time t and light moves in time t is $ct = \sqrt{(l_2^2 + x^2)}$.

Hence,

$$t = \frac{x}{v} = \frac{\sqrt{(l_2^2 + x^2)}}{c}.$$

Or,

$$\sqrt{(l_2^2 + x^2)} = \frac{l_2}{\sqrt{(1 - \frac{v^2}{c^2})}}. \tag{2.2}$$

On the back, it travels an equal distance, so the time for to and fro travel of the light beam (i.e. the path BM_2B) is

$$t_2 = \frac{(\frac{2l_2}{c})}{\sqrt{(1 - \frac{v^2}{c^2})}}. \tag{2.3}$$

Interference fringes are created by the difference between t_1 and t_2. The whole apparatus is now swung through $90°$, so the roles of the two arms are interchanged. Therefore, l_1 takes the place of l_2 and vice versa. We have

$$\bar{t}_1 = \frac{(\frac{2l_1}{c})}{\sqrt{(1 - \frac{v^2}{c^2})}} \tag{2.4}$$

$$\bar{t}_2 = \frac{(\frac{2l_2}{c})}{(1 - \frac{v^2}{c^2})} \tag{2.5}$$

The time differences between the two paths, before and after the apparatus is swung round, are

$$\Delta t = t_1 - t_2 = \frac{(\frac{2}{c})}{\sqrt{(1 - \frac{v^2}{c^2})}} \left(\frac{l_1}{\sqrt{(1 - \frac{v^2}{c^2})}} - l_2 \right) \tag{2.6}$$

$$\underline{\Delta t} = \underline{t}_1 - \underline{t}_2 = \frac{(\frac{2}{c})}{\sqrt{(1 - \frac{v^2}{c^2})}} \left(l_1 - \frac{l_2}{\sqrt{(1 - \frac{v^2}{c^2})}} \right). \tag{2.7}$$

The change in t brought about by rotating the apparatus is

$$\underline{\Delta t} - \Delta t = -\frac{(\frac{2}{c})(l_1 + l_2)}{\sqrt{(1 - \frac{v^2}{c^2})}} \left(\frac{1}{\sqrt{(1 - \frac{v^2}{c^2})}} - 1 \right). \tag{2.8}$$

We expand the right-hand side in a Binomial series, retaining terms up to the second power of (v/c) to yield

$$\underline{\Delta t} - \Delta t \approx -\frac{1}{c}(l_1 + l_2)\frac{v^2}{c^2}. \tag{2.9}$$

Hence, the shift in the number of fringes is

$$\Delta N = \frac{\underline{\Delta t} - \Delta t}{T} = -\frac{\gamma}{c}(l_1 + l_2)\frac{v^2}{c^2}. \tag{2.10}$$

[Period of one vibration is $T = (\frac{1}{\gamma}) = (\frac{\lambda}{c})$, where γ and λ are frequency and wave length of the light].

Earth's velocity is presumed to be 30 km/s = 3×10^6 cm/s, $(\frac{v}{c})^2 = 10^{-8}$ and $(\frac{\gamma}{c}) \approx 10^4$ cm^{-1} for visible light, so

$$\Delta N \approx -2(l_1 + l_2) \times 10^{-4}. \tag{2.11}$$

Since l_1 and l_2 were several metres, such a shift would be easily detectable and measurable. However, no fringe shift was observed. This experiment was repeated with multiple mirrors to increase the lengths l_1 and l_2, however, the experimental conclusion was that there was no fringe shift at all. In the year 1904, Trouton and Noble done again the experiment using electromagnetic waves instead of visible light, however, no fringe shift was observed. Since the velocity of earth relative to ether cannot always be zero, therefore, experiment does not depend solely on an absolute velocity of earth through an ether but also depends on changing velocity of earth with respect to 'ether'. Such a changing motion through the ether would be easily detected and measured by the precision of experiments, if there was an ether frame.

The null result seems to rule out an ether (absolute) frame. Thus, no preferred inertial frame exists. Also, one way to interpret the null result of M–M experiment is to conclude simply that the measured speed of light is the same, i.e. c for all directions in every inertial system. In the experiment, the downstream speed and upstream speed being c rather than $c + v$ or $c - v$ in any frame.

Note 1: In 1882, Fitzgerald and Lorentz independently proposed a hypothesis to explain null results of M–M experiment. Their hypothesis is 'Every body moving with a velocity v through the ether has its length in the direction of motion contracted by a factor $\frac{1}{\sqrt{1 - \frac{v^2}{c^2}}}$, while the dimensions in directions perpendicular to motion remain unchanged'.

Let us consider in M–M experiment, the arms are equal, i.e. $l_1 = l_2 = l$ at rest relative to ether and according to Fitzgerald and Lorentz hypothesis its length l when it is in motion with velocity v relative to ether will be $l\sqrt{(1 - \frac{v^2}{c^2})}$. By this, they interpreted the null result of M–M experiment as follows: From, Eqs. (2.1) and (2.3), we have

$$t_1 = \frac{(\frac{2l}{c})}{(1 - \frac{v^2}{c^2})} \approx \left(\frac{2l}{c}\right)\left(1 + \frac{v^2}{c^2}\right) \quad , \quad t_2 = \frac{(\frac{2l}{c})}{\sqrt{(1 - \frac{v^2}{c^2})}} \approx \left(\frac{2l}{c}\right)\left(1 + \frac{v^2}{2c^2}\right).$$

Using Fitzgerald and Lorentz hypothesis, we get by replacing l by $l\sqrt{(1 - \frac{v^2}{c^2})}$,

$$t_1 = \left(\frac{2l}{c}\right)\sqrt{1 - \frac{v^2}{c^2}}\left(1 + \frac{v^2}{c^2}\right).$$

Expanding binomially and neglecting $\frac{v^4}{c^4}$, we get

$$t_1 = \left(\frac{2l}{c}\right)\left(1 + \frac{v^2}{2c^2}\right) = t_2.$$

Therefore, according to Fitzgerald and Lorentz hypothesis, the times taken by reflected and transmitted beams are the same and hence no shift of fringe is observed. However, Fitzgerald and Lorentz hypothesis fails to give correct explanation of null results of M–M experiment when the two ears of the interferometer are not equal.

Note 2: If some instant, the velocity of earth were zero with respect to ether, no fringe shift would be expected. From, Eqs. (2.1) and (2.3), we have, by putting $v = 0$,

$$t_1 = \left(\frac{2l}{c}\right) = t_2.$$

Then, there is no relative velocity between earth and the ether, i.e. earth drags the ether with the same velocity as earth during its motion. However, this explanation is in contradiction with various experiments like Bradley's result of aberration, etc.

2.3 Phenomena of Aberration: Bradley's Observation

In, the year 1727, Bradley observed that the star, γ Draconis revolves in nearly circular orbit. He found that angular diameter of this circular orbit is about 41 s of arc and other stars appeared to move in elliptical orbit have almost same period. This phenomena of apparent displacement is known as aberration. This can be explained as follows. Let a light ray comes from a star at zenith. An observer together with

Fig. 2.3 Telescope will have
to tilt to see the star

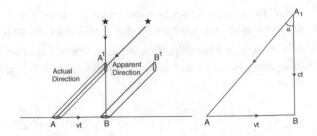

earth has a velocity v will see the direction of light ray will not be vertical. This is due to relative motion of the observer and light ray. To see the star, the telescope will have to tilt, i.e. direction slightly away from the vertical as shown in the Fig. 2.3.

The angle α between the actual and apparent directions of the star is called aberration.

Let v be the velocity of earth in the direction AB. The light ray enters the telescope tube at A^1 and reaches the observer's eye at B in time Δt. As the light reaches from A^1 to B, in the mean time, eyepiece moves from A to B. Hence, $A^1 B = c\Delta t$, $AB = v\Delta t$. The tilt of the telescope α is given by (see Fig. 2.3)

$$\tan \alpha = \frac{v}{c}$$

or

$$\alpha = \tan^{-1} \frac{v}{c}.$$

Using the velocities of earth and light as 30 km/s and 3×10^5 km/s, respectively, one gets,

$$\alpha = \tan^{-1} \frac{v}{c} = \tan^{-1} \frac{3 \times 10^4}{3 \times 10^8} = \tan^{-1}(10^{-4}) = 20.5 \text{ arc s.}$$

Since earth's motion is nearly circular, therefore every 6 months, the direction of aberration will reverse and telescope axis is tracing out a cone of aberration during the year. Hence, the angular diameter of the cone would be $2\alpha = 41$ arc s. In 1727, Bradley observed this motion and found the angular diameter was 41 arc s for the star γ Draconis. This experiment apparently showed the absolute velocity of earth is 30 km/s.

The outcome of this experiment is that ether is not dragged around with earth. If it were, then the phenomena of aberration would not be happened.

Fig. 2.4 Experimental set-up of Fizeau's experiment

2.4 Fizeau's Experiment

In 1817, Fresnel proposed that light would be partially dragged along by a moving medium. Using ether hypothesis, he provided the formula for the effect of moving medium on the velocity of light. In 1951, Fizeau confirmed this effect experimentally.

Figure 2.4 shows the Fizeau's experimental arrangement. Fizeau used water as the medium inside the tubes. Water flows through the tubes with velocity u.

The S is the source of light. The light ray from S falls on a semi-silver plate inclined $45°$ from the horizontal is divided into two parts. One is reflected in the direction $M_1 M_2$ and other part is transmitted in the direction $M_1 D$. Here, one beam of light is entering towards the velocity of water and other opposite to the motion of water. One can note that reflected beam travels the path $M_1 M_2 C D M_1$ in the direction of motion of water and the transmitted beam travels the path $M_1 E D C M_2 M_1$ opposite to the motion of water. Both beams finally enter the telescope.

Difference in time taken by both the beams travel equal distance d due to the motion of the water and velocity v of earth. It was assumed that space is filled uniformly with the ether and ether to be partially dragged. Now, the time difference is

$$\Delta t = \frac{d}{\frac{c}{\mu} + \left(\frac{1}{\mu^2}\right) v + u \left(1 - \frac{1}{\mu^2}\right)} - \frac{d}{\frac{c}{\mu} + \left(\frac{1}{\mu^2}\right) v - u \left(1 - \frac{1}{\mu^2}\right)}.$$

It was assumed that ether to be partially dragged and as a result component of the velocity in the tube $\left(\frac{1}{\mu^2}\right) v$. It follows also that if a material body is moving with velocity u through ether, the velocity of ether will be $u \left(1 - \frac{1}{\mu^2}\right)$. Here, $\left(\frac{1}{\mu^2}\right) =$ constant. Neglecting higher order terms, we get the time difference as

$$\Delta t = \frac{2 u d \mu^2}{c^2} \left(1 - \frac{1}{\mu^2}\right).$$

The phase difference is given by

$$\Delta f = \frac{\frac{2ud\mu^2}{c^2}\left(1 - \frac{1}{\mu^2}\right)}{T} = \frac{2udn\mu^2}{c^2}\left(1 - \frac{1}{\mu^2}\right),$$

where n be the frequency of the oscillation and T is time period of one oscillation ($nT = 1$). Fizeau observed the shift in the fringes and found exactly the same as given by the above equation. This result also confirms the result that the velocity of the light in any medium of refractive index μ is not simply $\frac{c}{\mu}$ but $\frac{c}{\mu} \pm \left(1 - \frac{1}{\mu^2}\right)$, u is the velocity of the medium. The factor $\left(1 - \frac{1}{\mu^2}\right)$ is called *Fresnel's ether dragging coefficient*. Thus, Fresnel's formula for ether drag was verified by Fizeau.

2.5   The Relativistic Concept of Space and Time

The null results of M–M Experiment ruled out the absolute concept of space and laid Einstein to formulate the new concept of space and time.

Consider the following two contradictory statements: [1] In classical mechanics, the velocity of any motion has different values for observers moving relative to each other. [2] The null results of M–M experiment concludes that the velocity of light is not affected by the motion of the frame of reference. The genius Einstein applied his intuition and realized that this contradiction comes from the imperfection of the classical ideas about measuring space and time. He ruled out the Newton's concept of absolute time by criticizing our preconceived ideas about simultaneity. He commented 'time is indeed doubtful'. His intuition may be supported as follows:

$$X - - - - - - - - - - - O - - - - - - - - - - - - Y.$$

Suppose an observer sitting at O where O is the middle point of the straight line XY. Suppose that two events occurring at X and Y which are fixed in an inertial frame S and the two clocks giving absolute time are kept at X and Y. In the inertial frame S, the events occurring at X and Y are said to be simultaneous, if the clocks placed at X and Y indicate the same time when the events occur. Let a light signal is giving at X and Y simultaneously. Suppose XOY is moving with v velocity in the direction XY. Then, one can show that these light signals will not reach simultaneously the observer at O which is moving with velocity v. Here, the moving O seems to lie in another inertial frame. The light signal takes t_1 time to reach from Y to O as $\frac{OY}{(c-v)}$ and the time t_2 taken by the light signal from X to O is $\frac{OX}{(c+v)}$.

[c is the velocity of light in all direction. So, according to Newtonian mechanics, downstream speed is $(c + v)$ and upstream speed is $(c - v)$].

One can note that

$$t_1 = \frac{OY}{(c-v)} \neq \frac{OX}{(c+v)} = t_2 \,, \text{ as } c+v \neq c-v \,, \quad OX = OY.$$

Therefore, time is not absolute. It varies from one inertial frame to another inertial frame, i.e. $t_1 \neq t_2$. In other words, every reference frame has its own particular time. Therefore, concept of absolute time was ruled out by Einstein.

The space coordinate is transformed due to Galilean Transformation as $x^1 = x - vt$. Since, time depends on frame of reference, therefore, distance between two points will also depend on the frame of reference. In other words, distance between two points is not absolute. Hence, Einstein also ruled out the concept of absolute space.

Einstein proposed his new relativistic concepts of space and time based on the fact that observers who are moving relative to each other measure the speed of light to be the same. This new concept of space and time is not only valid for mechanical phenomenon but also for all optical and electromagnetic phenomenon and known as theory of relativity. To develop these new concepts of space and time, Einstein replaced Galilean Transformation equations by a new type of transformation equations called *Lorentz transformation*, which was based on the invariant character of the speed of light.

The theory of relativity having new concept of space and time is divided into two parts. [1] *Special theory of relativity (STR)* [2] *General theory of relativity, (GTR)*.

The STR deals with inertial systems, i.e. the systems which move in uniform rectilinear motion relative to one another. GTR deals with non-inertial system, i.e. it is an extension of inertial system to accelerated systems, i.e. the systems moving with accelerated velocity relative to one another. The GTR is applied to the area of gravitational field and based on which laws of gravitation can be explained in a more perfect manner than given by Newton.

Chapter 3
Lorentz Transformations

3.1 Postulates of Special Theory of Relativity

To develop the special theory of relativity, Einstein used two fundamental postulates.

[1] *The principle of relativity or equivalence*: The laws of physics are the same in all inertial systems. No preferred inertial system exists.

[2] *The principle of constancy of the speed of light*: The velocity of light in an empty space is a constant, independent not only of the direction of propagation but also of the relative velocity between the source of the light and observer.

Note that the first principle has not changed its Newtonian form except the Galilean Transformation equations is replaced by a new type of transformation equations known as Lorentz Transformation. The second principle is not consistent with Galilean Transformation and it indicates a clear division between Newtonian theory and Einstein theory. Therefore, we have to replace Galilean Transformation equations by Lorentz Transformation equations which fulfil the above principles.

3.2 Lorentz Transformations

The principle of relativity states that the laws of nature are invariant under a particular group of space–time coordinate transformations. Newton's laws of motion are invariant under GT but Maxwell's equations are not, and that Einstein resolved this conflict by replacing Galilean Transformation with Lorentz Transformation.

F. Rahaman, *The Special Theory of Relativity*, La Matematica per il 3+2 136,
https://doi.org/10.1007/978-981-19-0497-4_3

3.2.1 Lorentz Transformation Between Two Inertial Frames of Reference (Non-axiomatic Approach)

Let us suppose that an event occurred at a point P in an inertial frame S specifying the coordinates x, y, z and t. In another inertial frame S_1, which is moving with a constant velocity v along x-axis, the same event will be specified by x_1, y_1, z_1 and t_1. Let the observers in the two systems be situated at the origin O and O_1 of S and S_1 frames. The rectangular axes x_1, y_1, z_1 are parallel to x, y, z, respectively. Therefore, the coordinates x_1, y_1, z_1 and t_1 are related to x, y, z and t by a functional relationship. Hence, we must have

$$x_1 = x_1(x, y, z, t); \quad y_1 = y_1(x, y, z, t); \quad z_1 = z_1(x, y, z, t); \quad t_1 = t_1(x, y, z, t).$$
(3.1)

Since S and S_1 are both inertial frames, therefore laws of mechanics remain the same in both the frames. As a result, a particle is moving with a constant velocity along a straight line relative to an observer in S, then an observer in S_1 will see also that that particle is moving with uniform rectilinear motion relative to him. Under these conditions, the transformation Eq. (3.1) must be linear and assume the most general form as

$$x_1 = a_{11}x + a_{12}y + a_{13}z + a_{14}t + a_{15}$$
$$y_1 = a_{21}x + a_{22}y + a_{23}z + a_{24}t + a_{25}$$
$$z_1 = a_{31}x + a_{32}y + a_{33}z + a_{34}t + a_{35}$$
$$t_1 = a_{41}x + a_{42}y + a_{43}z + a_{44}t + a_{45},$$
(3.2)

[where a_{ij} are constants]

Note that we have chosen the relative velocity v of the S and S_1 frames to be along a common $x - x_1$-axis to keep corresponding planes parallel. This assumption simplifies the algebra without any loss of generality. Also at the instant, when the origins O ($x = y = z = 0$) and O_1 ($x_1 = y_1 = z_1 = 0$) coincide, the clocks read $t = 0$ and $t_1 = 0$, respectively. These imply $a_{15} = a_{25} = a_{35} = a_{45} = 0$. Now, regarding the remaining 16 coefficients, it is expected that their values will depend on the relative velocity v of the two inertial frames. Now, we use the postulates of the STR (1. The principle of relativity 2. The principle of constancy of the speed of light) to determine the values of the 16 coefficients. Here, we note that x-axis coincides continuously with x_1-axis. Therefore, points on x-axis and x_1-axis are characterized by $y = z = 0$ and $y_1 = z_1 = 0$. As a result, the transformation formulas for y and z should be of the form

$$y_1 = a_{22}y + a_{23}z \quad \text{and} \quad z_1 = a_{32}y + a_{33}z.$$
(3.3)

That is, coefficients a_{21}, a_{24}, a_{31} and a_{34} must be zero. In a similar manner, the $x - y$- and $x - z$-planes (which are characterized by $z = 0$ and $y = 0$, respectively) should transform over to the $x_1 - y_1$- and $x_1 - z_1$-planes (which are characterized by $z_1 = 0$ and $y_1 = 0$, respectively). Therefore, in Eq. (3.3), the coefficients a_{23} and a_{32} are zero to yield

$$y_1 = a_{22}y \text{ and } z_1 = a_{33}z. \tag{3.4}$$

Now, we try to find the coefficients a_{22} and a_{33}. Suppose we put a rod of unit length (measured by observer in S frame) along y-axis. According to Eq. (3.4), the observer in S_1 frame measures the rod's length as a_{22}. Now, we put a rod of unit length (measured by observer in S_1 frame) along y_1-axis. The observer in S frame measures this rod as $\frac{1}{a_{22}}$. The first postulate of STR requires that these measurements must be same. Hence, we must have $a_{22} = \frac{1}{a_{22}}$, i.e. $a_{22} = 1$. Using similar argument, one can determine $a_{33} = 1$. Therefore, we get

$$y_1 = y \text{ and } z_1 = z. \tag{3.5}$$

For the reason of symmetry, we assume that t_1 does not depend on y and z. Hence, $a_{42} = 0$ and $a_{43} = 0$. Thus, t_1 takes the form as

$$t_1 = a_{41}x + a_{44}t. \tag{3.6}$$

We know points which are at rest relative to S_1 frame will move with velocity v relative to S frame in x-direction. Therefore, the statement $x_1 = 0$ must be identical to the statement $x = vt$. As a result, the correct transformation equation for x_1 will be

$$x_1 = a_{11}(x - vt). \tag{3.7}$$

Still we will have to determine the three coefficients, namely, a_{11}, a_{41}, a_{44}. These can be determined by using the principle of the constancy of the velocity of light. Let us assume that a light pulse produced at the time $t = 0$ and will propagate in all direction with same velocity c. Initially, it coincides with the origin of S_1. Due to the second postulate, the wave propagates with a speed c in all directions in each inertial frame. Hence, equations of motion of this wave with respect to both frames S and S_1 are given as

$$x^2 + y^2 + z^2 = c^2t^2 \tag{3.8}$$

$$x_1^2 + y_1^2 + z_1^2 = c^2t_1^2. \tag{3.9}$$

Now, we substitute the transformation equations for x_1, y_1, z_1 and t_1 into Eq. (3.9) and get

$$[a_{11}^2 - c^2a_{41}^2]x^2 + y^2 + z^2 - 2[va_{11}^2 + c^2a_{41}a_{44}]xt = [c^2a_{44}^2 - v^2a_{11}^2]t^2. \tag{3.10}$$

But this equation and Eq. (3.8) are identical, so we must have

$$[c^2 a_{44}^2 - v^2 a_{11}^2] = c^2; \quad [a_{11}^2 - c^2 a_{41}^2] = 1; \quad [v a_{11}^2 + c^2 a_{41} a_{44}] = 0.$$

Solving these equations, we get

$$a_{44} = \frac{1}{\sqrt{(1 - \frac{v^2}{c^2})}}; \quad a_{11} = \frac{1}{\sqrt{(1 - \frac{v^2}{c^2})}}; \quad a_{41} = -\frac{v}{c^2 \sqrt{(1 - \frac{v^2}{c^2})}}.$$

So finally we get, the Lorentz Transformation as

$$x_1 = a(x - vt); \quad y_1 = y; \quad z_1 = z; \quad t_1 = a\left[t - \left(\frac{vx}{c^2}\right)\right], \tag{3.11}$$

where

$$a = \frac{1}{\sqrt{(1 - \beta^2)}} \quad \text{and} \quad \beta = \frac{v}{c}.$$

The inverse relations can be obtained by solving x, y, z, t in terms of x_1, y_1, z_1, t_1 as

$$x = a(x_1 + vt_1); \quad y = y_1; \quad z = z_1; \quad t = a\left[t_1 + \left(\frac{vx_1}{c^2}\right)\right]. \tag{3.12}$$

Actually, Eq. (3.12) are obtained from Eq. (3.11) by replacing v by $-v$ and subscripted quantities by unsubscripted ones and vice versa. In particular, for speed small compared to c, i.e. $\frac{v}{c} \ll 1$, the Lorentz Transformations reduce to Galilean Transformations. Thus,

$$x_1 = x - vt; \quad y_1 = y; \quad z_1 = z; \quad t_1 = t.$$

Note 1: If the velocity of light would be infinity, then Lorentz Transformation will never exist and we have only Galilean Transformation.

Example 3.1 Show that the expressions

$$[i] \, x^2 + y^2 + z^2 - c^2 t^2$$

$$[ii] \, dx^2 + dy^2 + dz^2 - c^2 dt^2 \equiv dx_1^2 + dx_2^2 + dx_3^2 + dx_4^2,$$

where, $x_1 = x$, $x_2 = y$, $x_3 = z$, $x_4 = ict$, are invariant under Lorentz Transformations.

Fig. 3.1 The frame S_1 is moving with velocity v along x-direction

3.2.2 Axiomatic Derivation of Lorentz Transformation

Let there be two reference system S, S_1. S_1 is moving with velocity v with respect to S along x-axis (see Fig. 3.1). At time, $t = 0$, two origins O and O_1 coincide. If (x, y, z, t) be any event in S frame, then this event is represented by (x_1, y_1, z_1, t_1) in S_1 frame.

To find the relation between the coordinates of S and S_1 frames, we use the following **axioms**:

[i] The direct transformation (i.e. S to S_1) and reverse transformation (i.e. S_1 to S) should be symmetrical with respect to both systems, i.e. one should be derivable from the other by replacing v by $-v$ and subscripted quantities by unsubscripted ones and vice versa.

[ii] If (x, y, z, t) are finite, then (x_1, y_1, z_1, t_1) are also finite.

[iii] In the limit $v \to 0$, the transformations should be reduced to identity transformations, i.e. $x_1 = x$, $y_1 = y$, $z_1 = z$, $t_1 = t$.

[iv] In the limit $c \to \infty$, the Lorentz Transformations reduce to Galilean Transformations. Thus,

$$x_1 = x - vt; \quad y_1 = y; \quad z_1 = z; \quad t_1 = t.$$

[v] The velocity of light is same in two reference frames, i.e. $c_1 = c$.

Now, we search the possible transformations between the coordinates of two frames. If the transformation functions are quadratic or higher, then the inverse transformations should be irrational functions. Thus, transformations cannot be quadratic or higher. Another choice is linear fractional transformation function, $x_1 = \frac{ax+b}{cx+d}$. Then, the inverse transformation gives, $x = \frac{b-dx_1}{cx_1-a}$. But this function for $x_1 = \frac{a}{c}$ becomes infinite and this violates the condition (ii). Hence, linear transformation is the only acceptable one. Since the relative motion between the reference frames is along the x-axis, we get,

$$y_1 = y; \quad z_1 = z. \tag{3.13}$$

Now, we choose the linear transformations for x and t as

$$x_1 = \lambda x + \mu t + A \tag{3.14}$$

$$t_1 = \psi x + \phi t + B. \tag{3.15}$$

Since origins coincide initially, i.e. $x = 0, t = 0, x_1 = 0, t_1 = 0$, so the constants present in Eqs. (3.14) and (3.15), namely, $A = B = 0$ should be zero. Since the point O_1 (its coordinate $x_1 = 0$) moves with a velocity v relative to O, we get

$$x = vt.$$

Now, we put together $x_1 = 0$ and $x = vt$ in (3.14), we get,

$$\lambda v + \mu = 0. \tag{3.16}$$

Solving (3.14) and (3.15), we get

$$x = \frac{\phi x_1 - \mu t_1}{\lambda \phi - \mu \psi} \tag{3.17}$$

$$t = \frac{\lambda t_1 - \psi x_1}{\lambda \phi - \mu \psi}. \tag{3.18}$$

From condition (i), we can write the inverse transformation by replacing v by $-v$ and subscripted quantities by unsubscripted ones as

$$x = \lambda x_1 - \mu t_1 \tag{3.19}$$

$$t = -\psi x_1 + \phi t_1. \tag{3.20}$$

Since the coefficients μ and ψ are related with x- and t-coordinates and therefore changes its sign when $v \longrightarrow -v$ (one can also take λ and ϕ).

Since Eqs. (3.17) and (3.19) are identical, therefore, comparing the coefficients, we have

$$\lambda = \frac{\phi}{\lambda \phi - \mu \psi}; \quad -\mu = \frac{-\mu}{\lambda \phi - \mu \psi}.$$

These imply

$$\lambda = \phi \tag{3.21}$$

$$\lambda \phi - \mu \psi = 1. \tag{3.22}$$

From Eqs. (3.14) and (3.15), we have

$$\frac{x_1}{t_1} = \frac{\lambda(\frac{x}{t}) + \mu}{\psi(\frac{x}{t}) + \phi}. \tag{3.23}$$

Let a light signal emitted from the origin in S frame, then $\frac{x}{t} = c$. But condition (v) implies $\frac{x_1}{t_1} = c$ also. Hence, from Eq. (3.23), we have

$$c = \frac{\lambda c + \mu}{\psi c + \phi}. \tag{3.24}$$

From Eqs. (3.16), (3.21), (3.22) and (3.24), we get the following solutions as

$$\psi = -\frac{\lambda v}{c^2} \tag{3.25}$$

$$\lambda = \pm \frac{1}{\sqrt{1 - (\frac{v^2}{c^2})}}. \tag{3.26}$$

Condition (ii) indicates that we must get back $t_1 = t$ when $c \to 0$. Therefore, we should select $+$ sign.

Thus, finally we the Lorentz Transformation as,

$$x_1 = a(x - vt); \quad y_1 = y; \quad z_1 = z; \quad t_1 = a\left[t - \left(\frac{vx}{c^2}\right)\right], \tag{3.27}$$

where

$$a = \frac{1}{\sqrt{(1 - \beta^2)}} \quad \text{and} \quad \beta = \frac{v}{c}.$$

3.2.3 Lorentz Transformation Based on the Postulates of Special Theory of Relativity

Let us suppose that an event occurred at a point P in an inertial frame S specifying the coordinates x, y, z and t. In another inertial frame S_1, which is moving with a constant velocity v along x-axis, the same event will be specified by x_1, y_1, z_1 and t_1. Let the observers in the two systems be situated at the origin O and O_1 of S and S_1 frames. The Galilean Transformation between x and x^1 is $x^1 = x - vt$.

We know Galilean Transformation fails, so a reasonable guess about the nature of the correct relationship is

$$x^1 = k(x - vt). \tag{3.28}$$

Here, k is the factor that does not depend upon either x or t but may be a function of v.

The first postulate of Special Theory of Relativity: All laws of physics are the same on all inertial frames of reference. Therefore, Eq. (3.28) must have the same form on both S and S^1. We need only change the sign of v (in order to take into account the difference of the direction of relative motion) to write the corresponding equation for x in terms of x^1 and t^1. Therefore,

$$x = k(x^1 + vt^1). \tag{3.29}$$

It is obtained that y, y^1 and z, z^1 are equal to each other as they are perpendicular to velocity v. Therefore,

$$y^1 = y \tag{3.30}$$

and

$$z^1 = z. \tag{3.31}$$

Remember t^1 may not be equal to t. Putting the value of x^1 in (3.29), we get

$$x = k[k(x - vt) + vt^1] = k^2(x - vt) + kvt^1,$$

or

$$t^1 = kt + \left(\frac{1 - k^2}{kv}\right)x. \tag{3.32}$$

The second postulates of relativity gives us a way to evaluate k. At the instant $t = 0$, the origins of the two frames of reference S and S^1 are on the same place. Therefore, according to our initial condition $t^1 = 0$ when $t = 0$. Suppose that a light signal from a light source emits a pulse at the common origin of S and S^1 at time $t = t^1 = 0$ and the observers in each system measures the speed of the propagate of light wave. Both observes must find the same speed c, which means that in the S frame,

$$x = ct \tag{3.33}$$

and in S^1 frame,

$$x^1 = ct^1. \tag{3.34}$$

Now from (3.33), we get

$$k(x - vt) = c\left[kt + \left(\frac{1-k^2}{kv}\right)x.\right]$$

Therefore,

$$x = \frac{ct\left(1 + \frac{v}{c}\right)}{\left[1 - \left(\frac{1-k^2}{k^2}\right)\frac{c}{v}\right]},$$

i.e.

$$ct = \frac{ct\left(1 + \frac{v}{c}\right)}{1 - \left(\frac{1}{k^2} - 1\right)\frac{c}{v}} \quad \text{[since from (3.33), } x = ct\text{]}.$$

Hence,

$$k = \frac{1}{\sqrt{1 - \frac{v^2}{c^2}}}. \tag{3.35}$$

Using the value of k from (3.35) in Eqs. (3.28) and (3.32), we get the values of x^1 and t^1 as

$$x^1 = \frac{(x - vt)}{\sqrt{1 - \frac{v^2}{c^2}}} \tag{3.36}$$

$$t^1 = \frac{t - \frac{vx}{c^2}}{\sqrt{1 - \frac{v^2}{c^2}}}, \tag{3.37}$$

Thus, Eqs. (3.30), (3.31), (3.36) and (3.37) give the required Lorentz Transformations.

3.3   The General Lorentz Transformations

Now, we search the Lorentz Transformations for an arbitrary direction of the relative velocity $\vec{v}$ between S and S_1 frames. Let us consider the position vector $\vec{r}$ of a particle separate into two components, one parallel and another perpendicular to $\vec{v}$. That is,

$$\vec{r} = \vec{r}_\perp + \vec{r}_\parallel \tag{3.38}$$

Fig. 3.2 The frame S_1 is moving with velocity v in arbitrary direction

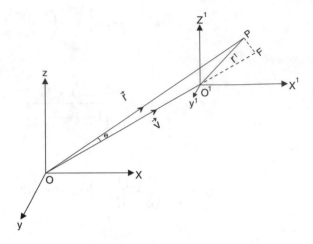

[here, $\vec{\mathbf{r}}_\perp \perp \vec{\mathbf{v}}$ and $\vec{\mathbf{r}}_\parallel \parallel \vec{\mathbf{v}}$].

The component $\vec{\mathbf{r}}_\parallel$ assumes the following value:

$$\vec{\mathbf{r}}_\parallel = \frac{(\vec{\mathbf{r}} \cdot \vec{\mathbf{v}})\vec{\mathbf{v}}}{v^2} \tag{3.39}$$

[from Fig. 3.2, we have $\vec{\mathbf{r}} \cdot \vec{\mathbf{v}} = r\,v\,\cos\theta$; $|\overrightarrow{OF}| = |\overrightarrow{OP}|\cos\theta$; $|\vec{\mathbf{r}}| = r$, $|\vec{\mathbf{v}}| = v$; $\overrightarrow{OF} = \vec{\mathbf{r}}_\parallel = |\overrightarrow{OF}|\tilde{n} = (\frac{\vec{r}.\vec{v}}{v})(\frac{\vec{v}}{v}) = (\frac{\vec{r}.\vec{v}}{v^2})\vec{\mathbf{v}}$]

The other component $\vec{\mathbf{r}}_\perp$ takes the value

$$\vec{\mathbf{r}}_\perp = \vec{\mathbf{r}} - \vec{\mathbf{r}}_\parallel = \vec{\mathbf{r}} - \frac{(\vec{\mathbf{r}} \cdot \vec{\mathbf{v}})\vec{\mathbf{v}}}{v^2}. \tag{3.40}$$

Since coordinates change in the direction of motion, therefore, the component $\vec{\mathbf{r}}_\parallel$ will transform like the x-coordinate. Hence,

$$\vec{\mathbf{r}}'_\parallel = a(\vec{\mathbf{r}}_\parallel - \vec{\mathbf{v}}\,t), \tag{3.41}$$

where

$$a = \frac{1}{\sqrt{(1 - \beta^2)}} \quad \text{and} \quad \beta = \frac{v}{c}.$$

The coordinates remain invariable in a direction perpendicular to motion as a result component $\vec{\mathbf{r}}_\perp$ will transform like y- or z-coordinates. Hence,

$$\vec{\mathbf{r}}'_\perp = \vec{\mathbf{r}}_\perp \tag{3.42}$$

Now, $\vec{\mathbf{r}}' = \vec{\mathbf{r}}'_\parallel + \vec{\mathbf{r}}'_\perp = a(\vec{\mathbf{r}}_\parallel - vt) + \vec{\mathbf{r}}_\perp$

$$= a\left(\frac{(\vec{\mathbf{r}} \cdot \vec{\mathbf{v}})\vec{\mathbf{v}}}{v^2} - \vec{\mathbf{v}}t\right) + \left(\vec{\mathbf{r}} - \frac{(\vec{\mathbf{r}} \cdot \vec{\mathbf{v}})\vec{\mathbf{v}}}{v^2}\right).$$

Rearranging the above expression, we get

$$\vec{\mathbf{r}}' = \vec{\mathbf{r}} - \vec{\mathbf{v}}t + (a-1)(\frac{\vec{\mathbf{v}}}{v^2})[(\vec{\mathbf{r}} \cdot \vec{\mathbf{v}}) - v^2 t]. \tag{3.43}$$

Similarly,

$$t' = a\left[t - \frac{(\vec{\mathbf{r}} \cdot \vec{\mathbf{v}})}{c^2}\right]. \tag{3.44}$$

The inverse transformations are

$$\vec{\mathbf{r}} = \vec{\mathbf{r}}' + \vec{\mathbf{v}}t' + (a-1)(\frac{\vec{\mathbf{v}}}{v^2})[(\vec{\mathbf{r}}' \cdot \vec{\mathbf{v}}) + v^2 t'].$$

$$t = a\left[t' + \frac{(\vec{\mathbf{r}}' \cdot \vec{\mathbf{v}})}{c^2}\right].$$

When $v \ll c$, $(a-1) \to 0$, then Eqs. (3.43) and (3.44) reduce to

$$\vec{\mathbf{r}}' = \vec{\mathbf{r}} - \vec{\mathbf{v}}t$$

$$t' = t.$$

This is the generalized form of the Galilean Transformation.

One can write the generalized Lorentz Transformation equations in component form:

$$x^1 = x\{1 + (a-1)\frac{v_x^2}{v^2}\} + y(a-1)\frac{v_x v_y}{v^2} + z(a-1)\frac{v_x v_z}{v^2} - av_x t \tag{3.45}$$

$$y^1 = x(a-1)\frac{v_x v_y}{v^2} + y\{1 + (a-1)\frac{v_y^2}{v^2}\} + z(a-1)\frac{v_y v_z}{v^2} - av_y t \tag{3.46}$$

$$z^1 = x(a-1)\frac{v_x v_z}{v^2} + y(a-1)\frac{v_y v_z}{v^2} + z\{1 + (a-1)\frac{v_z^2}{v^2}\} - av_z t \tag{3.47}$$

$$t^1 = a\left(t - \frac{xv_x + yv_y + zv_z}{c^2}\right). \tag{3.48}$$

[here, $v^2 = v_x^2 + v_y^2 + v_z^2$]

If the relative velocity v between two frames to be parallel to x-axis, then, $v_x = v$ and $v_y = v_z = 0$. Therefore, the above transformation equations reduce to

$$x_1 = a(x - v_x t); \quad y_1 = y; \quad z_1 = z; \quad t_1 = a\left[t - \left(\frac{xv_x}{c^2}\right)\right].$$

If the relative velocity v between two frames to be parallel to y-axis, then, $v_y = v$ and $v_x = v_z = 0$. Therefore, the above transformation equations reduce to

$$y_1 = a(y - v_y t); \quad x_1 = x; \quad z_1 = z; \quad t_1 = a\left[t - \left(\frac{yv_y}{c^2}\right)\right].$$

3.4 Thomas Precession

Consider three inertial frames S, S^1 and S^2. Let S^1 is moving with a constant velocity v_1 along x-axis of S and S^2 is moving with a constant velocity v_2 along y-axis of S^1. Then, the relative velocity between S and S^2 will not be parallel to any of the axes. Thus, there exists some rotation of axes of S^2 relative to those of S. The amount of rotation of axes is known as **Thomas Precession**.

The transformation equations between S and S^1 are given by

$$x^1 = a_1(x - v_1 t), \quad y^1 = y, \quad z^1 = z, \quad t^1 = a_1\left(t - \frac{v_1 x}{c^2}\right), \tag{3.49}$$

where $a_1 = \dfrac{1}{\sqrt{1 - \frac{v_1^2}{c^2}}}$.

Similarly, the transformation between S^1 and S^2 are given by

$$x^2 = x^1, \quad y^2 = a_2(y^1 - v_2 t^1), \quad z^2 = z^1, \quad t^2 = a_2\left(t^1 - \frac{v_2 y^1}{c^2}\right), \tag{3.50}$$

where, $a_2 = \dfrac{1}{\sqrt{1 - \frac{v_2^2}{c^2}}}$.

Using (3.49) in to (3.50), we get the required transformation between S and S^2 as

$$x^2 = a_1(x - v_1 t), \quad y^2 = a_2\left(y + \frac{v_1 v_2 a_1 x}{c^2} - v_2 a_1 t\right),$$
$$z^2 = z^1, \quad t^2 = a_2\left(a_1 t - \frac{v_1 a_1 x}{c^2} - \frac{v_2 y}{c^2}\right). \tag{3.51}$$

Let us assume that the relative velocity vector $\vec{u}$ between S and S^2 frames makes angles $90° - \theta_1$ and $90° - \theta_2$ with x and x^2 axes. This means normal to relative velocity vector $\vec{u}$ makes θ_1 and θ_1 angles with x- and x^2-axes. Therefore, Thomas Precession is the angle $\theta_2 - \theta_1$ (see Fig. 3.3). Due to Lorentz Transformation, the

Fig. 3.3 Thomas precession

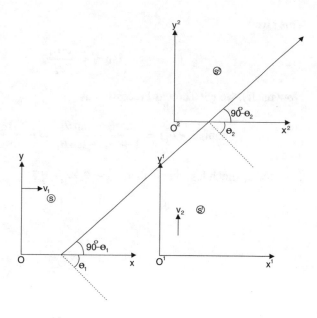

components of the position vectors $\overrightarrow{\mathbf{r}}$ and $\overrightarrow{\mathbf{r}^2}$ remain unchanged in a direction perpendicular to the relative velocity, therefore,

$$x \cos \theta_1 - y \sin \theta_1 = x^2 \cos \theta_2 - y^2 \sin \theta_2$$

$$= a_1 x \left(\cos \theta_2 - \frac{v_1 v_2 a_2 \sin \theta_2}{c^2} \right) - a_2 y \sin \theta_2 \qquad (3.52)$$

$$- a_1 t (v_1 \cos \theta_2 - v_2 a_2 \sin \theta_2).$$

[using the values of x^2 and y^2 from (3.51)]

Definitely, coefficient of t should vanish. Hence, we get

$$\tan \theta_2 = \frac{v_1}{v_2 a_2}. \qquad (3.53)$$

This gives

$$\sin \theta_2 = \frac{v_1}{D a_2}, \qquad \cos \theta_2 = \frac{v_2}{D}, \qquad (3.54)$$

where $D = v_1^2 + v_2^2 - \frac{v_1^2 v_2^2}{c^2}$.

From (3.54), we have

$$x \cos \theta_1 - y \sin \theta_1 = \frac{v_2 x}{D a_1} - \frac{v_1 y}{D}$$

This gives

$$\tan \theta_1 = \frac{v_1 a_1}{v_2}.$$
(3.55)

Now finally, we get Thomas Precession as

$$\tan(\theta_2 - \theta_1) = \frac{\tan \theta_2 - \tan \theta_1}{1 + \tan \theta_2 \tan \theta_1} = -\frac{v_1 v_2 (a_1 a_2 - 1)}{a_1 v_1^2 + v_2^2 a_2}.$$

If v_1 and v_2 much less that c, then, $\theta_2 - \theta_1 = -\frac{v_1 v_2}{2c^2}$.

Chapter 4
Mathematical Properties of Lorentz Transformations

The Lorentz Transformations have some interesting mathematical properties, particularly in the measurements of length and time. These new properties are not realized before.

4.1 Length Contraction (Lorentz–Fitzgerald Contraction)

Suppose there is a rod rest on the x-axis of S frame (see Fig. 4.1). Therefore, the coordinates of its end points are $A(x_1, 0, 0)$ and $B(x_2, 0, 0)$. Its length in this frame is $l = x_2 - x_1$. Let an observer fixed in the S^1 system which is moving uniformly along x-axis w.r.t. S system wants to measure the length of the given rod. He must find the coordinates of the two end points x_1^1 and x_2^1 in his system at the same time t^1. Thus, from the inverse Lorentz Transformation equations, we have

$$x_1 = a(x_1^1 + vt^1) \quad ; \quad x_2 = a(x_2^1 + vt^1)$$

so that

$$x_2 - x_1 = a(x_2^1 - x_1^1). \tag{4.1}$$

Therefore, the length of the rod in S^1 system is

$$l^1 = (x_1^1 - x_2^1) = l\sqrt{(1 - \beta^2)} \quad \text{[here, } a = \frac{1}{\sqrt{(1 - \beta^2)}} \quad , \quad \beta = \frac{v}{c}]. \tag{4.2}$$

Thus, to the moving observer, rod appears to be contracted. In other words, the length of the rod is greatest in the reference S, i.e. in the rest frame.

© The Author(s), under exclusive license to Springer Nature Singapore Pte Ltd. 2022
F. Rahaman, *The Special Theory of Relativity*, La Matematica per il 3+2 136,
https://doi.org/10.1007/978-981-19-0497-4_4

Fig. 4.1 Rod's length in S frame is $x_2 - x_1$ and in S^1 frame is $x_2^1 - x_1^1$

Alternative

Now, we place the rod at rest in S^1 frame with the space coordinates $(x_1^1, 0, 0)$ and $(x_2^1, 0, 0)$. If the observer on S frame measures the length of this rod, then he finds the coordinates of the two end points x_1 and x_2 in his system at the same time t. Now, using direct Lorentz Transformations equations, one gets

$$x_1^1 = a(x_1 - vt) \quad ; \quad x_2^1 = a(x_2 - vt)$$

so that

$$l^1 = x_1^1 - x_2^1 = a(x_1 - x_2) = al. \tag{4.3}$$

Therefore, the length of the rod in S system is

$$l = l^1\sqrt{(1 - \beta^2)}. \tag{4.4}$$

This proves that the length of a rod, i.e. rigid body appears to be contracted in the ratio, $\sqrt{(1 - \beta^2)} : 1$ in the direction of the relative motion in all other coordinate system, v being the relative velocity. This result is known as *Lorentz–Fitzgerald contraction* in special theory of relativity.

Proper length: The proper length of the rod is its length in a reference system in which it is at rest.

Note 1: The reader should note that we have used here inverse Lorentz Transformation equations. It would not be convenient to use direct equation since events (location of end points) that are simultaneous in primed system (ends are measured at the same time t^1) will not be simultaneous in unprimed system as they are at different points x_1 and x_2. As we are measuring from S^1 system, we do not know the value of x and t in S system. So, we have to use inverse Lorentz Transformation equations.

Note 2: We note that the contraction depends on v^2. Therefore, if one replaces v by $-v$, the contraction result remains the same. Therefore, the Lorentz–Fitzgerald contraction law is independent of the direction of the relative velocity between the systems.

Note 3: The proper length is the maximum length, i.e. every rigid body appears to have its maximum length in the coordinate system in which it is at rest.

Note 4: When the rod moves with a velocity relative to the observer, then rod's length remains unaffected in the perpendicular direction of motion.

Proof Let, S^1 is moving with velocity v along x-axis relative to a system S. Let a rod of length l is placed on y-axis. The coordinates of its ends are $(0, y_1, 0)$ and $(0, y_2, 0)$. Suppose, in S^1 system, the coordinates of its ends are $(0, y_1^1, 0)$ and $(0, y_2^1, 0)$. Now, using the Lorentz Transformation equations, one gets,

$$l^1 = y_2^1 - y_1^1 = y_2 - y_1 = l.$$

This means length of the rod remains unaltered when an observer in S^1 system measures it. In other words, the lengths perpendicular to the direction of motion remains unaffected.

4.2 Time Dilation

Suppose there is a clock A rest at point $(x, 0, 0)$ in S frame. Let this clock gives a signal at time t_1 in S frame. An observer fixed in the S^1 system which is moving uniformly along x-axis w.r.t. S system measures this time as t_1^1. According to the Lorentz Transformation equations,

$$t_1^1 = a\left[t_1 - \frac{vx}{c^2}\right]$$

Let the clock A fixed at the same point $(x, 0, 0)$ in S system gives another signal at time t_2. The observer in S^1 system notes this time as

$$t_2^1 = a\left[t_2 - \frac{vx}{c^2}\right].$$

Therefore, the apparent time interval is

$$t_2^1 - t_1^1 = a\left[t_2 - t_1\right]. \tag{4.5}$$

This shows that

$$(t_2^1 - t_1^1) > (t_2 - t_1). \tag{4.6}$$

Hence, it appears to the moving observer that the stationary clock is moving at a slow rate. This effect is called *time dilation*.

Alternative

Suppose there is a clock B that is rigidly attached at point $(x^1, 0, 0)$ in S^1 frame which is moving uniformly along x-axis w.r.t. S system. Let this clock gives a signal at time t_1^1 in S^1 frame. A clock in S system measures this time as t_1. Thus, from the inverse Lorentz Transformation equations, we have

$$t_1 = a\left[t_1^1 + \frac{vx^1}{c^2}\right].$$

Therefore, time interval $(t_2 - t_1)$ recorded by the clock at S frame is related to the interval $(t_2^1 - t_1^1)$, recorded by the clock in S^1 frame as follows:

$$(t_2 - t_1) = a(t_2^1 - t_1^1). \tag{4.7}$$

Obviously,

$$(t_2 - t_1) > (t_2^1 - t_1^1). \tag{4.8}$$

If $(t_2^1 - t_1^1)$ is 1 s, then the observer in S judges that more than 1 s have elapsed between two events. Thus, the moving clock appears to be slowed down or retarded.

Proper time: The time recorded by a clock moving with a given system is called *proper time*.

Note 5: Note that if we fix the clock in the moving (primed) system or rest (unprimed) system even then the conclusion is true, i.e. observer reports that the clock (in unprimed or primed system) is running slow as compared to his own. Therefore by recording the time on a clock fixed in a certain system by an observer fixed another system, it is difficult to distinguish which one of the systems is stationary or in motion; only a relative motion can be guessed.

Note 6: A clock will be found to run more and more slowly with the increase of the relative motion between clock and observer.

Note 7: (i) If moving observer measures the time of a stationary clock, then use

$$\Delta t^1 = a\Delta t.$$

(ii) If stationary observer measures the time of a moving clock, then use

$$\Delta t = a\Delta t^1.$$

In general, time dilation

$$t = \frac{t_0}{\sqrt{1 - \frac{v^2}{c^2}}}.$$

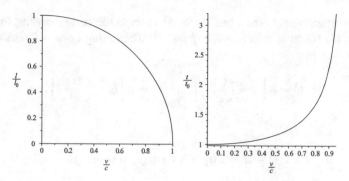

Fig. 4.2 Variations of length contraction (*left*) and time dilation (*right*) with respect to $\frac{v}{c}$ are shown in the figures

t_0 = time interval on clock at rest relative to an observer = proper time.
t = time interval on clock in motion relative to an observer.
v = relative speed.

Note 8: If the moving observer observes the rod, then

$$l = l_0\sqrt{1 - \frac{v^2}{c^2}}.$$

If the moving observer measures the time, then

$$t = \frac{t_0}{\sqrt{1 - \frac{v^2}{c^2}}}.$$

Note 9: Figure 4.2 indicates that for moving observer, the length is gradually decreasing with the increase of its velocity and when velocity is approaching to the velocity of light, the measurement of length is tending to zero. However, for time dilation, the nature is just opposite to the length contraction. That is, time interval is gradually increasing with the increase of observer's velocity and when velocity is approaching to the velocity of light, the time interval will be infinitely large.

4.3 Relativity of Simultaneity

Two events are said to be simultaneous if they occur at the same time. It can be shown that two events occur at a same time in one frame of reference S are not in general simultaneous w.r.t. another frame of reference S^1 which is in relative motion w.r.t. the previous frame. Let us consider two events happening at times t_1 and t_2 in S frame at rest. Suppose an observer in S^1 is observing these two events at two different points

x_1 and x_2, respectively. The observer in S^1 records the corresponding times with respect to his frame of reference are t_1^1 and t_2^1. Then using Lorentz Transformation equations, one gets

$$t_1^1 = a\left[t_1 - \left(\frac{vx_1}{c^2}\right)\right] \quad ; \quad t_2^1 = a\left[t_2 - \left(\frac{vx_2}{c^2}\right)\right]. \tag{4.9}$$

Thus,

$$t_2^1 - t_1^1 = a(t_2 - t_1) + a\left(\frac{v}{c^2}\right)(x_1 - x_2). \tag{4.10}$$

If the two events are simultaneous in the frame of reference S, then $t_2 = t_1$. Therefore, the above expression gives

$$t_2^1 - t_1^1 = a\left(\frac{v}{c^2}\right)(x_1 - x_2) \neq 0, \quad \text{since } x_1 \neq x_2. \tag{4.11}$$

Thus, if the events are simultaneous in S frame, then, in general, the same events are not simultaneous in S^1 frame which is moving with velocity v relative to S. This indicates that simultaneity is not absolute, but relative.

4.4   Twin Paradox in Special Theory of Relativity

Let Rina and Mina, each 20 years of age, are two twins. Suppose Mina travels within a spaceship with velocity, $v = \frac{3}{5}c$ for 4 years along positive direction of x-axis, whereas Rina stays in Kolkata. Therefore, one can imagine, Rina is staying in a rest frame, say S, and the spaceship containing Mina is moving as a moving frame S^1. Note that Mina counts 4 year in her reference frame, i.e. in S^1. Then, Mina returns with same speed and reaches the starting point after the end of another 4 years of her time. Then according to Mina, the age of Mina is $20 + 8 = 28$ years, but according to Rina (i.e. with respect to rest frame S), the age of Mina is

$$20 + \frac{8}{\sqrt{1 - (\frac{3c}{5c})^2}} = 30 \text{ years.}$$

On the other hand, the age of Rina according to her (i.e. S frame) would be 30 years, whereas the age of Mina according to her (i.e. in S^1 frame) would be 28 years. These two statements are contradict to each other. That is, one of the two twins appear younger than the other. This is known as *twin paradox* in special theory of relativity. According to special theory of relativity, Mina will be 28 years old, while Rina 30 years. That is Mina is younger than Rina.

Note 10: The two situations are not equivalent. Mina changed from one inertial frame S to another inertial frame S^1 of her journey, however, Rina remained in the same inertial frame during Mina's journey. Therefore, the time dilation formula applies to Mina's movement from Rina.

Note 11: After travelling 4 years, the velocity of Mina changes from v to $-v$ instantly to reach the Kolkata. However, the calculation of time remains the same.

Note 12: The distance Mina covers should be shorten as compare to Rina, i.e. compare to the distance measured by S frame. Let us consider Mina reaches to a star 4 light years away from earth (at the beginning of her journey origins of both frames coincided). Mina covers the distance which is shorten to

$$L = L_0 \sqrt{\left(1 - \frac{v^2}{c^2}\right)} = (4 \text{ light years}) \sqrt{\left(1 - \frac{9}{25}\right)} = 3.2 \text{ light years.}$$

Note 13: The clock could be substituted for any periodic natural phenomena, i.e. process of life such as heartbeat, respiration, etc. For, Rina who stays at rest, Mina's life (who is moving in a high speed) is slower than her by a factor

$$\sqrt{\left(1 - \frac{v^2}{c^2}\right)} = \sqrt{\left(1 - \frac{9}{25}\right)} = 80\%.$$

It follows that in a certain period, Mina's heart beats only four times, whereas Rina's heart beats five times within that period. Mina takes 4 breaths for every 5 of Rina. Mina thinks 4 thoughts for every 5 of Rina. Thus, during the travel, Mina's biological functions proceeding at the same slower rate as her physical clock.

4.5 Car–Garage Paradox in Special Theory of Relativity

Consider a car and a garage with same proper length, l_0. When at rest, car be parked exactly inside the garage. Now, a person driving the car towards the garage with speed v. For the gate man, the car appears to be of length

$$l = l_0 \sqrt{\left(1 - \frac{v^2}{c^2}\right)} < l_0.$$

He realizes that the car smoothly enters the garage and the driver does not need to stop the car before the garage. Now, we see what will happen for the driver. For him, the car length is l_0, while the garage is of length

$$l = l_0 \sqrt{\left(1 - \frac{v^2}{c^2}\right)} < l_0.$$

Therefore, he realizes that the garage length is smaller than the car and he stops the car before the garage. This is known as car–garage paradox.

4.6 Real Example of Time Dilation

The elementary particle Muons are found in cosmic rays. It has a mass 207 times that of the electron (e) and has a charge of either $+e$ or $-e$. These Muons collide with atomic nuclei in earth's atmosphere and decay into an electron, a muon neutrino and an anti-neutrino:

$$\mu^{\pm} \longrightarrow e^{\pm} + \nu_{\mu} + \tilde{\nu}_{\mu}.$$

The Muon has speed of about $0.998c$ and mean lifetime in the rest frame is

$$t_0 = 2.2 \times 10^{-6}\,\text{s} = 2.2\,\mu\text{s}$$

These unstable elementary particles muons collide with atomic nuclei in earth's atmosphere at high altitudes and destroy within 2.2 μ s, however, they can be seen in the sea level in profusion. But in $t_0 = 2.2$ μ s, muons can travel a distance of only

$$vt_0 = 0.998 \times 3 \times 10^8 \times 2.2 \times 10^{-6} = 0.66\,\text{km}$$

before decaying, whereas the collisions occurred at altitudes of 6 km or more.

Now, we have the following problems: (i) either Muon's mean lifetime is more than 2.2 μ s or (ii) its velocity is more than c, the velocity of light. The second possibility is impossible as it violates the principle of theory of relativity. Since the cosmic rays Muons are moving violently towards earth with a high speed, $0.998c$, their lifetimes are extended in our frame of reference (rest frame) by time dilation to

$$t = \left[\frac{t_0}{\sqrt{1 - \left(\frac{v}{c}\right)^2}}\right] = \left[\frac{2.2 \times 10^{-6}}{\sqrt{1 - \left(\frac{0.998c}{c}\right)^2}}\right] = 34.8 \times 10^{-6}\,\text{s} = 34.8\,\mu\text{s}.$$

We note that the moving muon's mean lifetime almost 16 times longer than the proper lifetime, i.e. that at rest. Thus, the Muon will traverse the distance

$$vt = 0.998 \times 3 \times 10^8 \times 34.8 \times 10^{-6} = 10.4\,\text{km}.$$

Fig. 4.3 Distance is contracted for muons

Therefore, we conclude that a muon can reach the ground from altitudes about 10.4 km in spite its lifetime is only 2.2 μ s.

Note 14: The lifetime of Muon is only 2.2 μ s in its own frame of reference, i.e. moving frame of reference. The lifetime of Muon is 34.8 μ s in the frame of reference where these altitudes are measured, i.e. with respect to the rest frame of reference.

Note 15: Let an observer moving with the muon. This means the observer and muon are now in the same reference frame. Since the observer is moving relative to earth, he will find the distance 10.4 km to be contracted as (see Fig. 4.3)

$$L = L_0\sqrt{1 - \left(\frac{v}{c}\right)^2} = (10.4\,\text{km})\sqrt{1 - \left(\frac{0.998c}{c}\right)^2} = 0.66\,\text{km}.$$

Therefore, a muon will travel with the speed of about 0.998c in 2.2 μ s.

4.7 Terrell Effects

In 1959, James Terrell observed that if one takes a snapshot of a rapidly moving object with respect to the film, then the photograph may give a distorted picture, i.e. the object has gone a rotation rather than contraction. This phenomena is known as Terrell effects. Terrell proved that the shape of the moving body would remain unchanged to the observer, however, it would appear only to be rotated a little. This phenomena is happened due to combination of two different effects: (i) the length contraction effect (ii) the retardation effect. Second effect will arise because the light ray from different parts of the object take different times in reaching the observer's eyes. Hence, when a body is in motion, a distorted image is formed due to the combination of the above two effects.

Let us suppose that a square body flies past an observer with a constant velocity v at a large distance from the observer. Let l be length of each side. By length contraction effect, the image of front side will have a length $l\sqrt{1 - \frac{v^2}{c^2}}$. By retardation effect, light rays from the far side must originate $\frac{l}{c}$ time earlier than the light rays originating from the front face so as to reach the observer at the same observer time. If one takes a photo of this moving body, the snapshot shows the near side and far side as it was $\frac{l}{c}$ earlier when it was a distance $\frac{vl}{c}$ to the left. This means side face of the square is seen inclined at an angle θ with the front side, where $\tan \theta = \frac{\frac{vl}{c}}{l} = \frac{v}{c}$.

Example 4.1 A particle with mean proper lifetime of 2 μ s, moves through the laboratory with a speed of $0.9c$. Calculate its lifetime as measured by an observer in laboratory. What will be the distance traversed by it before decaying? Also find the distance traversed without taking relativistic effect.

Hint: We use the result of time dilation, $\Delta t^1 = \frac{\Delta t}{\sqrt{1 - \frac{v^2}{c^2}}}$.

Given: $\Delta t = 2 \mu s = 2 \times 10^{-6}$, $v = 0.9c = 0.9 \times 3 \times 10^8$. Therefore, $\Delta t^1 = 4.58 \mu$ s.
Distance traversed $= 0.9 \times 3 \times 10^8 \times 4.58 \times 10^{-6}$.
Distance traversed without relativistic effect $= 0.9 \times 3 \times 10^8 \times 2 \times 10^{-6}$.

Example 4.2 The proper lifetime of π^+ mesons is $\alpha \mu$ s. If a beam of these mesons of velocity f c produced ($f < 1$), calculate the distance, the beam can travel before the flux of the mesons beam is reduced to $e^{-\beta}$ times the initial flux, a being a constant.

Hint: Given: $\Delta t = \alpha$, $v = fc$. Therefore, $\Delta t^1 = \frac{\alpha}{\sqrt{1 - \frac{f^2 c^2}{c^2}}}$, etc.

If N_0 is the initial flux and N is the flux after time t, then we have $N = N_0 e^{-\frac{t}{\tau}}$, τ being the mean lifetime.
Here, $N = e^{-\beta} N_0$, therefore, $e^{-\beta} N_0 = N_0 e^{-\frac{t}{\tau}} \implies t = \beta \tau = \beta \Delta t^1 = \frac{\alpha \beta}{\sqrt{1 - f^2}}$.
Hence, the distance traversed by the beam before the flux of mesons beam is reduced to $e^{-\beta}$ times the initial flux is $d = t \times fc$, etc.

Example 4.3 At what speed should a clock be moved so that it may appear to loose α seconds in each minute.

Hint: If the clock is to loose α seconds in each minute, then the moving clock with velocity v records $60 - \alpha$ s for each 60 s recorded by stationary clock.
We use the result of time dilation, $\Delta t^1 = \frac{\Delta t}{\sqrt{1 - \frac{v^2}{c^2}}}$.
Given: $\Delta t = 60$ s and $\Delta t^1 = 60 - \alpha$ s, etc.

Example 4.4 Calculate the length and orientation of a rod of length 1 metre in a frame of reference which is moving with velocity fc ($f < 1$) in a direction making an angle θ with the rod.

Hint: Let the rod is moving with velocity fc along x-axis and the rod is inclined at the θ angle with x-axis. We know contraction will take place only in x-direction and not in y-direction. Here, $l_x = l \cos \theta$ and $l_y = l \sin \theta$. Now, using the result of length contraction ($L = L_0\sqrt{1 - \frac{v^2}{c^2}}$), we have $l_x^1 = l_x\sqrt{1 - \frac{f^2c^2}{c^2}} = l\cos\theta\sqrt{1 - f^2}$, $l_y^1 = l_y = l\sin\theta$. Therefore, the length of the rod as observed from moving frame, $l^1 = \sqrt{(l_x^1)^2 + (l_y^1)^2}$, etc.

If θ^1 is the angle made by the rod with x-axis in moving frame, then $\tan\theta^1 = \frac{l_y^1}{l_x^1}$, etc.

Example 4.5 An astronaut wants to go to a star α light year away. The rocket accelerates quickly and then moves at a uniform velocity. Calculate with what velocity the rocket must move relative to earth if the astronaut is to reach therein β year as measured by clock at rest on the rocket.

Hint: The distance of the star from earth is $= \alpha yc$, where $y = 365 \times 24 \times 3,600$ s. In the frame of reference of the astronaut (i.e. S^1), the time required is given by

$$t_2^1 - t_1^1 = \frac{(t_2 - t_1) - \frac{v}{c^2}(x_2 - x_1)}{\sqrt{1 - \frac{v^2}{c^2}}}.$$

Here, $(t_2 - t_1)$ is the time difference measured by the clock in earth and $t_2^1 - t_1^1$ is the time difference measured by the clock in the rocket, i.e. moving frame. Let $v = fc$ ($f < 1$) be the required velocity of rocket to reach the distance $x_2 - x_1 = \alpha yc$ in $t_2^1 - t_1^1 = \beta y$ year. Therefore, $x_2 - x_1 = \alpha yc = v(t_2 - t_1)$. Now using these, we get

$$\beta y = \frac{\frac{\alpha yc}{v} - \frac{v}{c^2}(\alpha yc)}{\sqrt{1 - \frac{v^2}{c^2}}} = \frac{\frac{\alpha yc}{fc} - \frac{fc}{c^2}(\alpha yc)}{\sqrt{1 - \frac{f^2c^2}{c^2}}}, \quad etc.$$

Example 4.6 A flash of light is emitted at the position x_1 on x-axis and is absorbed at position $x_2 = x_1 + l$. In a reference frame S^1 moving with velocity $v = \beta c$ along x-axis, find the spatial separation between the point of emission and the point of absorption of light.

Hint:
S frame
$$O - - - - - - \bullet x_1 - - - - - \bullet x_2 - - - - - - - > x$$
Here, the frame S^1 is moving with velocity $v = \beta c$.

In S frame, let the light is emitted on the position x_1 at time $t_1 = 0$ and will absorb on the position $x_2 = x_1 + l$ at time t_2, where $t_2 = \frac{(x_1 + l) - x_1}{c} = \frac{l}{c}$.

In S^1 frame, the points x_1 and x_2 will be

$$x_1^1 = \frac{x_1 - vt_1}{\sqrt{1 - \beta^2}} = \frac{x_1}{\sqrt{1 - \beta^2}}, \quad x_2^1 = \frac{x_2 - vt_2}{\sqrt{1 - \beta^2}} = \frac{(x_1 + l) - \frac{vl}{c}}{\sqrt{1 - \beta^2}}.$$

Therefore, the spatial separation is $l^1 = x_2^1 - x_1^1$, etc.

Example 4.7 Consider that a frame S^1 is moving with velocity v relative to the frame S. A rod in S^1 frame makes an angle θ^1 with respect to the forward direction of motion. What is the angle θ as measured in S frame?

Hint: Here, x-component of the length will be contracted relative to an observer in S frame, etc., see Example 4.4.

Example 4.8 Two rods which are parallel to each other move relative to each other in their length directions. Explain the apparent paradox that either rod may appear longer than the other depending on the state of motion of the observer.

Hint: Let us place a rod of length l_0 at rest on the x-axis of S frame. Consider two frames of reference S^1 and S^2 which are moving with velocities v_1 and v_2, respectively, relative to S frame. To the observer in S^1 frame, the rod appears as

$$l^1 = l_0 \sqrt{1 - \frac{v_1^2}{c^2}}. \tag{1}$$

Similarly, to the observer in S^2 frame, the rod appears as

$$l^2 = l_0 \sqrt{1 - \frac{v_2^2}{c^2}}. \tag{2}$$

Case 1: If $v_2 > v_1$, then $\frac{v_2^2}{c^2} > \frac{v_1^2}{c^2}$. Since $\frac{v_2^2}{c^2} = 1 - \frac{l_2^2}{l_0^2}$ and $\frac{v_1^2}{c^2} = 1 - \frac{l_1^2}{l_0^2}$, therefore $l_1 > l_2$.

Case 2: If $v_1 > v_2$, then $\frac{v_1^2}{c^2} > \frac{v_2^2}{c^2}$. As a result, $l_2 > l_1$.

Thus, either rod may appear longer than the other depending on the state of motion of the observer.

Example 4.9 Four-dimensional volume element 'dxdydzdt' is invariant under Lorentz Transformation.

Hint: According to Lorentz–Fitzgerald contraction, we have

$$dx^1 = dx \sqrt{1 - \frac{v^2}{c^2}}, \quad dy^1 = dy, \quad dz^1 = dz. \tag{1}$$

According to time dilation,

$$dt^1 = \frac{dt}{\sqrt{1 - \frac{v^2}{c^2}}}. \tag{2}$$

The four-dimensional volume element in S^1 frame is $dx^1 dy^1 dz^1 dt^1$, etc.

Example 4.10 Two events are simultaneous, though not coincide in some inertial frame. Prove that there is no limit on time separation assigned to these events in different frames but the space separation varies from ∞ to a minimum which is the measurement in that inertial frame.

Hint: From Lorentz Transformation, we have

$$x_2 - x_1 = \frac{(x_2^1 - x_1^1) + v(t_2^1 - t_1^1)}{\sqrt{1 - \frac{v^2}{c^2}}} \tag{1}$$

$$t_2 - t_1 = \frac{(t_2^1 - t_1^1) + \frac{v}{c^2}(x_2^1 - x_1^1)}{\sqrt{1 - \frac{v^2}{c^2}}}. \tag{2}$$

Two events are simultaneous but not coincide, therefore, $t_2^1 - t_1^1 = 0$ and $x_2^1 - x_1^1 \neq 0$. Thus, (1) implies

$$x_2 - x_1 = \frac{(x_2^1 - x_1^1)}{\sqrt{1 - \frac{v^2}{c^2}}}. \tag{3}$$

Now, if $v = 0$, then $x_2 - x_1 = x_2^1 - x_1^1$ and $v = c, x_2 - t_1 = \infty$.

Thus, space separation varies from ∞ to a minimum which is the measurement in the rest frame.

Also, from (2), we get

$$t_2 - t_1 = \frac{\frac{v}{c^2}(x_2^1 - x_1^1)}{\sqrt{1 - \frac{v^2}{c^2}}}. \tag{4}$$

If $v = 0$, then $t_2 - t_1 = 0$ and $v = c, t_2 - t_1 = \infty$.

Thus, there is no limit for time separation.

Example 4.11 A circular ring moves parallel to its plane relative to an inertial frame S. Show that the shape of the ring relative to S is an ellipse.

Hint: Let us consider a frame S^1 containing the ring is moving with velocity v relative to S frame in x-direction. Obviously, S^1 frame is the rest frame with respect to the ring. Let $(X, Y, 0, T)$ and $(X^1, Y^1, 0, T^1)$, respectively, be the coordinates of the centre of the ring relative to S and S^1. Consider $(x, y, 0, T)$, $(x^1, y^1, 0, T^1)$, respectively, be the coordinates of any point on the circumference of the ring relative to S and S^1 at the same observer time T. (here, ring lies in xy-plane.)

From Lorentz Transformation, we have

$$X^1 = \frac{X - vT}{\sqrt{1 - \frac{v^2}{c^2}}}, \quad x^1 = \frac{x - vT}{\sqrt{1 - \frac{v^2}{c^2}}}, \quad Y^1 = Y, \quad y^1 = y. \tag{1}$$

Equation of the ring is in S^1 frame is

$$(x^1 - X^1)^2 + (y^1 - Y^1)^2 = a^2 \tag{2}$$

a being the radius of the circular ring.

Using the transformation (1), we get

$$\frac{(x - X)^2}{1 - \frac{v^2}{c^2}} + (y - Y)^2 = a^2. \tag{3}$$

This is an ellipse as observed in S frame. The centre of the ellipse moves along x-axis with velocity v.

Example 4.12 A reference frame S^1 containing a body has the dimensions represented by $(a\hat{i} + b\hat{j} + c\hat{k})$ is moving with velocity f c $(0 < f < 1)$ along x-axis relative to a reference system S. How these dimensions will be represented in system S. $(\hat{i}, \hat{j}, \hat{k})$ are unit vectors along respective axes.

Hint: The dimension of the body along x-axis as viewed from S system is $a^1 = a\sqrt{1 - f^2}$. However, there are no contractions in the directions of y- and z-axes, etc.

Example 4.13 In the usual set-up of frames S^1 and S having relative velocity v along $x - x^1$, the origins coinciding at $t = t^1 = 0$, we shall find a later time, t, there is only one plane S on which the clocks agree with those of S^1. Show that this plane is given by $x = \frac{c^2}{v}\left[1 - \sqrt{1 - \frac{v^2}{c^2}}\right] t$ and moves with velocity $u = \frac{c^2}{v}\left[1 - \sqrt{1 - \frac{v^2}{c^2}}\right]$. Show also, this velocity is less than v.

Hint: Here, the frame S^1 is moving with velocity v relative to S frame in x-direction. Let (x, y, z, t) and (x^1, y^1, z^1, t^1), respectively, be the coordinates of an event relative to S and S^1. Then from Lorentz Transformation, we have

$$t = \frac{t^1 + \frac{vx^1}{c^2}}{\sqrt{1 - \frac{v^2}{c^2}}}, \quad x = \frac{x^1 + vt^1}{\sqrt{1 - \frac{v^2}{c^2}}}. \tag{1}$$

Considering the situation that clock in S frame agrees with the clock in S^1 frame, i.e. when $t = t^1$, then from the second expression in (1), we get

$$x^1 = x\sqrt{1 - \frac{v^2}{c^2}} - vt. \tag{2}$$

Using (2), we get from the first expression in (1) as

$$t\sqrt{1 - \frac{v^2}{c^2}} = t + \frac{v}{c^2}\left[x\sqrt{1 - \frac{v^2}{c^2}} - vt\right].$$

This gives the equation of the plane $x = \frac{c^2}{v}\left[1 - \sqrt{1 - \frac{v^2}{c^2}}\right]t$ and moves with velocity $u = \frac{c^2}{v}\left[1 - \sqrt{1 - \frac{v^2}{c^2}}\right]$.

Expanding binomially and neglecting higher power of $\frac{v^2}{c^2}$, we get

$$u = \frac{c^2}{v}\left[1 - \left(1 - \frac{v^2}{2c^2}\right)\right], \quad \text{etc.}$$

Example 4.14 Is it true that two events which occur at the same place and at the same time for one observer will be simultaneous for all observers?

Hint: Use

$$t_2^1 - t_1^1 = \frac{(t_2 - t_1) - \frac{v}{c^2}(x_2 - x_1)}{\sqrt{1 - \frac{v^2}{c^2}}} \quad \text{etc.}$$

Example 4.15 The S and S^1 are two inertial frames of reference. Further, at any instant of time, the origin of S^1 is situated at the position $(vt, 0, 0)$. A particle describes a parabolic trajectory, $x = ut, y = \frac{1}{2}ft^2$ in S frame. Find the trajectory of the particle in S^1 frame. Show also that its acceleration in S^1 frame along y^1-axis is $\frac{f\left(1 - \frac{v^2}{c^2}\right)}{\left(1 - \frac{uv}{c^2}\right)^2}$.

Hint: Since S and S^1 are inertial frames, the relative velocity between them is constant. Also, at t, the origin of S^1 is situated at the position $(vt, 0, 0)$, therefore, S^1 must move with a velocity v along x-axis. From, Lorentz Transformation, we have

$$t = \frac{t^1 + \frac{vx^1}{c^2}}{\sqrt{1 - \frac{v^2}{c^2}}}, \quad x = \frac{x^1 + vt^1}{\sqrt{1 - \frac{v^2}{c^2}}}.$$

Now, $x = ut$ implies

$$\frac{x^1 + vt^1}{\sqrt{1 - \frac{v^2}{c^2}}} = u\left[\frac{t^1 + \frac{vx^1}{c^2}}{\sqrt{1 - \frac{v^2}{c^2}}}\right].$$

This gives

$$x^1 = \left[\frac{u - v}{1 - \frac{uv}{c^2}}\right]t^1.$$

Hence, the particle moves with a uniform velocity $\frac{u-v}{1-\frac{uv}{c^2}}$ along x^1-axis.

Again,

$$y = y^1 = \frac{1}{2}ft^2 = \frac{1}{2}f\left[\frac{t^1 + \frac{vx^1}{c^2}}{\sqrt{1 - \frac{v^2}{c^2}}}\right]^2 = \frac{1}{2}f\left[t^1 + \frac{v\left(\frac{u-v}{1-\frac{uv}{c^2}}t^1\right)}{c^2}\right]^2 \bigg/ \sqrt{1 - \frac{v^2}{c^2}}.$$

This gives

$$y^1 = \left[\frac{f\left(1 - \frac{v^2}{c^2}\right)}{2\left(1 - \frac{uv}{c^2}\right)^2}\right]t^{12}.$$

Hence, the acceleration along y^1 axis $f_y^1 = \frac{d^2 y^1}{d^2 t^1}$, etc.

Example 4.16 The S and S^1 are two inertial frames of reference. Further, at the instant of time, the origin of S^1 is situated at the position $(vt, 0, 0)$. Trajectory of a particle describes a straight line through the origin in S frame. Find the trajectory of the particle in S^1 frame. Show also that it has no acceleration in S^1 frame.

Hint: The parametric equations of the straight line passing through origin is, $x = t$, $y = mt$. See Example 4.15.

Example 4.17 A rocket of proper length L metres moves vertically away from earth with uniform velocity. A light pulse sent from earth and is reflected from the mirrors at the tail end and the front end of the rocket. If the first pulse is received at the ground $2T$ s after emission and second pulse is received $2f$ μ s later. Calculate (i) the distance of the rocket from earth (ii) the velocity of the rocket relative to earth and (iii) apparent length of the rocket.

Hint: (i) The pulse sent from earth reaches the back part of the rocket T s after its emission, hence at this instant, the distance of the back part of the rocket from earth $= Tc$ metre (see Fig. 4.4).

(ii) Let the tail end and the front end of the rocket be characterized by (x_1, t_1), (x_2, t_2) and (x_1^1, t_1^1), (x_2^1, t_2^1) in earth frame S and in the rocket frame S^1, respectively. From Lorentz Transformation, we have

$$t_2 - t_1 = \frac{(t_2^1 - t_1^1) + \frac{v}{c^2}(x_2^1 - x_1^1)}{\sqrt{1 - \frac{v^2}{c^2}}}. \tag{1}$$

Here, $x_2^1 - x_1^1 = L, t_2^1 - t_1^1 = \frac{x_2^1 - x_1^1}{c} = \frac{L}{c}, t_2 - t_1 = f \times 10^{-6}$. Therefore, from (1), one can get

Fig. 4.4 Distance of the
back part of the rocket from
Earth is Tc metre

$$f \times 10^{-6} = \frac{(\frac{L}{c}) + \frac{v}{c^2}(L)}{\sqrt{1 - \frac{v^2}{c^2}}} \implies v = \text{etc.} \tag{2}$$

(iii) The apparent length of the rocket $l = L\sqrt{1 - \frac{v^2}{c^2}}$, using the value of v from (2), one can get the required apparent length.

Example 4.18 Let, V_0 is the volume of a cube in reference system S. Find the volume of the cube as viewed from S^1 frame which is moving with velocity v along one edge of the cube.

Hint: Let L_0 be the length of one edge of the cube in reference system S. Then, $V_0 = L_0^3$. Let S^1 is moving along one edge of the cube parallel to x-axis. Therefore, the length of the edge is contracted by a factor $\sqrt{1 - \frac{v^2}{c^2}}$ in the direction of motion, i.e. parallel to the x-axis. The other edges remain unaffected. So, volume of the cube as viewed from S^1 frame is $V = L_0 L_0 L_0 \sqrt{1 - \frac{v^2}{c^2}} = V_0 \sqrt{1 - \frac{v^2}{c^2}}$.

Example 4.19 Show that Lorentz Transformation equations are symmetric in x and ct.

Hint: We know

$$x' = a(x - vt), \quad y' = y, \quad z' = z, \quad t' = a\left(t - \frac{vx}{c^2}\right), \quad a = \frac{1}{\sqrt{1 - \frac{v^2}{c^2}}}.$$

Here,

$$x' = a\left(x - \frac{v}{c}ct\right).$$

Now interchanging x and ct, we get

$$ct' = a\left(ct - \frac{v}{c}x\right), \qquad [as, \ x = ct, \ hence, \ x' = ct']$$

or,

$$t' = a\left(t - \frac{vx}{c^2}\right).$$

Again,

$$ct' = a\left(ct - \frac{vx}{c}\right)$$

So,

$$x' = a\left(x - \frac{v}{c}ct\right).$$

Example 4.20 Suppose radius of our galaxy is 3×10^4 light years. A person wants to make a trip from the centre to the edge of our galaxy is 10^{20} years. What will be the velocity of the person?

Hint: Suppose the velocity of the man is v. The actual radius of our galaxy is contracted during the trip of the man. Therefore,

$$v \times 20 \times 365 \times 24 \times 60 \times 60 = 3 \times 10^4 \times 365 \times 24 \times 60c\sqrt{1 - \frac{v^2}{c^2}}.$$

Example 4.21 An elliptical ring moves parallel to the major axis relative to an inertial frame S. Find the speed v for which the ring becomes circular relative to S.

Hint: Let $(X, Y, 0, T)$ and $(X', Y', 0, T_1')$ be the co-ordinates of the centre of the ellipse (assumed $X - Y$-plane as the plane of the ellipse) relative to frame S and S'.
Suppose $(x, y, 0, T)$ and $(x', y', 0, T_2')$ be the coordinates of any point on the elliptical ring relative to S and S' at the same observer time T. So,

$$X' = \nu(X - vT), \quad x' = \nu(x - vT), \quad y' = y, \quad Y' = Y,$$

where $\nu = \dfrac{1}{\sqrt{1 - \frac{v^2}{c^2}}}$.

Now, equation of the ellipse is

$$\frac{(x' - X')^2}{a^2} + \frac{(y' - Y')^2}{b^2} = 1.$$

Hence,

$$\nu^2 \frac{(x - X)^2}{a^2} + \frac{(y - Y)^2}{b^2} = 1.$$

If this represents a circle then,

$$\frac{a^2}{\nu^2} = b^2.$$

Example 4.22 A rod of length L_0 moves with speed v along x-direction with respect to S frame. The rod makes θ angle with x'-axis. Show that rod in S appears contracted and rotated. Find the length of the rod as measured by S frame. Also find the angle θ the rod makes with x'-axis.

Hint: Here, the rod lies in the S' frame and S' frame moves with velocity v along x-axis with respect to S frame.
The rod in S' frame makes θ angle with x'-axis. Therefore,

$$\Delta x' = L_0 cos\theta , \quad \Delta y' = L_0 \sin \theta.$$

In S frame, the observer sees that the rod is moving along x-axis with v velocity. Therefore,

$$\Delta y = \Delta y' = L_0 \sin \theta , \quad \Delta x = \Delta x' \left(1 - \frac{v^2}{c^2}\right)^{\frac{1}{2}}.$$

Here, an observer in S frame finds the length of the rod as

$$L = \sqrt{\Delta x^2 + \Delta y^2} = L_0 \left(1 - \frac{v^2}{c^2} \cos^2 \theta\right)^{\frac{1}{2}}.$$

Also rod makes ϕ angle with x-axis, where

$$\tan \phi = \frac{\Delta y}{\Delta x}.$$

Example 4.23 A rod of length l_0 in its rest frame. In another inertial frame S', which is oriented in a direction of the unit vector $\vec{e}$ and is moving with a velocity $\vec{v}$. Show that the length of the rod in the second frame is

$$l = \frac{l_0\sqrt{c^2 - v^2}}{\sqrt{c^2 - v^2 sin^2\theta}},$$

Fig. 4.5 At time
$t = 0$, $t' = 0$, origin of two
systems coincide

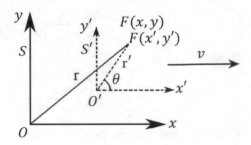

where $v = |\vec{v}|$ and θ is the angle between $\vec{e}$ and $\vec{v}$.

Hint: Let, $OF = r = l_0$ be the rod in S frame and the length of the rod as seen from S' frame is $O'F = r' = l$. At time $t = 0$, $t' = 0$, the origins O and O' coincide, then the two systems are related to each other by inverse transformation relations.

$$x = a(x' + vt') = ax_1; \quad y = y', \quad a = \frac{1}{\sqrt{1 - \beta^2}}, \beta = \frac{v}{c},$$

therefore,

$$r = \sqrt{x^2 + y^2} = \sqrt{a^2 x'^2 + y'^2}.$$

Figure 4.5 indicates that the length of the rod $O'F$ in S' system is

$$l = r' = \sqrt{x'^2 + y'^2}.$$

Also Fig. 4.5 yields

$$x' = r' \cos\theta.$$

Hence,

$$r'^2 = r'^2 \cos^2\theta + y'^2,$$

$$or, \ r'^2(1 - \cos^2\theta) = y'^2,$$

$$or. \ r'^2 \sin^2\theta = y'^2.$$

therefore,

$$a^2 x'^2 + r'^2 \sin^2\theta = a^2 x'^2 + y'^2 = r^2.$$

Hence,

$$a^2 r'^1 \cos^2\theta + r'^2 \sin^2\theta = r^2,$$

$$or. \ r'^2 \left[\frac{\cos^2\theta}{1 - \beta^2} + \sin^2\theta \right] = r^2,$$

$$or. \ r'^2(1 - \beta^2 sin^2\theta) = r^2(1 - \beta^2),$$

$$\therefore r' = \frac{r\sqrt{1 - \beta^2}}{\sqrt{1 - \beta^2 sin^2\theta}}.$$

Example 4.24 A rod with length $13\,m$ is kept fixed in the S frame and makes an angle $sin^{-1}(5/13)$ with the x-axis. Determine its length and inclination as observed from S', which is moving along the x-direction with a velocity $0.6c$ with respect to S.

Hint: Here, in S system,

$$l_x = l cos\theta \ \ and \ \ l_y = l sin\theta, \ \ l = 13\,m \ \ and \ \ \theta = sin^{-1}\frac{5}{13}.$$

Therefore,

$$l_x = 12\,m \ \ and \ \ l_y = 5\,m.$$

Since S' is moving parallel to the x-axis, so we have

$$l'_x = l_x\sqrt{1 - \frac{v^2}{c^2}} = 12\sqrt{1 - (0.6)^2} = 9.6\,m.$$

Note that

$$l'_y = l_y = 3m.$$

Let the inclination of the rod in the S' frame is θ', then

$$l'_x = l' cos\theta' \ \ and \ \ l'_y = l' sin\theta'$$

or,

$$l'^2 = \sqrt{l'^2_x + l'^2_y} = etc.,$$

and,

$$tan\,\theta' = \frac{l'_y}{l'_x} = etc.$$

Example 4.25 A cube of side length l_0 in it's rest frame. Now the cube is moving along the base diagonal with a velocity fc. Find the volume of the cube as measured by the observer in rest frame.

Hint:

The cube is moving along OA direction with velocity fc (see Fig. 4.6). So components of the velocity along x and y directions are

Fig. 4.6 At time
$t = 0$, $t' = 0$, origin of two
systems coincide

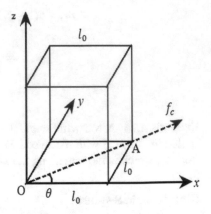

$$v_x = fc \cos \theta ; \quad v_y = fc \sin \theta.$$

Therefore, side of the cube contracted along x- and y-directions and side along z-direction will remain unchanged.

$$l'_x = l_0 \sqrt{1 - \left(\frac{v_x}{c}\right)^2} = l_0 \sqrt{1 - f^2 \cos^2 \theta},$$

$$l'_y = l_0 \sqrt{1 - \left(\frac{v_y}{c}\right)^2} = l_0 \sqrt{1 - f^2 \sin^2 \theta}.$$

So, the volume of the cube as measured by rest frame is

$$V' = l'_x l'_y l'_z = l_0^3 \sqrt{1 - f^2 \cos^2 \theta} \sqrt{1 - f^2 \sin^2 \theta}.$$

Example 4.26 Rahil is sitting in a moving car throws a stone with velocity u vertically upwards, Ayat is standing on a road. Find how she measures the path of the stone?

Hint: The motion of the stone as observed by Rahil is given by the set of equations:

$$x' = 0; \quad y' = ut' - \frac{1}{2}gt'^2.$$

Let the velocity of the car along the horizontal direction is v.

Then Ayat measures the motion of the stone by the equations

$$x = a(x' + vt'), \quad y = y';$$

$$t = a \left(t' + \frac{vx'}{c^2} \right), \quad a = \frac{1}{\sqrt{1 - \frac{v^2}{c^2}}}.$$

Eliminating prime terms, we get

$$y = y' - ut' - \frac{1}{2}gt'^2 = \frac{ut}{a} - \frac{gt^2}{2a^2}.$$

Now replacing t in terms of x, one gets,

$$y = \frac{u}{av} - \frac{g}{2v^2a^2}x^2.$$

Then Ayat observes the motion of the stone as a parabola.

Chapter 5
More Mathematical Properties
of Lorentz Transformations

5.1 Interval

Any event is described by two different things: One is the place where it occurred and another is time when it occurred. Therefore, an event is expressed by three spatial coordinates and one time coordinate. Such a space (four axes of which represent three space coordinates and one time coordinate) is called *Minkowski's Four-dimensional world*. Any event is represented by a point, called *World point* in this space and there corresponds to each particle a certain line called *World line*, i.e. A curve in the four-dimensional space is called *World line*. Let us consider two events which are represented by the coordinates (x_1, y_1, z_1, t_1) and (x_2, y_2, z_2, t_2) w.r.t. a frame of reference which is at rest. Then the quantity

$$\rho_{12} = [c_2(t_2 - t_1)^2 - (x_2 - x_1)^2 - (y_2 - y_1)^2 - (z_2 - z_1)^2]^{\frac{1}{2}} \qquad (5.1)$$

is called the *Interval* between two events.

5.2 The Interval Between Two Events Is Invariant Under Lorentz Transformation

Proof Let consider two events which are represented by the coordinates (x_1, y_1, z_1, t_1) and (x_2, y_2, z_2, t_2) w.r.t. a frame of reference which is at rest, i.e. in S frame. Suppose the corresponding coordinates for these events in another frame of reference S^1, which is moving with uniform velocity along x-axis relative to the previous frame, be $(x_1^1, y_1^1, z_1^1, t_1^1)$ and $(x_2^1, y_2^1, z_2^1, t_2^1)$. Then the interval in S^1 frame

$$\rho_{12}^1 = [c^2(t_2^1 - t_1^1)^2 - (x_2^1 - x_1^1)^2 - (y_2^1 - y_1^1)^2 - (z_2^1 - z_1^1)^2]^{\frac{1}{2}}. \qquad (5.2)$$

© The Author(s), under exclusive license to Springer Nature Singapore Pte Ltd. 2022
F. Rahaman, *The Special Theory of Relativity*, La Matematica per il 3+2 136,
https://doi.org/10.1007/978-981-19-0497-4_5

Using Lorentz Transformation, we have

$$(\rho_{12}^1)^2 = c^2 \left[a \left\{ t_2 - \left(\frac{vx_2}{c^2} \right) \right\} - a \left\{ t_1 - \left(\frac{vx_1}{c^2} \right) \right\} \right]^2$$
$$- [a(x_2 - vt_2) - a(x_1 - vt_1)]^2 - (y_2 - y_1)^2 - (z_2 - z_1)^2 = (\rho_{12})^2.$$

Thus, the interval between two events is invariant under Lorentz Transformation. We have shown that the interval between two events are invariant, i.e. $(\rho_{12}^1)^2 = (\rho_{12})^2$. This gives

$$[c^2(t_2^1 - t_1^1)^2 - (x_2^1 - x_1^1)^2 - (y_2^1 - y_1^1)^2 - (z_2^1 - z_1^1)^2]$$
$$= [c^2(t_2 - t_1)^2 - (x_2 - x_1)^2 - (y_2 - y_1)^2 - (z_2 - z_1)^2].$$

Now, if the two events occur at the same point in the system S^1, i.e.

$$(x_2^1 - x_1^1)^2 + (y_2^1 - y_1^1)^2 + (z_2^1 - z_1^1)^2 = 0.$$

Then

$$(\rho_{12})^2 = [c^2(t_2 - t_1)^2 - (x_2 - x_1)^2 - (y_2 - y_1)^2 - (z_2 - z_1)^2 = [c^2(t_2^1 - t_1^1)^2] > 0$$

This indicates the interval between two events is a real number.

Real intervals are said to be *Time like*. Thus, if the interval between two events is time like, there exists a system of reference in which the two events occur at one and the same place.

Now consider a case when the two events occur at one the same time, i.e. $(t_2^1 - t_1^1) = 0$. Then,

$$(\rho_{12})^2 = [c^2(t_2 - t_1)^2 - (x_2 - x_1)^2 - (y_2 - y_1)^2 - (z_2 - z_1)^2]$$
$$= -[(x_2^1 - x_1^1)^2 + (y_2^1 - y_1^1)^2 + (z_2^1 - z_1^1)^2] < 0.$$

Therefore, ρ_{12} is imaginary.

Imaginary intervals are said to be *space like*. Here, the space part dominates time part. Hence, if the interval between two events is space like, then there exists a system of reference in which the two events occur simultaneously.

If $\rho_{12} = 0$, then we call the interval to be *light like*. Then,

$$c^2(t_2 - t_1)^2 = (x_2 - x_1)^2 + (y_2 - y_1)^2 + (z_2 - z_1)^2$$

so that the events can be connected to the light signal. Obviously, this is a limiting case dividing space like from time like.

Note 1: The expression $x^2 + y^2 + z^2 - c^2t^2$ remains invariant under LT. In terms of differentials, the quantity

$$ds^2 = dx^2 + dy^2 + dz^2 - c^2 dt^2 \tag{5.3}$$

remains also invariant under Lorentz Transformation. Here, ds is known as line element.

Remember that in Euclidean geometry the square of the distance between two infinitely close points in three dimensions (i.e. in Cartesian coordinates) can be written as

$$ds^2 = dx_1^2 + dx_2^2 + dx_3^2 \tag{5.4}$$

For a four-dimensional Euclidean space, square of the distance between two infinitely close points will be written (generalizing the above notion)

$$ds^2 = dx_1^2 + dx_2^2 + dx_3^2 + dx_4^2, \tag{5.5}$$

Now, comparing (5.3) and (5.5), one can write

$$x_1 = x, \quad x_2 = y, \quad x_3 = z, \quad x_4 = ict. \tag{5.6}$$

Here, we have introduced the imaginary fourth coordinate for mathematical convenience and for giving our equation a more symmetrical form. ds is the invariant space–time interval between two events which are infinitely close to each other.

Note 2: Suppose a point moves from (x, y, z, t) to any neighbouring point, then we know the space–time interval

$$ds^2 = c^2 dt^2 - dx^2 - dy^2 - dz^2,$$

or

$$\left(\frac{ds}{dt}\right)^2 = c^2 - \left[\left(\frac{dx}{dt}\right)^2 + \left(\frac{dy}{dt}\right)^2 + \left(\frac{dz}{dt}\right)^2\right] = c^2 - v^2.$$

If $c^2 > v^2$, then $ds^2 > 0$, then the distance or interval is positive. It is time-like interval. If $c^2 < v^2$, then $ds^2 < 0$, then the distance or interval is negative. It is space-like interval. For photon particle $c^2 = v^2$, $ds^2 = 0$, then it is light-like or null interval. If $c^2 < v^2$, then particle does not obey the law of causality, i.e. if the particle is space-like then, the particle is causally disconnected.

Note 3: Suppose two events occur at the same place with respect to S^1 frame, then $\sum (x_i^1 - y_i^1)^2 = 0$. Since interval is invariant, we have $\rho_{12}^2 = (\rho_{12}^1)^2$. This implies

$$c^2 (t_1^1 - t_2^1)^2 = c^2 (t_1 - t_2)^2 - \Sigma (x_i - y_i)^2.$$

Fig. 5.1 x_1- and x_4-axes are
rotated through an angle θ

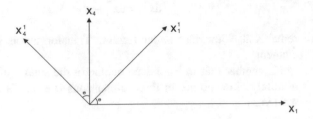

Writing differential notations, we have

$$c^2 (\mathrm{d}t^1)^2 = c^2 \mathrm{d}t^2 - \mathrm{d}x^2 - \mathrm{d}y^2 - \mathrm{d}z^2.$$

This yields

$$\mathrm{d}t^1 = \mathrm{d}t \sqrt{1 - \frac{v^2}{c^2}}.$$

Thus, we can re-derive time dilation from the invariant property of intervals.

Theorem 5.1 *Lorentz Transformation may be regarded as a rotation of axes through an imaginary angle.*

Proof Let us consider four-dimensional space with coordinates x_1, x_2, x_3 and x_4, where $x_4 = ict$. Suppose x_1- and x_4-axes are rotated through an angle θ so that new axes are x_1^1 and x_4^1; x_2, x_3 axes remain unchanged (see Fig. 5.1). Then obviously $x_2^1 = x_2$, $x_3^1 = x_3$ and

$$x_1^1 = x_1 \cos \theta + x_4 \sin \theta \qquad (5.7)$$

$$x_4^1 = -x_1 \sin \theta + x_4 \cos \theta. \qquad (5.8)$$

Let us take angle to be imaginary such that

$$\tan \theta = \frac{iv}{c} = i\beta, \qquad (5.9)$$

where $\beta = \frac{v}{c}$. From (5.9), we have

$$\cos \theta = \frac{1}{a}, \quad \sin \theta = \frac{i\beta}{a} \qquad (5.10)$$

where $a = \sqrt{1 - \beta^2}$. Now, putting the values of $\sin \theta$, $\cos \theta$ and x_4 in (5.7) and (5.8), we get

$$x_1^1 = x_1 \left(\frac{1}{a}\right) + ict \left(\frac{i\beta}{a}\right) = \left(\frac{1}{a}\right)[x_1 - vt],$$

$$ict^1 = -x_1 \left(\frac{i\beta}{a}\right) + ict \left(\frac{1}{a}\right)$$

or

$$t^1 = \left(\frac{1}{a}\right)\left[t - \frac{vx_1}{c^2}\right].$$

Thus, Eqs. (5.7) and (5.8) give L.T. Hence, we can conclude that LT are equivalent to rotation of axes in four-dimensional space through an imaginary angle $tan^{-1}(i\beta)$.

Theorem 5.2 *Lorentz Transformation may be expressed in hyperbolic form.*

Proof Let us consider,

$$\tanh \phi = \beta = \frac{v}{c}.$$

Then,

$$\cosh \phi = a = \frac{1}{\sqrt{1 - \beta^2}}, \qquad \sinh \phi = a\beta.$$

Let construct a matrix

$$A = \begin{bmatrix} \cosh\phi & 0 & 0 & -\sinh\phi \\ 0 & 1 & 0 & 0 \\ 0 & 0 & 1 & 0 \\ -\sinh\phi & 0 & 0 & \cosh\phi \end{bmatrix}$$

We will show that this matrix represents the Lorentz Transformation of (x, y, z, t).

Using $X^1 = AX$, where, $X = \begin{bmatrix} x \\ y \\ z \\ ct \end{bmatrix}$, one can write

$$\begin{bmatrix} x^1 \\ y^1 \\ z^1 \\ t^1 \end{bmatrix} = \begin{bmatrix} \cosh\phi & 0 & 0 & -\sinh\phi \\ 0 & 1 & 0 & 0 \\ 0 & 0 & 1 & 0 \\ -\sinh\phi & 0 & 0 & \cosh\phi \end{bmatrix} \begin{bmatrix} x \\ y \\ z \\ t \end{bmatrix}.$$

From these, we get

$$x^1 = x \cosh \phi - ct \sinh \phi, \quad y^1 = y, \quad z^1 = z, \quad ct^1 = -x \sinh \phi + ct \cosh \phi.$$

Now, putting the values of $\cosh \phi$ and $\sinh \phi$, we get

$$x^1 = a[x - vt], \quad y^1 = y, \quad z^1 = z, \quad t^1 = a\left[t - \frac{xv}{c^2}\right].$$

Hence, the matrix A expressed in hyperbolic form represents the Lorentz Transformation of (x, y, z, t).

Note 4: Lorentz Transformation of (x,y,z,t) can be written in the matrix form as

$$A = \{A_{ij}\} = \begin{bmatrix} a & 0 & 0 & ia\beta \\ 0 & 1 & 0 & 0 \\ 0 & 0 & 1 & 0 \\ -ia\beta & 0 & 0 & a \end{bmatrix},$$

where $\beta = \frac{v}{c}$ and $a = \frac{1}{\sqrt{1-\beta^2}}$. Here, v is the parameter and $\mid A_{ij} \mid = 1$.

Theorem 5.3 *Lorentz Transformations form a group.*

Proof We know Lorentz Transformation of (x, y, z, t) can be written in the matrix form as

$$L = \{A_{ij}\} = \begin{bmatrix} a & 0 & 0 & ia\beta \\ 0 & 1 & 0 & 0 \\ 0 & 0 & 1 & 0 \\ -ia\beta & 0 & 0 & a \end{bmatrix},$$

where $\beta = \frac{v}{c}$ and $a = \frac{1}{\sqrt{1-\beta^2}}$. Here, v is the parameter and $\mid A_{ij} \mid = 1$.

Since v is a parameter, therefore, for two different velocities, v_1 and v_2, we have L_1 and $L_2 \in L$, where

$$L_1 = \begin{bmatrix} \frac{1}{\sqrt{1-(\frac{v_1}{c})^2}} & 0 & 0 & \frac{i(\frac{v_1}{c})}{\sqrt{1-(\frac{v_1}{c})^2}} \\ 0 & 1 & 0 & 0 \\ 0 & 0 & 1 & 0 \\ -\frac{i(\frac{v_1}{c})}{\sqrt{1-(\frac{v_1}{c})^2}} & 0 & 0 & \frac{1}{\sqrt{1-(\frac{v_1}{c})^2}} \end{bmatrix}$$

$$L_2 = \begin{bmatrix} \frac{1}{\sqrt{1-(\frac{v_2}{c})^2}} & 0 & 0 & \frac{i(\frac{v_2}{c})}{\sqrt{1-(\frac{v_2}{c})^2}} \\ 0 & 1 & 0 & 0 \\ 0 & 0 & 1 & 0 \\ -\frac{i(\frac{v_2}{c})}{\sqrt{1-(\frac{v_2}{c})^2}} & 0 & 0 & \frac{1}{\sqrt{1-(\frac{v_2}{c})^2}} \end{bmatrix}$$

Now,

$$L_1 L_2 = \begin{bmatrix} \frac{1}{\sqrt{1-(\frac{w}{c})^2}} & 0 & 0 & \frac{i(\frac{w}{c})}{\sqrt{1-(\frac{w}{c})^2}} \\ 0 & 1 & 0 & 0 \\ 0 & 0 & 1 & 0 \\ -\frac{i(\frac{w}{c})}{\sqrt{1-(\frac{w}{c})^2}} & 0 & 0 & \frac{1}{\sqrt{1-(\frac{w}{c})^2}} \end{bmatrix} \in L$$

where

$$w = \frac{v_1 + v_2}{1 + \frac{v_1 v_2}{c^1}}.$$

Hence, closure property holds good.

We know matrix multiplication is associative.

For $v=0$, we get

$$L_0 = \begin{bmatrix} 1 & 0 & 0 & 0 \\ 0 & 1 & 0 & 0 \\ 0 & 0 & 1 & 0 \\ 0 & 0 & 0 & 1 \end{bmatrix} \in L.$$

Thus, L_0 is an identity element.

Now, consider L_2 is the inverse of L_1. Then $L_1 L_2 = L_0$ implies

$$\begin{bmatrix} \frac{1}{\sqrt{1-(\frac{w}{c})^2}} & 0 & 0 & \frac{i(\frac{w}{c})}{\sqrt{1-(\frac{w}{c})^2}} \\ 0 & 1 & 0 & 0 \\ 0 & 0 & 1 & 0 \\ -\frac{i(\frac{w}{c})}{\sqrt{1-(\frac{w}{c})^2}} & 0 & 0 & \frac{1}{\sqrt{1-(\frac{w}{c})^2}} \end{bmatrix} = \begin{bmatrix} 1 & 0 & 0 & 0 \\ 0 & 1 & 0 & 0 \\ 0 & 0 & 1 & 0 \\ 0 & 0 & 0 & 1 \end{bmatrix}.$$

Hence, we get $w = 0$, i.e.

$$\frac{v_1 + v_2}{1 + \frac{v_1 v_2}{c^1}} = 0 \Rightarrow v_2 = -v_1.$$

In other words,

$$L_2 = \begin{bmatrix} \frac{1}{\sqrt{1-(\frac{v_1}{c})^2}} & 0 & 0 & -\frac{i(\frac{v_1}{c})}{\sqrt{1-(\frac{v_1}{c})^2}} \\ 0 & 1 & 0 & 0 \\ 0 & 0 & 1 & 0 \\ \frac{i(\frac{v_1}{c})}{\sqrt{1-(\frac{v_1}{c})^2}} & 0 & 0 & \frac{1}{\sqrt{1-(\frac{v_1}{c})^2}} \end{bmatrix} \in L.$$

Thus, inverse element exists in L [here, v_1 is replaced by $-v_1$].

Therefore, Lorentz Transformations form a group.

Example 5.1 Show two successive Lorentz Transformation with velocity parameter $\beta_i = \frac{1}{\sqrt{1-v_i^2/c^2}}$, $i = 1, 2$ equivalent to a single Lorentz Transformation with velocity parameter $\beta = \beta_1 \beta_2 (1 + v_1 v_2/c^2)$, where v_1 and v_2 are two velocities with respect to which the two Lorentz Transformation are applied in the same direction.

Hint: The Lorentz Transformations with velocity parameter v_1 are

$$x^1 = \frac{x - v_1 t}{\sqrt{1 - \frac{v_1^2}{c^2}}}, \quad y^1 = y, \quad z^1 = z, \quad t^1 = \frac{t - \frac{xv_1}{c^2}}{\sqrt{1 - \frac{v_1^2}{c^2}}}.$$

Thus, the transformation matrix is

$$L(v_1) = \begin{bmatrix} \beta_1 & 0 & 0 & -v_1\beta_1 \\ 0 & 1 & 0 & 0 \\ 0 & 0 & 1 & 0 \\ -\frac{\beta_1 v_1}{c^2} & 0 & 0 & \beta_1 \end{bmatrix},$$

where $\beta_1 = \frac{1}{\sqrt{1-\frac{v_1^2}{c^2}}}$.

Similarly, Lorentz Transformation matrix with velocity parameter v_2

$$L(v_2) = \begin{bmatrix} \beta_2 & 0 & 0 & -v_2\beta_2 \\ 0 & 1 & 0 & 0 \\ 0 & 0 & 1 & 0 \\ -\frac{\beta_2 v_2}{c^2} & 0 & 0 & \beta_2 \end{bmatrix}$$

where $\beta_2 = \frac{1}{\sqrt{1-\frac{v_2^2}{c^2}}}$.

Two successive Lorentz Transformation is obtained by multiplying the transformation matrices of these transformation with each other. Hence,

$$L(v_1)L(v_2) = \begin{bmatrix} \beta_1 & 0 & 0 & -v_1\beta_1 \\ 0 & 1 & 0 & 0 \\ 0 & 0 & 1 & 0 \\ -\frac{\beta_1 v_1}{c^2} & 0 & 0 & \beta_1 \end{bmatrix} \begin{bmatrix} \beta_2 & 0 & 0 & -v_2\beta_2 \\ 0 & 1 & 0 & 0 \\ 0 & 0 & 1 & 0 \\ -\frac{\beta_2 v_2}{c^2} & 0 & 0 & \beta_2 \end{bmatrix}$$

$$= \begin{bmatrix} \beta_1\beta_2\left(1 + \frac{v_1 v_2}{c^2}\right) & 0 & 0 & -(v_1 + v_2)\beta_1\beta_2 \\ 0 & 1 & 0 & 0 \\ 0 & 0 & 1 & 0 \\ -\frac{(v_1+v_2)\beta_1\beta_2}{c^2} & 0 & 0 & \beta_1\beta_2\left(1 + \frac{v_1 v_2}{c^2}\right) \end{bmatrix}$$

$$= \begin{bmatrix} \beta & 0 & 0 & -V\beta \\ 0 & 1 & 0 & 0 \\ 0 & 0 & 1 & 0 \\ -\frac{V\beta}{c^2} & 0 & 0 & \beta \end{bmatrix} = L(V),$$

where $\beta = \beta_1\beta_2\left(1 + \frac{v_1 v_2}{c^2}\right)$ and $V = \frac{v_1 + v_2}{1 + \frac{v_1 v_2}{c^2}}$.

Example 5.2 Prove that the Lorentz Transformation of x and t can be written in the following forms:

$$ct^1 = -x \sinh\phi + ct \cosh\phi, \quad x^1 = x \cosh\phi - ct \sinh\phi,$$

where ϕ is defined by $\phi = tanh^{-1}\frac{v}{c}$.

Establish also the following version of these equations:

$$ct^1 + x^1 = e^{-\phi}(ct + x), \quad ct^1 - x^1 = e^{\phi}(ct - x), \quad e^{2\phi} = \frac{1 + \frac{v}{c}}{1 - \frac{v}{c}}.$$

Hint:

$$\phi = tanh^{-1}\frac{v}{c} \implies \tanh\phi = \frac{v}{c}.$$

Hence,

$$\cosh\phi = \frac{1}{\sqrt{1 - \frac{v^2}{c^2}}}, \quad \sinh\phi = \frac{\frac{v}{c}}{\sqrt{1 - \frac{v^2}{c^2}}}.$$

Now,

$$x^1 = \frac{x - vt}{\sqrt{1 - \frac{v^2}{c^2}}} = \frac{x}{\sqrt{1 - \frac{v^2}{c^2}}} - \frac{ct\left(\frac{v}{c}\right)}{\sqrt{1 - \frac{v^2}{c^2}}} = x \cosh\phi - ct \sinh\phi.$$

Similarly,

$$ct^1 = \frac{ct - \frac{vx}{c}}{\sqrt{1 - \frac{v^2}{c^2}}} = -x \sinh \phi + ct \cosh \phi.$$

Now,

$$ct^1 + x^1 = (ct + x)(\cosh \phi - \sinh \phi)$$

$$= (ct + x)\left[\frac{1}{2}(e^\phi + e^{-\phi}) - \frac{1}{2}(e^\phi - e^{-\phi})\right] = e^{-\phi}(ct + x)$$

$$ct^1 - x^1 = (ct - x)(\cosh \phi + \sinh \phi) = e^\phi(ct - x).$$

Also,

$$\tanh \phi = \frac{v}{c} \implies \frac{\sinh \phi}{\cosh \phi} = \frac{v}{c} \implies \frac{e^\phi - e^{-\phi}}{e^\phi + e^{-\phi}} = \frac{v}{c} \; etc.$$

Note 5: Here, the quantity ϕ is called **rapidity** or pseudo-velocity of the transformation. In terms of rapidity, the velocity addition formula takes the following form:

$$\phi(w) = \phi(u) + \phi(v). \qquad [\; see \; Example \; 7.4 \;].$$

Example 5.3 Show that the transformations

$$x^1 = x \cosh \phi - ct \sinh \phi, \quad y^1 = y, \quad z^1 = z, \quad ct^1 = -x \sinh \phi + ct \cosh \phi$$

preserve the line element.

Hint: We know the line element is

$$ds^{1^2} = -c^2 dt^{1^2} + dx^{1^2} + dy^{1^2} + dz^{1^2}$$
$$= -(-dx \sinh \phi + cdt \cosh \phi)^2 + (dx \cosh \phi - cdt \sinh \phi)^2 + (dy)^2 + (dz)^2$$
$$= -c^2 dt^2 + dx^2 + dy^2 + dz^2 = ds^2$$

Example 5.4 Express the law of addition of parallel velocities in terms of the parameter θ.

Hint: Suppose a particle is moving with speed V along x-direction. Let $\frac{V}{c} = \tanh \Theta$ The velocity of the particle as measured from S^1 frame which is moving with velocity v relative to S frame in x-direction is given by

$$V^1 = \frac{V - v}{1 - \frac{vV}{c^2}}.$$

Now, we put $\frac{V^1}{c} = \tanh \Theta^1$ and $\frac{v}{c} = \tanh \theta$ in the above expression, we find

$$\Theta^1 = \Theta - \theta.$$

So, the law of addition of parallel velocities becomes the addition of θ parameter.

Chapter 6
Geometric Interpretation of Space–Time

6.1 Space–Time Diagrams

Various types of interval as well as trajectories of the particles in Minkowski's four-dimensional world can be described by some diagrams which are known as space–time diagrams. To draw the diagram, one has to use some specific inertial frame S in which one axis indicates the space coordinate x and other effectively the time axis. However, the time axis is taken as ct instead of t so that the coordinates will have the same physical dimensions. We don't need to specify the second and third space coordinates axes, i.e. y- and z-axes. These can be imagined such as both coordinates axes are at right angles to ct and to x-axis. To visualize the space–time diagram, we draw the two 45° lines [$\tan \alpha = \frac{dx}{d(ct)} = \frac{u}{c}$ and we must have $u < c$ for a material particle; the world line for light signal, i.e. $u = c$, is a straight line making an angle 45° with the axes], which are trajectories of light signals through O (these light signals pass through the point $R = 0$ at time $t = 0$; $R = (x, y, z)$). If one includes another axis too, then the set of all such trajectories form a right circular double cone around time axis, with apex at O, called the *light cone* through O. However, in general, we just ignore y- and z-axes and the resulting light cone consists of two sloping lines.

One can signify any event by a point on the diagram. Time-like events, i.e. the events which are time-like separated from the apex O lie inside the light cone, like A and B (see Fig. 6.1). Since the sign of t is invariant, therefore, points in the upper half of the cone posses positive time coordinates and known as inside future light cone. Here, the event A with $t_A > 0$ lies inside future light cone. The point B with $t_B < 0$ is said to be inside past light cone. Light-like events lie on the light cone and as before one can distinguish between future and past; for instance, C lies on the future light cone and D lies on the past light cone. The region outside the cone consists of space-like events like E, F. Since time coordinate of space-like event depends on choice of the reference frame, it is usually immaterial whether space-like points lie above or below the x-axis. The world lines on the diagram are the trajectories of the particles. World lines of massless particle (light-like) lie on the light cone and world

© The Author(s), under exclusive license to Springer Nature Singapore Pte Ltd. 2022
F. Rahaman, *The Special Theory of Relativity*, La Matematica per il 3+2 136,
https://doi.org/10.1007/978-981-19-0497-4_6

Fig. 6.1 Light cone

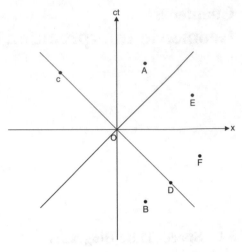

Fig. 6.2 World line of
accelerated particle

lines of the massive particles (time-like) lie within the light cone. They are restricted by the speed limit c.

Let us consider a particle is moving with constant velocity u in x-direction. Then its world line is a straight line making an angle α to the time axis where $\mid \tan \alpha \mid = \mid \frac{u}{c} \mid < 1$ so that $\mid \alpha \mid < 45°$. If the particle's velocity $u(t)$ changes during the motion, i.e. particle is moving with an acceleration, then its world line will be curved rather than a straight line (see Fig. 6.2). Here, the curve makes the angle with the time axis $\alpha(t) = \tan^{-1}\left(\frac{u(t)}{c}\right)$ that varies but always remains between $\pm 45°$. Therefore, it is possible for any signal from O to reach any event in or on the future light cone (like A or C). Likewise signals will be received by O from any event in or on the past light cone. Note that World lines passing through O are confined within the light cone through O. World lines can not leave the light cone because entry or exit would require $\alpha(t) > 45°$, i.e. speed should be greater than c, which is impossible. Hence, it is not possible to connect between O and any other event outside the light cone.

Fig. 6.3 Possible and impossible events

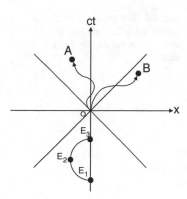

6.2 Some Possible and Impossible World Lines

Let us consider, Space–time diagram with the same origin, axes and the light cone as the previous one (see Fig. 6.3). The particle trajectory from O to A is possible because it is within the future light, moreover, its inclination to the time axis is always less than 45°. This indicates that the speed of the particle along the trajectory is always less than c. The trajectory from O to B is impossible as it crosses the light cone.

This means that the speed of the particle should exceed c. Again the trajectory $E_1E_2E_3O$ is not possible in spite it lies entirely within the light cone. The reason is that near E_1 and E_3 its slope indicates speeds greater than c (in fact, at the points E_1 and E_3 speed of the particle would be infinite).

Remember that in a different inertial frame S^1, the space–time diagram looks exactly same as in Fig. 6.3, except that its axes are labelled by ct^1 and x^1. Two or more events can be entered on both diagrams with different coordinates. However, space–time separations (intervals) ρ_{12} between two events remain invariant. In particular, if events lie inside the future light cone of one diagram, then these remain inside of the future light cone of the other.

Note 1: We know the quantity $s^2 = x^2 + y^2 + z^2 - c^2t^2$, called the square of the four-dimensional distance between the event (x^μ) and the origin $(0, 0, 0, 0)$, remains invariant under Lorentz Transformation. Now, the trajectories of the points whose distance from the origin is zero form a surface described by the equation

$$s^2 = x^2 + y^2 + z^2 - c^2t^2 = 0.$$

This surface is light cone. This equation describes the propagation of a spherical light wave from the origin $(0, 0, 0)$ at $t = 0$. The space–time, i.e. (3+1) space is divided into two invariant separated domains S_t and S_s by the light cone (see Fig. 6.4). The region S_t is defined by $s^2 = x^2 + y^2 + z^2 - c^2t^2 < 0$.

From this, we get

Fig. 6.4 The region S_t is within the light cone and the region S_s is outside the light cone

$$\frac{s^2}{t^2} = \left[\frac{\sqrt{x^2 + y^2 + z^2}}{t}\right]^2 - c^2 = v^2 - c^2 < 0.$$

This indicates trajectories of the massive particle lie always inside the light cone. Thus, world lines passing through O always be confined within the cone. They cannot enter in the region S_s, which is defined by $s^2 = x^2 + y^2 + z^2 - c^2 t^2 > 0$, because entry or exist would require speed greater than c ($S_s : \frac{s^2}{t^2} = v^2 - c^2 > 0$). Let an event occurs at $t = 0$, $x = 0$, then it can causally connect any point inside the positive light cone.

Note 2: Photon has a null trajectory as

$$ds^2 = c^2 dt^2 - dx^2 - dy^2 - dz^2$$

or

$$\left(\frac{ds}{dt}\right)^2 = c^2 - \left[\left(\frac{dx}{dt}\right)^2 + \left(\frac{dy}{dt}\right)^2 + \left(\frac{dz}{dt}\right)^2\right] = c^2 - v^2$$

For photon, $v = c$, $\left(\frac{ds}{dt}\right)^2 = 0$, i.e. *light-like*: $ds^2 = 0$.

Remember For time-like $ds^2 > 0$ and space-like: $ds^2 < 0$.

Note 3: Can an event at $t = 0$, $x = 0$ affect the event occurred at P in the future light cone region?

Yes, because, P is a point which is in future. So an event affects the event occurred at P. But the reverse is not, i.e. event at P can not affect the event at $t = 0$, $x = 0$, since t is always positive. All the events in the future light cone region occur after the event at t = 0, x = 0. Light signals emitted at $t = 0$, $x = 0$ travel out along the boundary of the light cone and affect the events on the light cone.

Note 4: A collision between two particles corresponds to an intersection of their world lines.

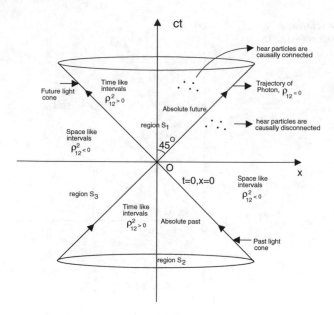

Fig. 6.5 Geometrical representation of space–time diagram

6.3 Importance of Light Cone

The light cone structure is important because inside the light cone particles are causally connected and outside the light cone particles are causally disconnected. Thus, it is very significant in the sense that geometrical representation of space–time diagram, i.e. light cone is completely consistent with the causality principle. For example, the events within the light cone have a cause-and-effect relationship between events and events outside the light cone have no cause-and-effect relationship between events (Fig. 6.5).

6.4 Relationship Between Space–Time Diagrams in S and S^1 Frames

Let us consider two frames of reference S and S^1 and S^1 is moving with constant velocity v along x-direction. Let S frame consists of coordinate axes ct and x so that a light ray has the slope $\frac{\pi}{4}$

$$\left[\frac{ct}{x} = \tan \theta \implies \cot \theta = \frac{1}{c} \left(\frac{x}{t} \right) = \frac{c}{c} = 1 \implies \theta = \frac{\pi}{4} \right].$$

Now, we try to draw the ct^1 and x^1 axes of S^1. Here, the equation of x^1 is $ct^1 = 0$. Using Lorentz Transformation, we get

Fig. 6.6 Two reference frames have the same origin

$$\frac{c\left(t - \frac{xv}{c^2}\right)}{\sqrt{1 - \frac{v^2}{c^2}}} = 0 \implies ct = \left(\frac{v}{c}\right)x.$$

Thus, x^1 axis is a straight line, $ct = \left(\frac{v}{c}\right)x$ with slope $\left(\frac{v}{c}\right) < 1$.

Similarly, equation of ct^1 axis, $x^1 = 0$ gives $ct = \left(\frac{c}{v}\right)x$. Hence, ct^1-axis is a straight line, $ct = \left(\frac{c}{v}\right)x$ having slope $\left(\frac{c}{v}\right) > 1$. The fixed points in S^1 frame lie in world line parallel to $O(ct^1)$-axis. The events with fixed time in S^1 frame lie in world line parallel to Ox^1-axis and are known as lines of simultaneity in S^1 frame. Let P be any event (see Fig. 6.6). Then, the coordinates of P with respect to S system are $(ct, x) = (OR, OQ)$ and $(ct^1, x^1) = (OV, OU)$ with respect to S^1 system. Suppose A be any point on x^1-axis having length from O is 1. Therefore, the coordinates of A with respect to S^1 system are $(ct^1, x^1) = (0, 1)$. The coordinates of A with respect to S system are $(ct, x) = (\frac{av}{c}, a)$, where $a = \frac{1}{\sqrt{1 - \frac{v^2}{c^2}}}$.

[by L.T : $ct^1 = \frac{c\left(t - \frac{xv}{c^2}\right)}{\sqrt{1 - \frac{v^2}{c^2}}} = 0$ and $x^1 = \frac{(x - vt)}{\sqrt{1 - \frac{v^2}{c^2}}} = 1 \implies x = a, ct = \frac{xv}{c}$]

Hence, the distance OA is given by

$$OA = \sqrt{(c^2t^2 + x^2)} = a\sqrt{1 + \frac{v^2}{c^2}}.$$

Thus, scale along the primed axes is greater than the scale along the unprimed by a factor

$$\left[\frac{1 + \frac{v^2}{c^2}}{\sqrt{1 - \frac{v^2}{c^2}}}\right].$$

Fig. 6.7 Simultaneous
events

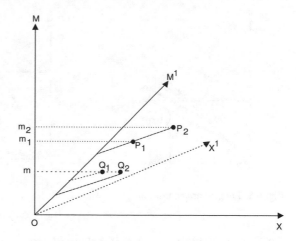

6.5   Geometrical Representation of Simultaneity, Space Contraction and Time Dilation

Let us consider two frames of reference S and S^1 and S^1 is moving with constant velocity v along x-direction. Let S frame consists of coordinate axes $ct \equiv M$ & x and S^1 frame consists of coordinate axes $ct^1 \equiv M^1$ & x^1.

6.5.1   Simultaneity

In S^1 frame, two events will be simultaneous if they have the same time coordinate m^1. Thus, simultaneous events lie on a line parallel to x^1-axis. In Fig. 6.7, P_1 and P_2 are simultaneous in S^1 system but not simultaneous in S system as they will be occurring at different times m_1 and m_2 in S system (i.e. they have different time coordinates). Similarly, Q_1 and Q_2 are two simultaneous events in S frame but they are separated in times in S^1 frame (i.e. they have different time coordinates m_1^1 and m_2^1 in S^1 frame).

6.5.2   Space Contraction

Consider a rod having end points at $x = x_1$ and $x = x_2$ rest in S frame. As time elapses, the world line of each end points traces out a vertical line parallel to the M-axis. The length of the rod (l_0) is the distance between its end points measured simultaneously, i.e. $l_0 = x_2 - x_1$. In S system, the length is the distance in S between the intersections of the world lines (these are parallel to M axis) with the x-axis or

Fig. 6.8 Lengths are contracted

any line parallel to x-axis. Actually, these intersecting points represent simultaneous events in S. Like in S frame, we measure the rod in S^1 frame as the distance between end points measured simultaneously which is the separation of the intersections of the world lines with x^1-axis or any line parallel to the x^1-axis, i.e. $l = x_2^1 - x_1^1$. Here, these intersecting points represent simultaneous events in S^1 (see Fig. 6.8).

The length of rod in S^1 system (moving) is $l = l_0\sqrt{1 - \frac{v^2}{c^2}}$, which is clearly less than l_0. Note that if events are simultaneous in one reference frame, then they are not simultaneous in other frame. To measure the length, observer in S uses P_1 and P_2, while the observer in S^1 system uses either P_2 and P_4 or P_1 and P_3.

Alternatively, let us consider a rod at rest in S^1 frame having end points x_1^1 and x_2^1. The world lines of its end points are parallel to M^1-axis. Its rest length is $l_0 = x_2^1 - x_1^1$. Observer in S system, finds the rod is in motion and to him, the measured length is the distance in S system between intersections of these world lines with x axis or any line parallel to x axis. Clearly, the length of rod in motion is observed by the observer in S system as $l = l_0\sqrt{1 - \frac{v^2}{c^2}}$, which is less than l_0.

6.5.3   Time Dilation

Let us consider a clock at rest at A in S system which ticks off units of time. Since time interval in system S between ticks being unity, let T_1 and T_2 be two events of ticking at m = 2 and m = 3. The solid vertical line through A, i.e. $T_1 T_2$ is the world line corresponding to the clock in S system. In S^1 system, clock ticks in different places. Therefore, to measure the times interval between T_1 and T_2 in S^1 system, two clocks must be used. The difference in reading of these clocks in S^1 is the difference in times between T_1 and T_2 as measured in S^1 system. Figure 6.9 confirms that the time interval between T_1 and T_2 in S^1 system is greater than unity. Hence, relative to S^1 the clock at rest in system S appears slowed down. Here, S clock registered unit time, however, S^1 registered a time greater than one unit.

Fig. 6.9 Time intervals are increased

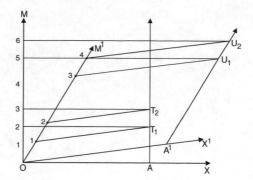

Alternatively, a clock at rest at A^1 in S^1 system which ticks off units of time at U_1 and U_2. Here, the line parallel to M^1-axis through A^1, i.e. the line $U_1 U_2$ is the world line corresponding to the clock in system S^1. The time interval to the observer in S frame (relative to which the clock is in motion) is greater than unity.

Chapter 7
Relativistic Velocity and Acceleration

Keywords Relativistic velocity · Relativistic acceleration · Velocity addition law

7.1 Relativistic Velocity Addition

Let us consider three frames of reference S, S^1 and S^2. S^1 is moving with constant velocity u relative to S in x-direction and S_2 is moving with constant velocity v relative to S^1 also in x-direction (see Fig. 7.1). We want to find the transformation equations between S_2 and S. In Newtonian mechanics, the relative velocity is

$$w = u + v.$$

However, we try to see what will happen in special theory of relativity. The transformation equations among S, S^1 and S^1, S^2 are given by

$$x^1 = \frac{1}{\sqrt{1 - (u/c)^2}}[x - ut], \ y^1 = y, \ z^1 = z, \ t^1 = \frac{1}{\sqrt{1 - (u/c)^2}}\left[t - \frac{ux}{c^2}\right],$$
(7.1)

$$x^2 = \frac{1}{\sqrt{1 - (v/c)^2}}[x^1 - vt^1], \ y^2 = y^1, \ z^2 = z^1, \ t^2 = \frac{1}{\sqrt{1 - (v/c)^2}}\left[t^1 - \frac{vx^1}{c^2}\right].$$
(7.2)

Now, putting the values of x^1, t^1 in the first expression of Eq. (7.2), one gets

Fig. 7.1 Addition of velocities

$$x^2 = \frac{1}{\sqrt{1-(v/c)^2}} \frac{1}{\sqrt{1-(u/c)^2}} \left[1 + \frac{uv}{c^2}\right]\left[x - t\left\{\frac{u+v}{1+uv/c^2}\right\}\right]. \qquad (7.3)$$

Let us choose,

$$w = \frac{u+v}{1+\frac{uv}{c^2}}. \qquad (7.4)$$

Using the expression of w in (7.3), we get

$$x^2 = \frac{x - wt}{\sqrt{1-(w/c)^2}}. \qquad (7.5)$$

Obviously,

$$y^2 = y, \qquad z^2 = z$$

Again, putting the values of x^1, t^1 in the last expression of Eq. (7.2), one gets as before,

$$t^2 = \frac{t - \frac{wx}{c^2}}{\sqrt{1-(w/c)^2}}. \qquad (7.6)$$

Eq. (7.4) is known as the Einstein law of addition of velocities.
From Eq. (7.4), we have

$$w = \frac{u+v}{1+\frac{uv}{c^2}} = c\left[1 - \frac{(1-\frac{u}{c})(1-\frac{v}{c})}{1+\frac{uv}{c^2}}\right].$$

If $u < c$ and $v < c$, then $w < c$. This shows that the composition of velocities which are separately less than c cannot be added to give a velocity greater than c. If we put $v = c$, i.e. if we consider that a photon travelling with velocity c along x-direction, then, $w = c$. This indicates that if a particle is moving with velocity c, it will be observed c in all inertial frames whatever their relative velocity is. This confirms the validity of the second postulate of special theory of relativity that the velocity of light is an absolute quantity independent of the motion of the reference frame. Even when, $v = u = c$, then, $w = c$.

The relativistic law of addition of velocities reduces to the law of Newtonian mechanics for small values of u and v in comparison to the velocity of light, i.e. if $uv << c^2$, then $w = u + v$.

Note 1: If one takes, $u = v = \frac{4c}{5}$, then relativistic law of addition of velocities gives, $w = \frac{40c}{41} < c$. However, according to Newtonian mechanics law of addition of velocities gives, $w = \frac{8c}{5} > c$.

Note 2: The general relativistic addition velocity w can be written as

$$w = \frac{c\sqrt{c^2(\vec{u} + \vec{v}).(\vec{u} + \vec{v}) - u^2v^2 + (\vec{u}.\vec{v})^2}}{c^2 - (\vec{u}.\vec{v})}.$$

This reduces to the classical formula $w = |\vec{u} + \vec{v}|$ when $u, v << c$.

7.2 Relativistic Velocity Transformations

Let a frame of reference S^1 is moving with constant velocity v relative to S frame along x-direction. Assume a particle at point P is moving with velocity u^1 relative to S^1 frame and u its velocity relative to S frame. The space and time coordinates of P in S^1 and S frames are (x^1, y^1, z^1, t^1) and (x, y, z, t), respectively. The component of velocity vector $\vec{u}$ relative to S frame are

$$u_x = \frac{dx}{dt}, \quad u_y = \frac{dy}{dt}, \quad u_z = \frac{dz}{dt} \tag{7.7}$$

and

$$u = (u_x^2 + u_y^2 + u_z^2)^{\frac{1}{2}}.$$

Similarly, the components of velocity vector $\vec{u}^1$ relative to S^1 frame are

$$u_x^1 = \frac{dx^1}{dt^1}, \quad u_y^1 = \frac{dy^1}{dt^1}, \quad u_z^1 = \frac{dz^1}{dt^1} \tag{7.8}$$

and

$$u^1 = [(u_x^1)^2 + (u_y^1)^2 + (u_z^1)^2]^{\frac{1}{2}}.$$

Now, from Lorentz Transformations, we get

$$dx^1 = a(dx - vdt), \quad dy^1 = dy, \quad dz^1 = dz, \quad dt^1 = a\left(dt - \frac{vdx}{c^2}\right), \qquad (7.9)$$

where $a = \dfrac{1}{\sqrt{1-v^2/c^2}}$. Hence,

$$u_x^1 = \frac{dx^1}{dt^1} = \frac{u_x - v}{1 - vu_x/c^2} \qquad (7.10)$$

$$u_y^1 = \frac{dy^1}{dt^1} = \frac{u_y\sqrt{1 - v^2/c^2}}{1 - vu_x/c^2} \qquad (7.11)$$

$$u_z^1 = \frac{dz^1}{dt^1} = \frac{u_z\sqrt{1 - v^2/c^2}}{1 - vu_x/c^2} \qquad (7.12)$$

$$(u^1)^2 = [(u_x^1)^2 + (u_y^1)^2 + (u_z^1)^2] = \frac{(u_x - v)^2 + (u^2 - u_x^2)(1 - v^2/c^2)}{(1 - vu_x/c^2)^2}. \qquad (7.13)$$

Note 3: The general Lorentz Transformation of velocity vector $\overrightarrow{u}^1$ can be written as

$$\overrightarrow{u}^1 = \frac{\sqrt{1 - v^2/c^2}\,\overrightarrow{u} + \frac{(\overrightarrow{u}.\overrightarrow{v})\overrightarrow{v}}{v^2}\left[1 - \sqrt{1 - v^2/c^2}\right] - \overrightarrow{v}}{1 - \frac{(\overrightarrow{u}.\overrightarrow{v})}{c^2}}.$$

Since, $\overrightarrow{v}$ parallel to x-axis, i.e. $\overrightarrow{v} = (v, 0, 0)$, therefore, $(\overrightarrow{u}.\overrightarrow{v}) = u_x v$.

Note 4: The inverse transformations can be obtained by replacing v by $-v$ and interchanging primed by unprimed in Eqs. (7.10)–(7.12).

$$u_x = \frac{u_x^1 + v}{1 + \frac{vu_x^1}{c^2}}, \quad u_y = \frac{u_y^1\sqrt{1 - v^2/c^2}}{1 + \frac{vu_x^1}{c^2}}, \quad u_z = \frac{u_z^1\sqrt{1 - v^2/c^2}}{1 + \frac{vu_x^1}{c^2}}. \qquad (7.14)$$

Note 5: Einstein's law of addition of velocities is a special case of Relativistic velocity. Putting, $u_x = u, u_y = u_z = 0$ in (7.14), one obtains

$$u = \frac{u^1 + v}{1 + \frac{vu^1}{c^2}}.$$

Note 6: The transformation of Lorentz Transformation factor can be found from Eq. (7.13) as

$$\sqrt{1 - \left(u^1/c\right)^2} = \frac{\sqrt{1 - (v/c)^2}\sqrt{1 - (u/c)^2}}{(1 - \frac{u_x v}{c^2})}.$$

7.3 Relativistic Acceleration Transformations

Let acceleration of the particle relative to S and S^1 frames are $\vec{f}$ and $\vec{f}^{\,1}$, respectively. The components of the acceleration vector relative to S and S^1 frames are, respectively, as

$$f_x = \frac{du_x}{dt}, f_y = \frac{du_y}{dt}, f_z = \frac{du_z}{dt}$$

and

$$f_x^1 = \frac{du_x^1}{dt^1}, f_y^1 = \frac{du_y^1}{dt^1}, f_z^1 = \frac{du_z^1}{dt^1}.$$

Using (7.10), we can get

$$f_x^1 = \frac{du_x^1}{dt^1} = \frac{f_x \left(1 - \frac{v^2}{c^2}\right)^{\frac{3}{2}}}{(1 - \frac{vu_x}{c^2})^3}. \tag{7.15}$$

This acceleration is known as *Longitudinal Acceleration*.

Similarly, from (7.11) and (7.12), we get

$$f_y^1 = \frac{du_y^1}{dt^1} = \left(1 - \frac{v^2}{c^2}\right)\left[\frac{f_y}{(1 - \frac{vu_x}{c^2})^2} + \frac{vf_x u_y}{c^2(1 - \frac{vu_x}{c^2})^3}\right], \tag{7.16}$$

$$f_z^1 = \frac{du_z^1}{dt^1} = \left(1 - \frac{v^2}{c^2}\right)\left[\frac{f_z}{(1 - \frac{vu_x}{c^2})^2} + \frac{vf_x u_z}{c^2(1 - \frac{vu_x}{c^2})^3}\right]. \tag{7.17}$$

These are known as *Transverse Acceleration*.

Note 7: The general Lorentz Transformation of acceleration vector $\vec{f}^{\,1}$ can be written as

$$\vec{f}^{\,1} = \frac{\vec{f}}{a^2 \left(1 - \frac{\vec{u}.\vec{v}}{c^2}\right)^2} + \frac{(\vec{f}.\vec{v})\left[\frac{a\vec{u}}{c^2} - (a - 1)\frac{\vec{v}}{v^2}\right]}{a^3 \left(1 - \frac{\vec{u}.\vec{v}}{c^2}\right)^3}.$$

Note 8: For a particle instantaneously rest in S frame, i.e. $u_x = u_y = u_z = 0$, we get

$$f_x^1 = \frac{du_x^1}{dt^1} = f_x \left(1 - \frac{v^2}{c^2}\right)^{\frac{3}{2}} , f_y^1 = \left(1 - \frac{v^2}{c^2}\right) f_y, \quad f_z^1 = \left(1 - \frac{v^2}{c^2}\right) f_z.$$

The transformation formula of acceleration indicates that the acceleration is not invariant in special theory of relativity. In other words, Newton's second law of motion is not invariant under Lorentz Transformation. But, it is evident that acceleration is an absolute quantity, i.e. all observers agree whether a body is accelerating or not. In other words, the acceleration of a particle is zero in one frame, it is necessarily zero in all other frames.

7.4   Uniform Acceleration

In special theory of relativity, an acceleration is said to be *uniform* if for any co-moving frame its value remains the same. Let S and S^1 be two inertial frames with constant relative velocity v along x-direction. Let velocity of a particle relative to S is u. If S^1 is a co-moving frame, then $v = u$. This means relative velocity of the particle and S^1 frame is zero, $u^1 = 0$. Now, if the acceleration is a constant $f^1 = \frac{du^1}{dt^1} = b$ (say, a constant). From Eq. (7.15), one can write,

$$f = \frac{du}{dt} = \frac{f^1 \left(1 - \frac{u^2}{c^2}\right)^{\frac{3}{2}}}{\left(1 + \frac{vu^1}{c^2}\right)^3}. \tag{7.18}$$

Now, using $u^1 = 0$ and $\frac{du^1}{dt^1} = b$, we get

$$\frac{du}{dt} = b \left(1 - \frac{u^2}{c^2}\right)^{\frac{3}{2}}. \tag{7.19}$$

Solving this equation, one gets

$$u = \frac{dx}{dt} = \frac{b(t - t_0)}{\left[1 + \frac{b^2(t-t_0)^2}{c^2}\right]}. \tag{7.20}$$

[initially, particle starts from rest at $t = t_0$]

However, in Newtonian mechanics, if a particle is moving with uniform acceleration, then

$$\frac{du}{dt} = \text{constant}.$$

The path of this particle is parabola.

Further integration of Eq. (7.20) yields (using $x = x_0$ at $t = t_0$)

$$\frac{(x - x_0 + \frac{c^2}{b})^2}{(\frac{c^2}{b})^2} - \frac{(ct - ct_0)^2}{(\frac{c^2}{b})^2} = 1. \tag{7.21}$$

This is an equation of a hyperbola in (x, ct) space.

7.5 Relativistic Transformations of the Direction Cosines

Let us consider the direction cosines of the motion of a particle in S frame be (m, n, l) and corresponding quantities be (m^1, n^1, l^1) in S^1 which is moving with constant velocity v along x-direction. Then

$$(l, m, n) = \frac{(u_x, u_y, u_z)}{\sqrt{u_x^2 + u_y^2 + u_z^2}}. \tag{7.22}$$

Similarly,

$$(l^1, m^1, n^1) = \frac{(u_x^1, u_y^1, u_z^1)}{\sqrt{(u_x^1)^2 + (u_y^1)^2 + (u_z^1)^2}}. \tag{7.23}$$

Using Eqs. (7.4)–(7.7), one will get

$$(l^1, m^1, n^1) = \frac{(l - \frac{v}{u}), \; m\sqrt{1 - (v/c)^2}, \; n\sqrt{1 - (v/c)^2}}{D}, \tag{7.24}$$

where

$$D = \frac{u^1}{u}\left(1 - \frac{u_x v}{c^2}\right) = \sqrt{\left[1 - 2lv/u + \frac{v^2}{u^2} + (1 - l^2)\frac{v^2}{c^2}\right]}.$$

7.6 Application of Relativistic Velocity and Velocity Addition Law

7.6.1 The Fizeau Effect: The Fresnel's Coefficient of Drag

Now, we will try to give explanation of index of refraction of moving bodies (Fizeau effect). Fizeau performed the experiment by employing an interferometer to measure the velocity of light propagating in water moving at the velocity v. Here, we consider two inertial frames: one is S as laboratory and other is S^1, the moving water. Since the water has the velocity v, therefore, one can think, S^1 frame is moving with velocity v relative to S frame which is at rest. Suppose μ be the refractive index of water; then, $\frac{c}{\mu}$ is the velocity of light in water relative to S^1. If V is the velocity of a light relative to S (i.e. laboratoty), then using Relativistic Velocity Addition law, we get

$$V = \frac{u_x + v}{1 + \frac{u_x v}{c^2}} = \frac{\frac{c}{\mu} + v}{1 + \frac{\frac{c}{\mu} v}{c^2}} = \frac{c}{\mu}\left(1 + \frac{v\mu}{c}\right)\left(1 + \frac{v}{c\mu}\right)^{-1}.$$

Expanding binomially and neglecting higher powers of $\frac{v}{c}$, we get

$$V = \frac{c}{\mu} + v\left(1 - \frac{1}{\mu^2}\right).$$

Coefficient of drag α is given by

$$\alpha = \frac{v\left(1 - \frac{1}{\mu^2}\right)}{v} = \left(1 - \frac{1}{\mu^2}\right).$$

At first, Fresnel obtained theoretically this expression and is known as Fresnel's coefficient of drag. Fizeau verified this result experimentally.

7.6.2 Aberration of Light

Bradley discovered the phenomena of aberration in 1927. The phenomena of aberration results: The speed of light does not depend on the medium of transmission. However, direction of rays depends on the relative motion of the source and observer. Now, we measure the directions of the light ray emitted from a star from two reference frames S and S^1, where S^1 frame is moving with velocity v relative to S frame in x-direction. Since earth revolves around the sun about its orbit, therefore, we assume, the frame S fixed in the sun and S^1 in earth. Let a light from a star P be observed by the observers O and O^1 in systems S and S^1, respectively. Suppose the observers notice that the directions of light make an angle α and α^1 with x-axis, respectively (see Fig. 7.2). Here, the light ray lies on the xy-plane.
    Then,

$$u_x = c\cos\alpha, \quad u_y = c\sin\alpha, \quad u_z = 0$$

$$u_x^1 = c\cos\alpha^1, \quad u_y^1 = c\sin\alpha^1, \quad u_z^1 = 0.$$

Fig. 7.2 Directions of light make angle α and α^1 with x axis

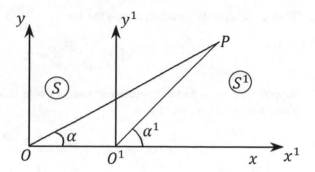

Using, relativistic velocity transformation law (7.10) and (7.11), we get the *relativistic equation of aberration* as

$$\tan \alpha^1 = \frac{u_y}{a(u_x - v)} = \frac{c \sin \alpha}{a(c \cos \alpha - v)} = \frac{\tan \alpha \sqrt{1 - (v/c)^2}}{1 - \frac{v}{c} \sec \alpha}.$$

Classical formula from relativistic result:
   For $v \ll c$ and using $\alpha^1 = \alpha + \delta\alpha$, where $\delta\alpha$ is very small, we get

$$\tan(\alpha + \delta\alpha) = \tan \alpha \left(1 - \frac{v}{c} \sec \alpha\right)^{-1} = \tan \alpha \left(1 + \frac{v}{c} \sec \alpha\right)$$

or

$$\tan(\alpha + \delta\alpha) - \tan \alpha = \frac{\sin \delta\alpha}{\cos(\alpha + \delta\alpha)} = \frac{v \sin \alpha}{c \cos \alpha}.$$

Since $\delta\alpha$ is very small, we finally get the *classical value of aberration* as

$$\delta\alpha = \frac{v}{c} \sin \alpha.$$

7.6.3 Relativistic Doppler Effect

Let a plane wave travelling in the direction (l, m, n) in S frame is given by

$$\psi = A \exp 2\pi i \left(\nu t - \frac{lx + my + nz}{\lambda}\right). \tag{7.25}$$

This ψ satisfies the invariant wave equation

$$\nabla^2 \psi - \frac{1}{c^2} \frac{\partial^2 \psi}{\partial t^2} = 0.$$

Suppose another frame of reference S^1 moving with velocity U in x-direction. This plane wave in S^1 is given by

$$\psi^1 = A \exp 2\pi i \left(\nu^1 t^1 - \frac{l^1 x^1 + m^1 y^1 + n^1 z^1}{\lambda^1} \right). \tag{7.26}$$

Now, using the transformation of (x^1, y^1, z^1, t^1), ψ^1 takes the following form:

$$\psi^1 = A \exp 2\pi i \left[\nu^1 a \left(t - \frac{Ux}{c^2} \right) - \frac{l^1 a(x - Ut) + m^1 y + n^1 z}{\lambda^1} \right], \tag{7.27}$$

where $a = \frac{1}{\sqrt{1 - U^2/c^2}}$.

Comparing the coefficients of x, y, z and t between Eqs. (7.25) and (7.27), we get

$$\nu = a \left(\nu^1 + \frac{U l^1}{\lambda^1} \right), \quad \frac{l}{\lambda} = a \left(\frac{l^1}{\lambda^1} + \frac{U \nu^1}{c^2} \right), \quad \frac{m}{\lambda} = \frac{m^1}{\lambda^1}, \quad \frac{n}{\lambda} = \frac{n^1}{\lambda^1}. \tag{7.28}$$

For electromagnetic waves, we have

$$\nu \lambda = \nu^1 \lambda^1 = c. \tag{7.29}$$

From Eqs. (7.28) and (7.29), one can get

$$\frac{l}{c} = \frac{l^1 + \frac{U}{c}}{c + U l^1}. \tag{7.30}$$

If the direction of propagation of the wave makes angles θ and θ^1 with the x- and x^1-axes, therefore,

$$l^1 = \cos \theta^1, \quad l = \cos \theta. \tag{7.31}$$

Now, expressing it in terms of angles, we get

$$\cos \theta = \frac{\cos \theta^1 + \frac{U}{c}}{1 + \frac{U \cos \theta^1}{c}}. \tag{7.32}$$

Again, we get

$$\frac{\lambda^1}{\lambda} = \frac{\nu}{\nu^1} = \frac{1 + \frac{U \cos \theta^1}{c}}{\sqrt{1 - U^2/c^2}} = \frac{\sqrt{1 - U^2/c^2}}{1 - \frac{U \cos \theta}{c}}. \tag{7.33}$$

Equation (7.32) gives the *relativistic aberration* and Eq. (7.33) gives *relativistic Doppler effect of light*.

Note 9: If the velocity of the source is along the x axis, then $\theta = \theta^1 = 0$, then one gets

$$\nu = \nu^1 \sqrt{\frac{1 + U/c}{1 - U/c}}, \quad \lambda^1 = \lambda \sqrt{\frac{1 + U/c}{1 - U/c}}.$$

This is known as *relativistic longitudinal Doppler effect*.

Note 10: If the velocity of the source is perpendicular to the direction of propagation, then $\theta^1 = 90°$, then one gets

$$\frac{\lambda^1}{\lambda} = \frac{\nu}{\nu^1} = \frac{1}{\sqrt{1 - U^2/c^2}}.$$

This is known as *relativistic transverse Doppler effect*. This indicates that a change of frequency occurs even when the observer is perpendicular to the direction of motion of the source. This effect is absent in the classical physics.

Note 11: Using the relativistic aberration formula, we can also obtain the transformation formula for the *solid angle* $d\Omega$ of a pencil of rays. From Eq. (7.32), we can write

$$1 + \frac{U \cos \theta^1}{c} = \frac{1 - \frac{U^2}{c^2}}{1 - \frac{U \cos \theta}{c}}.$$

We know,

$$\frac{d\Omega^1}{d\Omega} = \frac{\sin \theta^1 d\theta^1}{\sin \theta d\theta} = \frac{d(\cos \theta^1)}{d(\cos \theta)} = \frac{1 - \frac{U^2}{c^2}}{(1 - \frac{U \cos \theta}{c})^2}.$$

Example 7.1 An electron is moving with speed fc in a direction opposite to that of a moving photon. Find the relative velocity of the electron and photon.

Hint: We know the speed of photon is c. Let speed of electron is f c ($0 < f < 1$). Let the photon and electron are moving along positive and negative directions of x-axis, respectively. Let the electron moving with velocity $-fc$ be at rest in system S. Then we assume that system S^1 or laboratory is moving with velocity fc relative to system S (electron). Therefore, we may write, $v = fc, u_x^1 = c$. Using the theorem of composition of velocities, we get the relative velocity u_x of the electron and photon as

$$u_x = \frac{u_x^1 + v}{1 + \frac{u_x^1 v}{c^2}} = \frac{c + fc}{1 + \frac{cfc}{c^2}} = c.$$

Example 7.2 An observer notices two particles moving away from him in two opposite directions with velocity fc. What is the velocity of one particle with respect to other?

Hint: We may regard one particle as S frame, the observer as the S^1 frame and the other particle as the object whose speed is to be determined in the S frame. Then, $u^1 = fc$ and $v = fc$. Using the theorem of composition of velocities, we get the relative velocity u of one particle with respect to other as

$$u = \frac{u^1 + v}{1 + \frac{u^1 v}{c^2}} = \frac{fc + fc}{1 + \frac{fcfc}{c^2}} = \text{etc.}.$$

Example 7.3 Two particles come towards each other with speed fc with respect to laboratory. What is their relative speed?

Hint: Consider a system S in which the particle having velocity $-fc$ is a rest. Then, the system S^1, i.e. laboratory is moving with velocity fc relative to the system S, i.e. $v = fc$ (see Fig. 7.3). Now, we are to find the velocity of the particle u in system S which is moving with velocity $u^1 = fc$ relative to S^1, i.e. laboratory. Using the theorem of composition of velocities, we get the relative velocity u of the particles with respect to other as

$$u = \frac{u^1 + v}{1 + \frac{u^1 v}{c^2}} = \frac{fc + fc}{1 + \frac{fcfc}{c^2}} = \text{etc.}.$$

Example 7.4 If u and v are velocities in the same direction and V is their resultant velocity given by $\tanh^{-1} \frac{V}{c} = \tanh^{-1} \frac{u}{c} + \tanh^{-1} \frac{v}{c}$, then deduce the law of composition of velocities.

Hint: We know,

$$\tanh^{-1} x = \frac{1}{2} \ln \left(\frac{1+x}{1-x} \right).$$

Let, $\frac{V}{c} = z, \frac{u}{c} = x, \frac{v}{c} = y$, then the given condition implies

Fig. 7.3 The system S^1 is moving with velocity fc relative to the system S

$$\ln\left(\frac{1+x}{1-x}\right)^{\frac{1}{2}} + \ln\left(\frac{1+y}{1-y}\right)^{\frac{1}{2}} = \ln\left(\frac{1+z}{1-z}\right)^{\frac{1}{2}}$$

or

$$\left(\frac{(1+x)(1+y)}{(1-x)(1-y)}\right) = \left(\frac{1+z}{1-z}\right)$$

or

$$z = \frac{x+y}{1+xy}$$

or

$$\frac{V}{c} = \frac{\frac{u}{c}+\frac{v}{c}}{1+\frac{u}{c}\cdot\frac{v}{c}}$$

or

$$V = \frac{u+v}{1+\frac{uv}{c^2}}.$$

This is the law of composition of velocities or Einstein's law of addition of velocities.

Example 7.5 If v and v^1 be the velocities of a particle in two inertial frames S and S^1. Initially, two system coincided. If S^1 is moving with velocity V relative to S in x-direction. Show that $v^2 = \frac{v^{12}+V^2+2v^1 V \cos\theta^1 - (\frac{v^1 V \sin\theta^1}{c})^2}{(1+\frac{Vv^1\cos\theta^1}{c^2})^2}$, where θ^1 is the angle which v^1 makes with x^1-axis. Find also the direction of the velocity in S system.

Hint: According to the problem (see Fig. 7.4),

$$v^2 = v_x^2 + v_y^2, \quad v^{12} = v_x^{12} + v_y^{12} \quad \text{and} \quad v_x^1 = v^1\cos\theta^1, \quad v_y^1 = v^1\sin\theta^1.$$

Fig. 7.4 Direction of the
motion of the particle makes
θ and θ^1 angles with x- and
x^1-axes, respectively

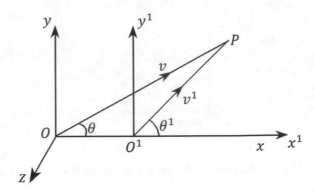

Also, we know

$$v_x^1 = \frac{v_x - V}{1 - \frac{v_x V}{c^2}}, \text{ and } v_y^1 = v_y \frac{\sqrt{1 - V^2/c^2}}{1 - \frac{v_x V}{c^2}}.$$

Thus,

$$v_x = \frac{v_x^1 + V}{1 + \frac{v_x^1 V}{c^2}}, \text{ and } v_y = v_y^1 \frac{\sqrt{1 - V^2/c^2}}{1 + \frac{v_x^1 V}{c^2}}.$$

Hence,

$$v^2 = \frac{(v_x^1 + V)^2 + v_y^{1^2}(1 - \frac{V^2}{c^2})}{(1 + \frac{v_x^1 V}{c^2})^2} = \frac{(v^1 \cos \theta^1 + V)^2 + (v^1 \sin \theta^1)^2(1 - \frac{V^2}{c^2})}{(1 + \frac{v^1 \cos \theta^1 V}{c^2})^2} \text{ etc..}$$

Now, the direction of the velocity in S system is given by

$$\tan \theta = \frac{v_y}{v_x} = \frac{\left(v_y^1 \frac{\sqrt{1-V^2/c^2}}{1+\frac{v_x^1 V}{c^2}}\right)}{\left(\frac{v_x^1 + V}{1+\frac{v_x^1 V}{c^2}}\right)} = \frac{\left(v^1 \sin \theta^1 \frac{\sqrt{1-V^2/c^2}}{1+\frac{v^1 \cos \theta^1 V}{c^2}}\right)}{\left(\frac{v^1 \cos \theta^1 + V}{1+\frac{v^1 \cos \theta^1 V}{c^2}}\right)} \text{ etc..}$$

Example 7.6 If a photon travels the path in such a way that it moves in $x^1 - y^1$
plane and makes an angle ϕ with x-axis of the system S^1, then prove that for the
frame S, $u_x^2 + u_y^2 = c^2$. Here, S and S^1 are two inertial frames.

Hint: We know the velocity of photon is c. Its components in S^1 system along x-
and y-axes, respectively, are (see Fig. 7.5)

$$u_x^1 = c \cos \phi \text{ and } u_y^1 = c \sin \phi.$$

Fig. 7.5 The path of the photon makes ϕ angle with x-axis

Let the frame S^1 is moving with velocity v relative to S frame, then from the theorem of composition of velocities, we have

$$u_x = \frac{u_x^1 + v}{1 + \frac{u_x^1 v}{c^2}}, \quad u_y = \frac{u_y^1 \sqrt{1 - v^2/c^2}}{1 + \frac{u_x^1 v}{c^2}}.$$

Now,

$$u_x^2 + u_y^2 = \left[\frac{c \cos \phi + v}{1 + \frac{c \cos \phi v}{c^2}} \right]^2 + \left[\frac{c \sin \phi \sqrt{1 - v^2/c^2}}{1 + \frac{c \cos \phi v}{c^2}} \right]^2 = c^2.$$

Example 7.7 An observer moving along x axis of S frame at a velocity v observes a body of proper volume V_0 moving at a velocity u along the x-axis of S. Show that the observer measures the volume to be equal to $V_0 \frac{\sqrt{c^2 - u^2} \sqrt{c^2 - v^2}}{c^2 - uv}$.

Hint: Let S^1 be the frame containing the observer (see Fig. 7.6). The relative velocity between S^1 and S is v along x-direction. The velocity of the body relative to S is u along x-direction. Therefore, its velocity relative to S^1 is

$$u^1 = \frac{u - v}{1 - \frac{uv}{c^2}}.$$

Therefore, an observer measures the volume of the body relative to him as

$$V = V_0 \sqrt{1 - \frac{u^{1^2}}{c^2}} = V_0 \sqrt{1 - \frac{\left(\frac{u - v}{1 - uv/c^2} \right)^2}{c^2}}.$$

Example 7.8 A particle has velocity $\vec{u} = (u_1, u_2, u_3)$ in S and $\vec{u^1} = (u_1^1, u_2^1, u_3^1)$ in S^1. Show that $c^2 - u^2 = \frac{c^2(c^2 - u^{1^2})(c^2 - v^2)}{(c^2 + u_1^1 v)^2}$.

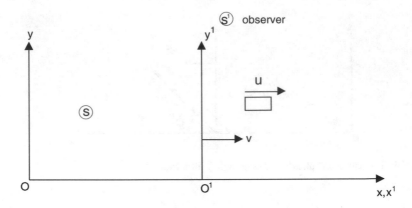

Fig. 7.6 A body is moving with velocity u along x-axis

Hint: We know, inverse transformations are obtained from by replacing v by $-v$ and subscripted quantities by unsubscripted ones and vice versa. Therefore, from note 6, we can write

$$\sqrt{1 - (u/c)^2} = \frac{\sqrt{1 - (v/c)^2}\sqrt{1 - (u^1/c)^2}}{(1 + \frac{u_1^1 v}{c^2})}$$

or

$$c^2 - u^2 = \frac{c^2(c^2 - u^{12})(c^2 - v^2)}{(c^2 + u_1^1 v)^2}.$$

Example 7.9 How fast would we need to drive towards a red traffic light for the light to appear green?

Hint: We know that the wavelengths of red light and green light are $\lambda_{\text{red}} \approx 7 \times 10^{-5}$ cm and $\lambda_{\text{green}} \approx 5 \times 10^{-5}$ cm, respectively. As velocity U of the car towards the signal, we use the relativistic formula for Doppler effect as $\lambda = \lambda^1 \sqrt{\frac{1 + U/c}{1 - \frac{U}{c}}}$. Thus, we have

$$5 \times 10^{-5} = 7 \times 10^{-5} \sqrt{\frac{1 - U/c}{1 + \frac{U}{c}}} \implies U = \text{etc..}$$

Example 7.10 Calculate the Doppler shift in wavelength for light of wavelength a Angstrom (i) when the source approaches to the observer at a velocity fc. (ii) when the source moves transversely to the line of sight at the same velocity fc.

Hint:

$$\text{(i) } \lambda = a\sqrt{\frac{1 - fc/c}{1 + fc/c}} \text{ etc. (ii) } \lambda = a\frac{1}{\sqrt{1 - (fc)^2/c^2}} \text{ etc..}$$

Example 7.11 If u and u^1 are the velocities of a particle in the frames S and S^1, which are moving with velocity relative to each other. Prove that $\sqrt{1 - \left(u^1/c\right)^2} = \frac{\sqrt{1-(v/c)^2}\sqrt{1-(u/c)^2}}{(1-\frac{u_x v}{c^2})}$.

Hint: Use Eq. (7.13).

Example 7.12 Show that velocity of light is the same in all inertial frame.

Hint: Let velocity in S^1 system is c^1, then

$$c^1 = \frac{dx^1}{dt^1} = \frac{a(dx - vdt)}{a\left(dt - \frac{vdx}{c^2}\right)} = \frac{\frac{dx}{dt} - v}{1 - \frac{v}{c^2}\frac{dx}{dt}}.$$

Now, velocity in S system is $\frac{dx}{dt} = c$, then using $\frac{dx}{dt} = c$, we get

$$c^1 = \frac{c - v}{1 - \frac{v}{c^2}c} = c.$$

Example 7.13 Let a reference frame moves with velocity $0.8c$ relative to S frame in $x-$ direction. A particle rest in S' frame experiences an acceleration in it represented by $\vec{f'} = 2i + 3j + 4k$. What is the acceleration measured from S?

Hint: Since the particle is at rest in S' frame, therefore

$$f_x = (1 - \frac{v^2}{c^2})^{\frac{3}{2}} f'_x,$$

$$f_y = (1 - \frac{v^2}{c^2}) f'_y,$$

$$f_z = (1 - \frac{v^2}{c^2}) f'_z.$$

Hence, $f = if_x + jf_y + kf_z$.

Chapter 8
Four-Dimensional World

Keywords Proper time · World velocity · Properties of four vectors

8.1   Four-Dimensional Space–Time

If any physical law may be expressed in a invariant four-dimensional form, then the law will be invariant under Lorentz Transformation. Minkowski said that the external world is formed by four-dimensional space–time continuum known as *Minkowski or World space*. The event in world space must be represented by three spatial coordinates and one time coordinate. The events in the world space are represented by points known as *world points*. In this world space there corresponds to each particle a certain line known as *world line*. In other words, a curve in the four dimensional space is called world line. Newton's laws retain their form under Galilean Transformation which is found to be incorrect and hence Newton's laws are inaccurate representation of experimental phenomena. We want to extend the three-vector law

$$F_i = \frac{d}{dt}\left[m\left(\frac{dx_i}{dt}\right)\right]$$

to four-vector form because above three-vector form does not satisfy the principle of relativity, i.e. its form changes under Lorentz Transformation.

The basic representation for such an extension are that:

1. A fourth component of force must be introduced.
2. The time is no more a scalar invariant and changes under Lorentz Transformation.

Note 1: In Newtonian mechanics, the physical parameters like position, velocity, momentum, acceleration, force have three-vector forms. We want to extend the three-vector form to four-vector form in the relativistic mechanics in order to satisfy the principle of relativity. Therefore, a fourth component of all physical parameters should be introduced.

8.2 Proper Time

The time recorded by a clock moving with a given system is called **proper time**.

Let S be a reference system which is at rest. Let a clock is moving with velocity v relative to S. Then at each instant we can introduce a coordinate system, rigidly linked to the moving clock, which constitutes an inertial reference system S^1. Suppose at any instant, a clock in S system indicates a time dt and at this time, the clock travels a distance $\sqrt{(dx^2 + dy^2 + dz^2)}$. Obviously, the relative velocity between the clock and S^1 system is zero, i.e. the clock is at rest in the system linked with them, i.e. in reference system S^1. Thus, $dx^1 = dy^1 = dz^1 = 0$. The interval

$$d\rho^2 = c^2 dt^2 - dx^2 - dy^2 - dz^2 = c^2 dt^{1^2} - dx^{1^2} - dy^{1^2} - dz^{1^2} = c^2 dt^{1^2}.$$

Let the moving clock in S^1 system reads dt^1 corresponding to the time dt. Then

$$dt^{1^2} = \frac{d\rho^2}{c^2} = \left(\frac{1}{c^2}\right)[dt^2 - dx^2 - dy^2 - dz^2]$$

or

$$dt^1 = dt\left[1 - \frac{1}{c^2}\left\{\left(\frac{dy}{dt}\right)^2 + \left(\frac{dx}{dt}\right)^2 + \left(\frac{dz}{dt}\right)^2\right\}\right]^{\frac{1}{2}} = dt\sqrt{1 - \frac{v^2}{c^2}} \qquad (8.1)$$

[v is the velocity of the moving clock]

Here, dt^1 is the proper read by the moving clock corresponding to dt read by the clock at rest. Important to note that $dt^1 < dt$. Thus proper time of a moving objects is always smaller than the corresponding time in the clock at rest.

Alternative
Let S be a reference system which is at rest. Let a rocket constitutes another reference system S^1 which is moving with velocity v relative to S along +ve direction of x-axis. Suppose at any instant, a clock in S system indicates a time t. Then an observer in rocket, i.e. in S^1 frame measures this time as t^1 which is given according to Lorentz Transformation as

$$t^1 = a\left[t - \frac{vx}{c^2}\right]$$

where $a = \dfrac{1}{\sqrt{1 - \frac{v^2}{c^2}}}$.

Since t is the time recorded by an observer at rest and v is the velocity of the rocket, therefore, at this time, the clock, i.e. the rocket travels a distance $x = vt$. Putting this value in the above equation, we get

$$t^1 = a\left[t - \frac{vx}{c^2}\right] = at\left[1 - \frac{v^2}{c^2}\right] = t\sqrt{1 - \frac{v^2}{c^2}}. \tag{8.2}$$

So that $t > t^1$. Thus the time recorded by the clock in rocket, i.e. in moving frame differs the time recorded by the clock at rest for the same journey. Hence, two clocks in the two systems S and S^1 run at different rate. Also we see that the time recorded in the rest frame to be greater than that the moving system. Thus a moving clock runs more slow than a stationary one.

Theorem 8.1 *Proper time is invariant under Lorentz Transformation.*

Proof We know the proper time,

$$d\tau = dt\sqrt{1 - \frac{v^2}{c^2}}$$

or,

$$d\tau = dt\left[1 - \frac{1}{c^2}\left\{\left(\frac{dy}{dt}\right)^2 + \left(\frac{dx}{dt}\right)^2 + \left(\frac{dz}{dt}\right)^2\right\}\right]^{\frac{1}{2}}.$$

This implies

$$d\tau^2 = \frac{1}{c^2}[c^2dt^2 - dx^2 - dy^2 - dz^2].$$

Now,

$$d\tau^{1^2} = \frac{1}{c^2}[c^2dt^{1^2} - dx^{1^2} - dy^{1^2} - dz^{1^2}].$$

Using the Lorentz Transformation

$$x^1 = a(x - vt); \quad y^1 = y; \quad z^1 = z; \quad t^1 = a\left[t - \left(\frac{xv}{c^2}\right)\right],$$

we get

$$d\tau^{1^2} = \frac{1}{c^2}\left[c^2a^2\left\{dt - \left(\frac{vdx}{c^2}\right)\right\}^2 - a^2(dx - vdt)^2 - dy^2 - dz^2\right]$$

$$= \frac{1}{c^2}[c^2dt^2 - dx^2 - dy^2 - dz^2] = d\tau^2.$$

Hence proper time is invariant under Lorentz Transformation.

Alternative:
We know

$$ds^2 = c^2 dt^2 - dx^2 - dy^2 - dz^2$$

$$= c^2 dt^2 \left[1 - \frac{1}{c^2} \left\{ \left(\frac{dy}{dt} \right)^2 + \left(\frac{dx}{dt} \right)^2 + \left(\frac{dz}{dt} \right)^2 \right\} \right]$$

$$= c^2 dt^2 \left[1 - \frac{v^2}{c^2} \right].$$

Therefore,

$$\left(\frac{ds}{c} \right) = dt \sqrt{1 - \frac{v^2}{c^2}} = d\tau$$

Since ds is invariant under Lorentz Transformation and c is also an invariant quantity, therefore, $d\tau$ is invariant under Lorentz Transformation.

8.3   World Velocity or Four Velocities

We know the expression

$$dx^2 + dy^2 + dz^2 - c^2 dt^2$$

remains invariant under Lorentz Transformation, i.e.

$$dx^2 + dy^2 + dz^2 - c^2 dt^2 = dx'^2 + dy'^2 + dz'^2 - c^2 dt'^2.$$

In terms of differential, the quantity

$$ds^2 = dx^2 + dy^2 + dz^2 - c^2 dt^2$$

remains invariant.

It can be written as

$$ds^2 = dx_1^2 + dx_2^2 + dx_3^2 + dx_4^2,$$
$$\text{where,} \quad x_1 = x, \quad x_2 = y, \quad x_3 = z \quad \text{and} \quad x_4 = ict.$$

Here ds is called *invariant space–time interval*.

The position four vectors, i.e. *world point* takes the form

$$x_\mu = [x_1, x_2, x_3, ict].\tag{8.3}$$

World velocity u_μ is defined as the rate of change of the position vector of a particle with respect to its proper time, i.e.

$$u_\mu = \frac{dx_\mu}{d\tau}.\tag{8.4}$$

The space components of u_μ are

$$\frac{dx_j}{d\tau} = \frac{dx_j}{dt\sqrt{1 - \frac{v^2}{c^2}}} = \frac{v_j}{\sqrt{1 - \frac{v^2}{c^2}}}\tag{8.5}$$

($j = 1, 2, 3$ and $v^2 = v_1^2 + v_2^2 + v_3^2$ where $v_j = \frac{dx_j}{dt}$ is the three-dimensional velocity)

The time component of u_μ is

$$u_4 = \frac{icdt}{d\tau} = \frac{ic}{\sqrt{1 - \frac{v^2}{c^2}}}.\tag{8.6}$$

Thus world velocity is given by

$$u_\mu = \frac{dx_\mu}{d\tau} = a\left[\frac{dx_1}{dt}, \frac{dx_2}{dt}, \frac{dx_3}{dt}, ic\right] = a\,[v_1, v_2, v_3, ic]\,,\tag{8.7}$$

where $a = \frac{1}{\sqrt{1 - \frac{v^2}{c^2}}}$.

In Euclidean geometry a vector has three components. In relativity, a vector has four components. Let us consider the transformations of coordinates $x \longrightarrow x^1$. Then we have

$$dx^1 = \left(\frac{\partial x^1}{\partial x}\right) dx.$$

[we use summation convention, therefore ignore Σ sign]

If the transformation of any vector A^μ is given by

$$A^{\mu 1} = \left(\frac{\partial x^{\mu 1}}{\partial x^\nu}\right) A^\nu,$$

then A^μ is called *contra-variant vector*.

If the transformation of any vector B_μ is given by

$$B_\mu^1 = \left(\frac{\partial x^\nu}{\partial x^{\mu 1}}\right) B_\nu,$$

then B_μ is called *covariant vector*.

In Minkowski four-dimensional space, the distance between two neighbouring points is defined by the metric

$$ds^2 = c^2 dt^2 - dx^2 - dy^2 - dz^2 = g_{\mu\nu} dx^\mu dx^\nu.$$

Here $g_{\mu\nu}$ is known as fundamental tensor,

$$g_{11} = g_{22} = g_{33} = -1; \quad g_{44} = 1.$$

If A^μ, i.e. contra-variant vector is known, then covariant component can be found with the help of fundamental tensor as $A_\mu = g_{\mu\nu} A^\nu$.

If A_μ, i.e. covariant vector is known, then contra-variant component is given by $A^\mu = g^{\mu\nu} A_\nu$. Here $g^{\mu\nu}$ is the inverse of $g_{\mu\nu}$.

$A^\mu A_\mu$ is called *norm of the vector* A^μ.

Note 1: For any two four vectors a^μ and b^μ, the four-dimensional dot product

$$a^\mu \cdot b^\mu \equiv a_0 b_0 - \vec{a}.\vec{b},$$

is an **invariant quantity**, i.e. independent of the reference frame. In particular, the invariant length of any $four - vector$ v^μ is $v^\mu \cdot v^\mu = v^\mu v_\mu$, which is known as **norm**.

This four-dimensional dot product can be written in matrix form as

$$a^\mu \cdot b^\mu = \begin{pmatrix} a_0 & a_1 & a_2 & a_3 \end{pmatrix} \begin{pmatrix} 1 & 0 & 0 & 0 \\ 0 & -1 & 0 & 0 \\ 0 & 0 & -1 & 0 \\ 0 & 0 & 0 & -1 \end{pmatrix} \begin{pmatrix} b_0 \\ b_1 \\ b_2 \\ b_3 \end{pmatrix} \qquad (i)$$

or,

$$a^\mu \cdot b^\mu = a^{\mu T} \eta b^\mu \quad \text{where} \quad \eta = \begin{pmatrix} 1 & 0 & 0 & 0 \\ 0 & -1 & 0 & 0 \\ 0 & 0 & -1 & 0 \\ 0 & 0 & 0 & -1 \end{pmatrix}. \tag{ii}$$

Here a^T denotes the transpose of the column matrix a.

Let L denotes the Lorentz Transformation matrix (see Eq. (8.13)) that transforms the *four vectors* a^μ and b^μ in S to the corresponding *four vectors* in S'. In matrix notation this means

$$a'^\mu = La^\mu \text{ and } b'^\mu = Lb^\mu.$$

The invariance of the four-dimensional dot product implies that

$$a'^{\mu T} \eta b'^\mu = a^{\mu T} \eta b^\mu.$$

Now,

$$a'^{\mu T} \eta b'^\mu = (La^\mu)^T \eta L b^\mu$$

$$= a^{\mu T} L^T \eta L b^\mu = a^{\mu T} \eta b^\mu,$$

if

$$L^T \eta L = \eta,$$

for arbitrary a^μ and b^μ.

Therefore, the Lorentz Transformation matrix should follow the defining property,

$$L^T \eta L = \eta.$$

Note 2: The theory of special relativity is established on Minkowski space which is a four-dimensional flat linear space endowed with the Lorentz inner product or metric. A matrix L satisfies the equation

$$L^T \eta L = \eta, \quad \eta = diag(1, -1, -1, -1)$$

is a Lorentz Transformation.

Check whether the matrices represent the Lorentz Transformations or not:

$$L = \begin{bmatrix} \gamma & -\gamma\beta & 0 & 0 \\ -\gamma\beta & \gamma & 0 & 0 \\ 0 & 0 & 1 & 0 \\ 0 & 0 & 0 & 1 \end{bmatrix}, \quad \left(\gamma = \frac{1}{\sqrt{1 - \beta^2}} \right); \quad L = \begin{bmatrix} \cosh\phi & -\sinh\phi & 0 & 0 \\ -\sinh\phi & \cosh\phi & 0 & 0 \\ 0 & 0 & 1 & 0 \\ 0 & 0 & 0 & 1 \end{bmatrix}.$$

(here, $\sinh\phi = \beta\gamma$ and $\beta = \frac{v}{c}$)

Hints: Check the relation $L^T \eta L = \eta$.

Theorem 8.2 *Norm is an invariant quantity.*

Proof

$$A^{\mu 1} A^1_\mu = \left[\left(\frac{\partial x^{\mu 1}}{\partial x^\nu} \right) A^\nu \right] \left[\left(\frac{\partial x^\sigma}{\partial x^{\mu 1}} \right) A_\sigma \right]$$

$$= \left(\frac{\partial x^{\mu 1}}{\partial x^\nu} \right) \left(\frac{\partial x^\sigma}{\partial x^{\mu 1}} \right) A^\nu A_\sigma$$

$$= \delta^\sigma_\nu A^\nu A_\sigma = A^\sigma A_\sigma = A^\mu A_\mu.$$

Thus norm is invariant and it is a scalar.

$A^\mu A_\mu = g_{\mu\nu} A^\mu A^\nu$ if A^μ is known.

$A^\mu A_\mu = g^{\mu\nu} A_\mu A_\nu$ if A_μ is known.

If $A^\mu A_\mu > 0$, then A^μ is called *time like vector* with signature $(+, -, -, -)$.
Alternatively,

If $A^\mu A_\mu < 0$, then A^μ is called *time like vector* with signature $(-, +, +, +)$.

If $A^\mu A_\mu < 0$, then A^μ is called *space like vector* with signature $(+, -, -, -)$.
Alternatively,

If $A^\mu A_\mu > 0$, then A^μ is called *space like vector* with signature $(-, +, +, +)$.

If $A^\mu A_\mu = 0$, then A^μ is called *null or light like vector.*

Theorem 8.3 *Any four vectors orthogonal to a time like vector is space like.*

Proof Let A^μ is a time like vector, therefore,

$$A^\mu A_\mu = [(A^0)^2 - (A^1)^2 - (A^2)^2 - (A^3)^2]$$

[A^0 = time component and A^1, A^2, A^3 are spatial components]

Now we transform the coordinate axis in such a way that $A^1 = A^2 = A^3 = 0$.
Then $A^\mu A_\mu = (A^0)^2 > 0$, i.e. $A^0 \neq 0$.

If B^μ be orthogonal vector to A^μ, then

$$A^\mu B_\mu = [A^0 B_0 + A^1 B_1 + A^2 B_2 + A^3 B_3] = 0,$$

i.e.

$$A^0 B^0 - (A^1 B^1 + A^2 B^2 + A^3 B^3) = 0$$

Therefore, $A^0 B^0 = 0$ [as $A^1 = A^2 = A^3 = 0$].
Hence, $B^0 = 0$ [as $A^0 \neq 0$].
Now,

$$B^\mu B_\mu = [(B^0)^2 - (B^1)^2 - (B^2)^2 - (B^3)^2] = -[(B^1)^2 + (B^2)^2 + (B^3)^2] < 0.$$

Therefore, B^μ is a space like vector.

Theorem 8.4 *Two-time-like vectors cannot be orthogonal to each other.*

Proof Let A^μ and B^μ be two-time-like vectors, therefore,

$$A^\mu A_\mu = [(A^0)^2 - (A^1)^2 - (A^2)^2 - (A^3)^2] > 0 \tag{8.8}$$

and

$$B^\mu B_\mu = [(B^0)^2 - (B^1)^2 - (B^2)^2 - (B^3)^2] > 0 \tag{8.9}$$

[A^0, B_0 are time components and A^1, A^2, A^3 & B^1, B^2, B^3 are spatial components]
If possible, let two vectors are orthogonal, then

$$A^0 B^0 - (A^1 B^1 + A^2 B^2 + A^3 B^3) = 0$$

or

$$(A^0 B^0)^2 = (A^1 B^1 + A^2 B^2 + A^3 B^3)^2. \tag{8.10}$$

Multiplying (8.8) and (8.9), we get

$$(A^0 B^0)^2 > [(A^1)^2 + (A^2)^2 + (A^3)^2][(B^1)^2 + (B^2)^2 + (B^3)^2]. \tag{8.11}$$

Using (8.10), we get

$$(A^1 B^1 + A^2 B^2 + A^3 B^3)^2 > [(A^1)^2 + (A^2)^2 + (A^3)^2][(B^1)^2 + (B^2)^2 + (B^3)^2]).$$

This implies

$$0 > (A^1 B^2 - A^2 B^1)^2 + (A^1 B^3 - A^3 B^1)^2 + (A^2 B^3 - A^3 B^2)^2.$$

Since sum of three squares of real quantities in the right-hand side is always positive, therefore, the above result is impossible. Hence two-time-like vectors cannot be orthogonal to each other.

Theorem 8.5 *Any four vectors orthogonal to a space like vector may be time like, space like or null.*

Proof Let A^μ is a space like vector with A^0 = time component and A^1, A^2, A^3 are spatial components.
Now we choose the inertial frame in such a way that $A^0 = A^2 = A^3 = 0$, i.e. $A^\mu = (0, A^1, 0, 0)$. Then $A^\mu A_\mu = -(A^1)^2 < 0$, i.e. $A^1 \neq 0$.
If B^μ be orthogonal vector to A^μ, then

$$A^\mu B_\mu = [A^0 B_0 + A^1 B_1 + A^2 B_2 + A^3 B_3] = 0,$$

i.e.

$$A^0 B^0 - (A^1 B^1 + A^2 B^2 + A^3 B^3) = 0$$

Therefore, $A^1 B^1 = 0$ [as $A^0 = A^2 = A^3 = 0$].
Hence, $B^1 = 0$ [as $A^1 \neq 0$].
Now,

$$B^\mu B_\mu = [(B^0)^2 - (B^2)^2 - (B^3)^2]$$

The right-hand side may be positive, negative or zero. Therefore, B^μ may be time like, space like or null.

Theorem 8.6 *Any four vectors orthogonal to a null vector is either space like or null.*

Proof Let A^μ is a null vector with A^0 = time component and A^1, A^2, A^3 are spatial components.

Now we transform the space axes in such a way that $A^2 = A^3 = 0$, i.e. $A^\mu = (A^0, A^1, 0, 0)$. Then $A^\mu A_\mu = (A^0)^2 - (A^1)^2 = 0$ implies $(A^1)^2 = (A^0)^2$. If B^μ be orthogonal vector to A^μ, then

$$A^\mu B_\mu = [A^0 B_0 + A^1 B_1 + A^2 B_2 + A^3 B_3] = 0,$$

i.e.

$$A^0 B^0 - (A^1 B^1) = 0$$

Therefore, $(B^1)^2 = (B^0)^2$ [as $(A^1)^2 = (A^0)^2$].
Now,

$$B^\mu B_\mu = [(B^0)^2 - (B^1)^2 - (B^2)^2 - (B^3)^2] = -[(B^2)^2 + (B^3)^2] \leq 0.$$

We use the equality sign since, both B^2 and B^3 happen to vanish. Therefore, B^μ is either space like or null.

The *magnitude or norm* of the world velocity is given by

$$u_\mu \cdot u_\mu \equiv u_1 u_1 + u_2 u_2 + u_3 u_3 + u_4 u_4$$
$$= a^2(v_1^2 + v_2^2 + v_3^2 - c^2) = a^2(v^2 - c^2) = -c^2.$$

[This is scalar product of given four vectors]

The square of world velocity, i.e. norm of the world velocity has a constant magnitude. So it is invariant under Lorentz Transformation. Here we see the usefulness of the four-dimensional language in relativity that all equations expressed in four-vector language will automatically retain their mathematical form under Lorentz Transformation and hence satisfy the postulates of special theory of relativity.

8.4 Lorentz Transformation of Space and Time in Four-Vector Form

Consider two systems S and S^1, the latter moving with velocity v relative to former along +ve direction of x-axis. In Minkowski space, let the coordinates of an event be represented by (x, y, z, ict) or x^μ ($\mu = 1, 2, 3, 4$) where $x_1 = x, x_2 = y, x_3 = z$ and $x_4 = ict$. The Lorentz Transformation

$$x^1 = a(x - vt), \quad y^1 = y, \quad z^1 = z, \quad t^1 = a\left[t - \left(\frac{vx}{c^2}\right)\right]$$

can be written as

$$x_1^1 = ax_1 + 0.x_2 + 0.x_3 + ia\beta x_4$$
$$x_2^1 = 0.x_1 + 1.x_2 + 0.x_3 + 0.x_4$$
$$x_3^1 = 0.x_1 + 0.x_2 + 1.x_3 + 0.x_4$$
$$x_4^1 = -ia\beta x_1 + 0.x_2 + 0.x_3 + ax_4,$$

where $a = \frac{1}{\sqrt{1-\beta^2}}$ and $\beta = \frac{v}{c}$.
This implies

$$x_\mu^1 = a_{\mu\nu}x_\nu, \tag{8.12}$$

where $x_\mu^1 = [x_1^1, x_2^1, x_3^1, x_4^1]^T$, $x_\mu = [x_1, x_2, x_3, x_4]^T$ and

$$a_{\mu\nu} = \begin{bmatrix} a & 0 & 0 & ia\beta \\ 0 & 1 & 0 & 0 \\ 0 & 0 & 1 & 0 \\ -ia\beta & 0 & 0 & a \end{bmatrix}. \tag{8.13}$$

Equation (8.12) represents the Lorentz Transformation of space and time in four-vector form.

Similarly, *Lorentz Transformation of any four vectors A_μ* ($\mu = 1, 2, 3, 4$) may be expressed as

$$A_\mu^1 = a_{\mu\nu}A_\nu, \tag{8.14}$$

where $a_{\mu\nu}$ is given in (8.13). One can verify $\mid a_{\mu\nu} \mid = 1$. Also, the Transformation matrix $a_{\mu\nu}$ is orthogonal since $[a_{\mu\nu}]^{-1} = [a_{\mu\nu}]^T$. Thus the *Transformation matrix $a_{\mu\nu}$ is a unitary orthogonal matrix.* The inverse Lorentz Transformation matrix $a_{\mu\nu}^1$ is given by

$$a^1_{\mu\nu} = [a_{\mu\nu}]^{-1} = \begin{bmatrix} a & 0 & 0 & -i a\beta \\ 0 & 1 & 0 & 0 \\ 0 & 0 & 1 & 0 \\ i a\beta & 0 & 0 & a \end{bmatrix}. \tag{8.15}$$

The inverse Lorentz Transformation of any four vectors can be obtained as

$$A_\mu = a^1_{\mu\nu} A^1_\nu. \tag{8.16}$$

Four vectors are four tensors of rank 1. A four tensor of rank 2 has 4^2 components $T_{\mu\nu}$. Here, we can also use the Lorentz Transformation matrix to get the required Lorentz Transformation of the tensor $T_{\mu\nu}$ as

$$T^1_{\mu\nu} = a_{\mu\sigma} a_{\nu\lambda} T_{\sigma\lambda}. \tag{8.17}$$

In general, a four tensor of rank n has 4^n components, which are written with n indices and they transform as above

$$T^1_{\mu\nu\rho...\omega} = a_{\mu\sigma} a_{\nu\lambda} a_{\rho\alpha} \ldots a_{\omega\beta} T_{\sigma\lambda\alpha...\beta}. \tag{8.18}$$

Note 3: Let L be a 4×4 matrix that represents a Lorentz Transformation (here, $L^T \eta L = \eta$). The following transformations of the type

$$x' = Lx + a$$

are known as **Poincaré transformations** (L, a), where x and x' are column vectors (4 vectors) representing spacetime coordinates of events as seen from S and S' frames, and a is a 4×1 column vector (translation vector) . This transformation is actually a combination of a Lorentz Transformation and Translation of the space and time coordinates. It is also known as **inhomogeneous Lorentz Transformation**. The Poincaré transformation leaves the interval invariant, $ds^2 = ds'^2$.

Two Poincaré transformations (L_1, a_1) and (L_2, a_2) , applied successively, is the Poincaré transformation, i.e. obey the composition law.

Hint: Here

$$x'' = L_2 x' + a_2 = L_2(L_1 x + a_1) + a_2 = (L_2 L_1) x + (L_2 a_1 + a_2).$$

One can see that

$$(L_2 L_1)^T \eta (L_2 L_1) = L_1^T (L_2^T \eta L_2) L_1 = L_1^T \eta L_1 = \eta.$$

So, two Poincaré transformations applied successively are a Poincaré transformation $(L_2 L_1, L_2 a_1 + a_2)$.

These transformations also form a group, known as the **Poincaré group**.

Example 8.1 Consider three inertial frames S, S^1 and S^2. Let S^1 is moving with a constant velocity v_1 along x-axis of S and S^2 is moving with a constant velocity v_2 along y-axis of S^1. Find the transformation between S and S^2. Again, S^1 is moving with a constant velocity v_2 along y-axis of S and S^2 is moving with a constant velocity v_1 along x-axis of S^1. Find the transformation between S and S^2. Show also these two transformations are not equal.

Hint: The transformation matrix from S to S^1 is

$$L_1 = \begin{bmatrix} a_1 & 0 & 0 & ia_1\beta_1 \\ 0 & 1 & 0 & 0 \\ 0 & 0 & 1 & 0 \\ -ia_1\beta_1 & 0 & 0 & a_1 \end{bmatrix}$$

where $a_1 = \frac{1}{\sqrt{1-\beta_1^2}}$ and $\beta_1 = \frac{v_1}{c}$.

The transformation matrix from S^1 to S^2 is

$$L_2 = \begin{bmatrix} 1 & 0 & 0 & 0 \\ 0 & a_2 & 0 & ia_2\beta_2 \\ 0 & 0 & 1 & 0 \\ 0 & -ia_2\beta_2 & 0 & a_2 \end{bmatrix},$$

where $a_2 = \frac{1}{\sqrt{1-\beta_2^2}}$ and $\beta_2 = \frac{v_2}{c}$.

Hence the transformation matrix from S to S^2 is

$$L_1 L_2 = \begin{bmatrix} a_1 & 0 & 0 & ia_1\beta_1 \\ 0 & 1 & 0 & 0 \\ 0 & 0 & 1 & 0 \\ -ia_1\beta_1 & 0 & 0 & a_1 \end{bmatrix} \begin{bmatrix} 1 & 0 & 0 & 0 \\ 0 & a_2 & 0 & ia_2\beta_2 \\ 0 & 0 & 1 & 0 \\ 0 & -ia_2\beta_2 & 0 & a_2 \end{bmatrix} = \text{etc.}$$

When the order is reversed then the transformation matrix from S to S^2 is

$$L_2 L_1 = \begin{bmatrix} 1 & 0 & 0 & 0 \\ 0 & a_2 & 0 & ia_2\beta_2 \\ 0 & 0 & 1 & 0 \\ 0 & -ia_2\beta_2 & 0 & a_2 \end{bmatrix} \begin{bmatrix} a_1 & 0 & 0 & ia_1\beta_1 \\ 0 & 1 & 0 & 0 \\ 0 & 0 & 1 & 0 \\ -ia_1\beta_1 & 0 & 0 & a_1 \end{bmatrix} = \text{etc.}$$

Obviously,

$$L_1 L_2 \neq L_2 L_1.$$

Example 8.2 Find the Lorentz Transformation of four velocities. Using it obtain Einstein's law of addition of velocities.

Hint: We know four velocities takes the form

$$U_\mu = [U_1, U_2, U_3, U_4] = \frac{1}{\sqrt{1 - \frac{u^2}{c^2}}} [u_1, u_2, u_3, ic],$$

where $u^2 = u_1^2 + u_2^2 + u_3^2$. Now, we use the Lorentz Transformation matrix to get the required Lorentz Transformation of U_μ as

$$\begin{bmatrix} U_1^1 \\ U_2^1 \\ U_3^1 \\ U_4^1 \end{bmatrix} = \begin{bmatrix} a & 0 & 0 & ia\beta \\ 0 & 1 & 0 & 0 \\ 0 & 0 & 1 & 0 \\ -ia\beta & 0 & 0 & a \end{bmatrix} \begin{bmatrix} U_1 \\ U_2 \\ U_3 \\ U_4 \end{bmatrix}$$

where $\beta = \frac{v}{c}$ and $a = \frac{1}{\sqrt{1 - \beta^2}}$.

Writing explicitly, we have

$$U_1^1 = aU_1 + ia\beta U_4$$

$$U_2^1 = U_2$$

$$U_3^1 = U_3$$

$$U_4^1 = aU_4 - ia\beta U_1.$$

Thus we get

$$\frac{u_1^1}{\sqrt{1 - \frac{u^{1^2}}{c^2}}} = \frac{au_1}{\sqrt{1 - \frac{u^2}{c^2}}} + iva\frac{ic}{c\sqrt{1 - \frac{u^2}{c^2}}} = \frac{a(u_1 - v)}{\sqrt{1 - \frac{u^2}{c^2}}}$$

$$\frac{u_2^1}{\sqrt{1 - \frac{u^{1^2}}{c^2}}} = \frac{u_2}{\sqrt{1 - \frac{u^2}{c^2}}}, \quad \frac{u_3^1}{\sqrt{1 - \frac{u^{1^2}}{c^2}}} = \frac{u_3}{\sqrt{1 - \frac{u^2}{c^2}}}$$

$$\frac{ic}{\sqrt{1 - \frac{u^{1^2}}{c^2}}} = \frac{aic}{\sqrt{1 - \frac{u^2}{c^2}}} - iva\frac{u_1}{c\sqrt{1 - \frac{u^2}{c^2}}}.$$

Last equation gives

$$a\left(1 - \frac{u_1 v}{c^2}\right) = \frac{\sqrt{1 - \frac{u^2}{c^2}}}{\sqrt{1 - \frac{u^{1^2}}{c^2}}}.$$

Using this result, we get the required transformation as

$$u_1^1 = \frac{u_1 - v}{1 - \frac{vu_1}{c^2}}, \quad u_2^1 = \frac{u_2\sqrt{1 - \frac{v^2}{c^2}}}{1 - \frac{vu_1}{c^2}}, \quad u_3^1 = \frac{u_3\sqrt{1 - \frac{v^2}{c^2}}}{1 - \frac{vu_1}{c^2}}.$$

The inverse transformations are given by

$$u_1 = \frac{u_1^1 + v}{1 + \frac{vu_1^1}{c^2}}, \quad u_2 = \frac{u_2^1\sqrt{1 - \frac{v^2}{c^2}}}{1 + \frac{vu_1^1}{c^2}}, \quad u_3 = \frac{u_3^1\sqrt{1 - \frac{v^2}{c^2}}}{1 + \frac{vu_1^1}{c^2}}.$$

If we take $u_2^1 = u_3^1 = 0$, $u_1^1 = u$ and $u_1 = w$, then we get

$$w = \frac{u + v}{1 + \frac{uv}{c^2}}.$$

This is Einstein's law of addition of velocities.

Example 8.3 Show that derivative of four velocities is perpendicular to it.

Hint: We know the magnitude of four velocities is constant. Hence

$$\frac{d}{d\tau}(u_\mu \cdot u_\mu) = \frac{d}{d\tau}(-c^2) = 0.$$

Hence,

$$u_\mu \cdot \frac{du_\mu}{d\tau} + \frac{du_\mu}{d\tau} \cdot u_\mu = 0 \text{ etc.}$$

Example 8.4 Show that four accelerations is perpendicular to four velocities.

Hint: Same as Example 8.3.

Example 8.5 Show that derivative of four vectors with constant magnitude is perpendicular to it.

Hint: Same as Example 8.3.

Example 8.6 Consider the four vectors a^μ, b^μ, c^μ whose components are given by

$$a^\mu = (-3, 0, 0, 2), \quad b^\mu = (10, 0, 6, 8), \quad c^\mu = (2, 0, 0, 3).$$

Check the nature of the four vectors.

Hint: Here,

$$a^\mu a_\mu = a_0^2 - a_1^2 - a_2^2 - a_3^2 = (-3)^2 - 2^2 > 0, \quad \text{hence,} \quad a^\mu \text{ is time like etc..}$$

Example 8.7 Consider a particle moving along x-axis whose velocity as a function of time is $\frac{dx}{dt} = \frac{at}{\sqrt{b^2 + a^2 t^2}}$, where a and b are constants and velocity of light c is assumed to be unity. (i) Does the particle's speed exceed the speed of light? (ii) Calculate the component of particle's four velocities. (iii) Express x and t in terms of proper time along the trajectory.

Hint: (i)

$$\text{Here,} \quad v = \frac{dx}{dt} = \frac{at}{\sqrt{b^2 + a^2 t^2}} < 1$$

since, $b^2 + a^2 t^2 > a^2 t^2$, therefore less than the velocity of light.

(ii) The particle's four velocities is given by

$$u_\mu = a[u_x, u_y, u_z, ic], \quad \text{where,} \quad a = \frac{1}{\sqrt{1 - \frac{v^2}{c^2}}}.$$

Here, $c = 1$ and $u_x = v = \frac{at}{\sqrt{b^2 + a^2 t^2}}, u_y = 0, u_z = 0$, etc.

(iii) The clock of an observer sitting on the particle reads proper time. The proper time elapsed from $t = 0$ to t is

$$\tau = \int_0^t dt \sqrt{1 - v^2} = \int_0^t \frac{b \, dt}{\sqrt{b^2 + a^2 t^2}} = \text{etc.}$$

Particle's trajectory is

$$x(t) - x_0 = \int_0^t \frac{at \, dt}{\sqrt{b^2 + a^2 t^2}} = \text{etc.}$$

Now, express x and t in terms of proper time τ.

Example 8.8 A π meson is moving with a speed $v = fc (0 < f < 1)$ in a direction $45°$ to the x-axis. Calculate the component of π meson's four velocities.

Hint: Suppose the π meson is moving in x–y plane. Therefore,

$$v_x = fc \cos 45°, \quad v_y = fc \sin 45°, \quad v_z = 0.$$

The π meson's four velocities is given by

$$u_\mu = a[v_x, v_y, v_z, ic], \quad \text{where, } a = \frac{1}{\sqrt{1 - \frac{v^2}{c^2}}} = \frac{1}{\sqrt{1 - f^2}}$$

Example 8.9 What do you mean by co-moving frame? What should be the components of a four-velocity vector in its co-moving frame?

Hint: A particle is moving in a frame. The observer sat on the particle. So it is also moving with the particle's velocity. This type of reference frame is known as co-moving frame.

The particle is not undergoing any spatial separation from the observer. That is $dx = dy = dz = 0$. However, the fourth component which is time cannot be stopped. Thus $dt \neq 0$.

Hence space components of u_μ are $\frac{dx_j}{dt} = 0$.

Now, $u_4 = \frac{dx_4}{dt} = \frac{d(ict)}{dt} = ic$. Therefore, $u_\mu = (0, 0, 0, ic)$.

Example 8.10 A particle of rest mass m_0 describes the trajectory $x = f(t)$, $y = g(t)$, $z = 0$ in an inertial frame S. Find the four-velocity components. Also show that norm of the four velocities is $-c^2$.

Hint: Four velocities is given by

$$u_\mu = \frac{dx_\mu}{d\tau} = \frac{1}{\sqrt{1 - \frac{v^2}{c^2}}} \frac{dx_\mu}{dt}.$$

For spatial components are

$$u_i = \frac{1}{\sqrt{1 - \frac{v^2}{c^2}}} \frac{dx_i}{dt},$$

where

$$v_1 = \dot{x} = \dot{f}(t), \quad v_2 = \dot{y} = \dot{g}(t), \quad v_3 = \dot{z} = 0, \quad v^2 = v_1^2 + v_2^2 + v_3^2 = \dot{f}^2 + \dot{g}^2.$$

Here,

$$\gamma = \frac{1}{\sqrt{1 - \frac{v^2}{c^2}}} = \frac{1}{\sqrt{1 - \frac{(\dot{f}^2 + \dot{g}^2)}{c^2}}}.$$

Now,

$$u_1 = \frac{1}{\sqrt{1 - \frac{v^2}{c^2}}} \frac{dx}{dt} = \frac{\dot{f}}{\sqrt{1 - \frac{(\dot{f}^2 + \dot{g}^2)}{c^2}}}$$

$$u_2 = \frac{1}{\sqrt{1 - \frac{v^2}{c^2}}} \frac{dy}{dt} = \frac{\dot{g}}{\sqrt{1 - \frac{(\dot{f}^2 + \dot{g}^2)}{c^2}}}$$

$$u_3 = 0$$

$$u_4 = \frac{1}{\sqrt{1 - \frac{v^2}{c^2}}} \frac{d(ict)}{dt} = \frac{ic}{\sqrt{1 - \frac{(\dot{f}^2 + \dot{g}^2)}{c^2}}}.$$

Now norm of the four velocities is given by

$$u_\mu \cdot u_\mu = u_1 u_1 + u_2 u_2 + u_3 u_3 + u_4 u_4$$

$$= \frac{\dot{f}^2}{1 - \frac{(\dot{f}^2 + \dot{g}^2)}{c^2}} + \frac{\dot{g}^2}{1 - \frac{(\dot{f}^2 + \dot{g}^2)}{c^2}} - \frac{c^2}{1 - \frac{(\dot{f}^2 + \dot{g}^2)}{c^2}}$$

$$= -c^2.$$

Example 8.11 A particle of rest mass m_0 describes the circular path $x = a \cos t$, $y = a \sin t$, $z = 0$ in an inertial frame S. Find the four-velocity components. Also show that norm of the four velocities is $-c^2$.

Hint: Use $f(t) = a \cos t$ and $g(t) = a \sin t$ in Example 8.10.

Example 8.12 A particle of rest mass m_0 describes the parabolic trajectory $x = at$, $y = bt^2$, $z = 0$ in an inertial frame S. Find the four-velocity components. Also show that norm of the four velocities is $-c^2$.

Hint: Use $f(t) = at$ and $g(t) = bt^2$ in Example 8.10.

Example 8.13 Show that $A^\mu = (3\frac{1}{2}, 2\frac{1}{2}, 0, 0)$ is a unit time like vector in special relativity. Show also the angle between A^μ and $e^\mu = (1, 0, 0, 0)$ is not real.

Hint: Norm of A^μ is

$$A^\mu A_\mu = g_{\mu\nu} A^\mu A^\nu = 1.$$

$[here\, g_{\mu\nu} = (1, -1, -1, -1)]$
If θ be the angle between A^μ and e^μ, then

$$\cos \theta = \frac{A^\mu e_\mu}{\sqrt{A^\mu A_\mu} \sqrt{e^\mu e_\mu}} = 3^{\frac{1}{2}} > 1.$$

Thus no real θ.

Example 8.14 Let Rahil travels in a spaceship on a parabolic path given by $x(t) = a\left(t - \frac{t^2}{b}\right)$ in Ayat's coordinate system S where a and b are constants. Find the time taken by Rahil to meet Ayat.

Hint: Note that $x(0) = 0$, $x(b) = 0$. This means Rahil leaves Ayat at $t = 0$ and meets again at time $t = b$. Here Rahil's speed is

$$\frac{dx}{dt} = a\left(1 - \frac{2t}{b}\right).$$

Now, integrating ds=cdτ we get $s = c\tau$ where τ is the time Rahil's time(proper time). Therefore,

$$c\tau = \int_0^b ds = \int_0^b \sqrt{(cdt)^2 - dx^2},$$

or,

$$\tau = \int_0^b \sqrt{1 - \left(\frac{dx}{cdt}\right)^2} \, dt = \int_0^b \left[1 - \frac{a^2}{c^2}\left(1 - \frac{2t}{b}\right)^2\right]^{\frac{1}{2}} dt.$$

Integrating it, one can get

$$\tau = \frac{b\left[sin^{-1}\frac{a}{c} + \frac{a}{c}\sqrt{1 - \frac{a^2}{c^2}}\right]}{2a}.$$

Example 8.15 $(1, 0, 0, -1)$ is a null vector whereas $(1, 0, 0, \sqrt{2})$ is a unit vector in Minkowski space

$$ds^2 = dt^2 - dx^2 - dy^2 - dz^2.$$

Hint: Here in the first case ,

$$g_{00} = 1, \ g_{11} = -1, \ g_{22} = -1, \ g_{33} = -1,$$

and

$$A^0 = 1 , A^1 = 0 , A^2 = 0 , A^3 = -1 .$$

Now,

$$l^2 = g_{ij} A^i A^j = g_{00} A^0 A^0 + g_{11} A^1 A^1 + g_{22} A^2 A^2 + g_{33} A^3 A^3 = 0 , \ etc.$$

Example 8.16 $(1, 1, 0, -1)$ and $(1, 0, 1, -1)$ are orthogonal vectors in Minkowski space

$$ds^2 = dt^2 - dx^2 - dy^2 - dz^2.$$

Hint: Here

$$g_{ij} A^i B^j = g_{00} A^0 B^0 + g_{11} A^1 B^1 + g_{22} A^2 B^2 + g_{33} A^3 B^3 = 0, \quad etc.$$

Example 8.17 Prove that the curve

$$x = \int r \cos\theta \cos\phi d\lambda ; \quad y = \int r \cos\theta \sin\phi d\lambda ; \quad z = \int r \sin\theta d\lambda ; \quad t = \int r d\lambda ,$$

represents a null curve in special relativity, where r, θ, ϕ are functions of λ.

Hint: We have to calculate

$$dx^2 + dy^2 + dz^2 - dt^2$$

along the curve.
    Now,

$$dx^2 + dy^2 + dz^2 - dt^2 = (r^2 \cos^2\theta \cos^2\phi + r^2 \cos^2\theta \sin^2\phi + r^2 \sin^2\theta - r^2)d\lambda^2 = 0.$$

So it is null curve.

Example 8.18 Let a particle is moving with velocity $\vec{v} = (\frac{c}{2}, \frac{c}{2}, \frac{c}{2})$. Show that its four-velocity vector is timelike. Show also that this particle is not massless.

Hint:
    The speed of the particle is

$$v = |\vec{v}| = \sqrt{\left(\frac{c}{2}\right)^2 + \left(\frac{c}{2}\right)^2 + \left(\frac{c}{2}\right)^2} = \frac{\sqrt{3}c}{2}.$$

Now, the four velocities of this particle is

$$u_\mu = \frac{1}{\sqrt{1 - \frac{v^2}{c^2}}} \left(\frac{c}{2}, \frac{c}{2}, \frac{c}{2}, ic\right) = (c, c, c, 2ic).$$

The norm-squared is

$$u_\mu \cdot u_\mu = c^2 + c^2 + c^2 + (2ic)^2 = -c^2 < 0.$$

Hence the four velocities is a timelike four vectors. Here actually, we have used the signature $(+, +, +, -)$. We know, massless particles move at the speed of light. Here the particle's speed, $\frac{\sqrt{3}c}{2} < c$, so the particle is not massless.

Example 8.19 Show that a two-dimensional Lorentz Transformation L connecting any two inertial frames satisfies the following conditions:

(i)

$$L^T \eta L = \eta, \quad where \quad \eta = \begin{bmatrix} 1 & 0 \\ 0 & -1 \end{bmatrix}.$$

(ii)

$$det\, L = \pm 1.$$

Prove that Lorentz Transformations form a group under matrix multiplication. Show also that if a two-dimensional linear transformation L connecting inertial frames satisfies above conditions then it is a Lorentz Transformation.

Hint: We knew

$$x' = (x - vt)a \text{ and } t' = a\left(t - \frac{vx}{c^2}\right),$$

where $a = \frac{1}{\sqrt{1 - \frac{v^2}{c^2}}}$.

Rewrite the above transformation as

$$x' = ax - a\frac{v}{c}ct = ax - a\beta(ct),$$

$$ct' = -a\frac{v}{c}x + a(ct) = -a\beta x + a(ct).$$

So, the transformation matrix is

$$L = \begin{bmatrix} a & -a\beta \\ -a\beta & a \end{bmatrix}.$$

Now

$$L^T \eta L = \begin{bmatrix} a & -a\beta \\ -a\beta & a \end{bmatrix} \cdot \begin{bmatrix} 1 & 0 \\ 0 & -1 \end{bmatrix} \cdot \begin{bmatrix} a & -a\beta \\ -a\beta & a \end{bmatrix} = \begin{bmatrix} 1 & 0 \\ 0 & -1 \end{bmatrix} = \eta.$$

Also ,

$$det(L^T \eta L^T) = det(\eta),$$

or,

$$det(L).1.det(L) = 1,$$

or,
$$det(L) = \pm 1.$$

Let L_1, L_2 be the Lorentz Transformations. Assume, A = $L_1 L_2$.

Now,
$$A^T \eta A = (L_1 L_2)^T \eta (L_1 L_2),$$

or,
$$A^T \eta A = L_2^T L_1^T \eta L_1 L_2,$$

or,
$$A^T \eta A = L_2^T (L_1^T \eta L_1) L_2,$$

or,
$$A^T \eta A = L_2^T \eta L_2,$$

$$A^T \eta A = \eta.$$

Hence L is closed with respect to matrix multiplication. We know matrix multiplication is associative. Identity matrix is a Lorentz matrix since

$$I^T \eta I = \eta.$$

We have proved that det (L) = $\pm 1 \neq 0$ for any Lorentz matrix. Then every Lorentz matrix has an inverse. Hence Lorentz Transformations form a group.

Let L = $\begin{bmatrix} a_{11} & a_{12} \\ a_{21} & a_{22} \end{bmatrix}$ be any matrix which satisfy condition (i), then

$$\begin{bmatrix} a_{11} & a_{12} \\ a_{21} & a_{22} \end{bmatrix} \cdot \begin{bmatrix} 1 & 0 \\ 0 & -1 \end{bmatrix} \cdot \begin{bmatrix} a_{11} & a_{12} \\ a_{21} & a_{22} \end{bmatrix} = \begin{bmatrix} 1 & 0 \\ 0 & -1 \end{bmatrix}.$$

Hence,
$$a_{11}^2 - a_{21}^2 = 1 ; \quad a_{11}a_{12} = a_{21}a_{22} ; \quad a_{12}^2 - a_{22}^2 = -1 .$$

One can solve these equations in terms of a_{11} as

$$a_{11}^2 = a_{22}^2 ; \quad a_{12}^2 = a_{21}^2 ; \quad a_{12}^2 = -1 + a_{11}^2. \tag{i}$$

To find a_{11}, consider the two vectors ($c\Delta t$, Δx), the difference between time and position of two events in S. In S' frame, these two vectors are represented as ($c\Delta t'$, $\Delta x'$).

Now by our transformation,

$$\begin{bmatrix} c\Delta t' \\ \Delta x' \end{bmatrix} = \begin{bmatrix} a_{11} & a_{12} \\ a_{21} & a_{22} \end{bmatrix} \cdot \begin{bmatrix} c\Delta t \\ \Delta x \end{bmatrix}.$$

Let two events occur at the same location, then $\Delta x = 0$. Then S is measuring proper time so that $\Delta t = \Delta \tau$.

Hence,

$$c\Delta t' = a_{11}c \quad i.e. \quad \Delta t' = a_{11}\Delta \tau. \tag{ii}$$

In the Michelson–Morley experiment, we have

$$\Delta t' = \frac{\Delta \tau}{\sqrt{1 - \beta^2}}, \tag{iii}$$

which is time dilation formula. Now comparing equations (ii) and (iii) we have

$$a_{11} = \frac{1}{\sqrt{1 - \beta^2}}..$$

Thus, finally we get,

$$a_{12} = \pm \beta a, \quad a_{22} = a.$$

Hence,

$$L = \begin{bmatrix} a & -a\beta \\ -a\beta & a \end{bmatrix},$$

(taking negative sign), is a Lorentz Transformation.

Example 8.20 Show that the four-dimensional space time gradient operators $(\frac{\partial}{\partial x}, \frac{\partial}{\partial y}, \frac{\partial}{\partial z}, \frac{1}{c}\frac{\partial}{\partial t})$ are invariant under Lorentz Transformation.

Hint: We know

$$x_1 = a(x_1' + ut'); \quad x_2 = x_2', \quad y_2 = y_2', \quad t = a\left(t' + \frac{ux_1'}{c^2}\right).$$

$[\beta = \frac{u}{c}, a = \frac{1}{\sqrt{1-\beta^2}}].$

Applying chain rule, we have

$$\frac{\partial}{\partial x_1'} = \frac{\partial}{\partial x_1}\frac{\partial x_1}{\partial x_1'} + \frac{\partial}{\partial x_2}\frac{\partial x_2}{\partial x_1'} + \frac{\partial}{\partial x_3}\frac{\partial x_3}{\partial x_1'} + \frac{\partial}{\partial t}\frac{\partial t}{\partial x_1'}$$

$$\Rightarrow \frac{\partial}{\partial x_1'} = a\left(\frac{\partial}{\partial x_1} + \frac{u}{c^2}\frac{\partial}{\partial t}\right).$$

Similarly,

$$\frac{\partial}{\partial x_2'} = \frac{\partial}{\partial x_2}; \quad \frac{\partial}{\partial x_3'} = \frac{\partial}{\partial x_3}$$

$$\frac{\partial}{\partial t'} = a\left(u\frac{\partial}{\partial x_1} + \frac{\partial}{\partial t}\right).$$

These can be written as (suppressing x_2, x_3)

$$\begin{pmatrix} \frac{\partial}{\partial t'} \\ \frac{\partial}{\partial x_1'} \end{pmatrix} = \begin{pmatrix} a & a\beta \\ a\beta & a \end{pmatrix} \begin{pmatrix} \frac{\partial}{\partial t} \\ \frac{\partial}{\partial x_1} \end{pmatrix}.$$

Now we prove that the matrix, $L = \begin{pmatrix} a & a\beta \\ a\beta & a \end{pmatrix}$ represents a Lorentz Transformation matrix by showing that $L^T \eta L = \eta$, where $\eta = diag(1, -1)$.

One can check

$$L^T \eta L = \begin{pmatrix} a & a\beta \\ a\beta & a \end{pmatrix} \begin{pmatrix} 1 & 0 \\ 0 & -1 \end{pmatrix} \begin{pmatrix} a & a\beta \\ a\beta & a \end{pmatrix} = \begin{pmatrix} 1 & 0 \\ 0 & -1 \end{pmatrix} = \eta.$$

Chapter 9
Mass in Relativity

9.1 Relativistic Mass

We know the fundamental quantities, length and time are dependent on the observer. Therefore, it is expected that mass would be an observer-dependent quantity.

9.1.1 First Method Based on Hypothetical Experiment of Tolman and Lews

Let us consider an elastic collision between two identical perfectly elastic particles P and Q as seen by different inertial frame S and S^1. Let m_0 be mass of each particle at rest. Here, P is in S system and Q is in S^1 system. We assume that the relative velocity between S and S^1 is v in which S^1 is approaching to S. Let P and Q have initial velocities along y-axis that are equal in magnitude but opposite in direction, i.e. $u_y^1 = -u_y$. Since the collision is elastic, the final velocities have the same magnitude at the initial velocities but in opposite directions. Now, the velocity components as measured by observer in S according to law of transformation of velocities are

$$\left[u_x = \frac{u_x^1 + v}{1 + \frac{vu_x^1}{c^2}}, \quad u_y = \frac{u_y^1 \sqrt{1 - \frac{v^2}{c^2}}}{1 + \frac{vu_x^1}{c^2}}, \quad u_z = \frac{u_z^1 \sqrt{1 - \frac{v^2}{c^2}}}{1 + \frac{vu_z^1}{c^2}} \right].$$

For particle P, we have

$$u_{x_P} = 0, \quad u_{y_P} = u_y, \quad u_{z_P} = 0.$$

F. Rahaman, *The Special Theory of Relativity*, La Matematica per il 3+2 136, https://doi.org/10.1007/978-981-19-0497-4_9

For particle Q, we have

$$u_{xQ} = v, \quad u_{yQ} = -u_y\sqrt{1 - \frac{v^2}{c^2}}, \quad u_{zQ} = 0$$

[since, $u_x^1 =, u_y^1 = -u_y, u_z^1 = 0$].

Thus, the resultant momentum of the whole system before collision as seen from S system is

$$= m_0 u_y \vec{j} + mv\,\vec{i} - mu_y\sqrt{1 - \frac{v^2}{c^2}}\,\vec{j}, \tag{9.1}$$

where m is the mass of the particle moving with velocity v and $\vec{i}$ & $\vec{j}$ are unit vectors along x- and y-axes, respectively.

Now after elastic collision, the velocities of the particles P and Q as observed from S system are given by

$$w_{xP} = 0, \quad w_{yP} = -u_y, \quad w_{zP} = 0 \quad \text{and} \quad w_{xQ} = v, \quad w_{yQ} = u_y\sqrt{1 - \frac{v^2}{c^2}}, \quad w_{zQ} = 0.$$

Thus, the resultant momentum of the whole system after the collision as seen from S system is

$$= -m_0 u_y \vec{j} + mv\,\vec{i} + mu_y\sqrt{1 - \frac{v^2}{c^2}}\,\vec{j}. \tag{9.2}$$

Using the principle of conservation of momentum, i.e. momentum before impact = momentum after impact, we have

$$m_0 u_y \vec{j} + mv\,\vec{i} - mu_y\sqrt{1 - \frac{v^2}{c^2}}\,\vec{j} = -m_0 u_y \vec{j} + mv\,\vec{i} + mu_y\sqrt{1 - \frac{v^2}{c^2}}\,\vec{j}.$$

This implies

$$m = \frac{m_0}{\sqrt{1 - \frac{v^2}{c^2}}}. \tag{9.3}$$

This mass m of the particle moving with velocity v is known as *relativistic mass* of the particle whose mass was m_0 at rest. Note that mass of the particle increases with increase of velocity.

Fig. 9.1 Collisions of two particles as seen from two-reference frames

9.1.2 Second Method Based on a Thought Experiment

As length and time are changed with the motion of the observer, we assume that mass of a particle which is moving with a velocity u with respect to an inertial frame must be a function of u, i.e.

$$m = m(u). \tag{9.4}$$

Let two particles, one be at rest and the other is moving with velocity u collide (inelastically) and coalesce and combined objects move with velocity U. The masses of two particles are denoted by m_0 (rest mass) and $m(u)$ (relativistic mass). The mass of the combined particles be $M(U)$. Let us consider the reference system S^1 to be the centre of mass frame. Therefore, it is evident that relative to S^1 frame, collision takes place for the two equal particles coming with same speeds with opposite directions leaving the combined particle with mass M_0 at rest. It is obvious that S^1 has the velocity U relative to S (Fig. 9.1).

Now, using the principles of conservation of mass and linear momentum in S frame, we get

$$m(u) + m_0 = M(U) \tag{9.5}$$

$$u\, m(u) + 0 = U M(U). \tag{9.6}$$

These give

$$m(u) = \frac{m_0 U}{u - U}. \tag{9.7}$$

The left-hand particle has a velocity U relative to S. Now, using Einstein's law of addition of velocities, we get

$$u = \frac{U + U}{1 + \frac{UU}{c^2}}.$$

This equation yields the solution of U in terms of u as

$$U = \frac{c^2}{u}\left[1 \pm \sqrt{1 - \frac{u^2}{c^2}}\right].$$ (9.8)

We should take negative sign as positive sign gives U to be greater than c. Using the value of U in (9.7), we get

$$m(u) = \frac{m_0}{\sqrt{1 - \frac{u^2}{c^2}}}.$$ (9.9)

9.1.3   Third Method

Let a particle of mass m is moving with velocity u in an inertial frame S. Then conservation of linear momentum in S frame implies

$$m\frac{\mathrm{d}x_\mu}{\mathrm{d}t} = \text{consant.}$$ (9.10)

We rewrite this equation in terms of proper time as

$$m\frac{\mathrm{d}x_\mu}{\mathrm{d}\tau}\sqrt{1 - \frac{u^2}{c^2}} = \text{consant,}$$ (9.11)

where velocity of the particle, $u = \left[\left(\frac{\mathrm{d}x_1}{\mathrm{d}t}\right)^2 + \left(\frac{\mathrm{d}x_2}{\mathrm{d}t}\right)^2 + \left(\frac{\mathrm{d}x_3}{\mathrm{d}t}\right)^2\right]^{\frac{1}{2}}$. Let us consider another reference system S^1 which is moving with velocity v relative to S frame. Now, we write the Eq. (9.11) with respect to S^1 frame as

$$ma_{\mu\nu}^1\frac{\mathrm{d}x_\nu^1}{\mathrm{d}\tau}\sqrt{1 - \frac{u^2}{c^2}} = \text{consant,}$$ (9.12)

where $a_{\mu\nu}^1$ is the inverse Lorentz Transformation matrix whose determinant is non-zero constant. Therefore, Eq. (9.12) implies

$$m\frac{dx_\nu^1}{d\tau}\sqrt{1-\frac{u^2}{c^2}} = \text{consant.} \qquad (9.13)$$

Since the principle of conservation of linear momentum holds good in S^1 frame also, we have

$$m^1\frac{dx_\mu^1}{d\tau}\sqrt{1-\frac{u^{1^2}}{c^2}} = \text{consant,} \qquad (9.14)$$

here, m^1 is the value of mass in S^1 system. Since the constants appearing in right-hand sides of Eqs. (9.13) and (9.14) may not be equal, so we multiply (9.14) by a constant k to make them equal. Hence

$$\left[m\sqrt{1-\frac{u^2}{c^2}} - km^1\sqrt{1-\frac{u^{1^2}}{c^2}} \right]\frac{dx_\mu^1}{d\tau} = 0.$$

This equation is valid for any arbitrary value of $\frac{dx_\mu^1}{d\tau}$, we can write

$$m\sqrt{1-\frac{u^2}{c^2}} = km^1\sqrt{1-\frac{u^{1^2}}{c^2}} = m_0 \qquad (9.15)$$

Here, m_0 is the mass of the particle in rest frame. In S frame, rest mass of the particle is m_0 whereas in S^1 frame rest mass of the particle is $\frac{m_0}{k}$. However, rest mass in all inertial frame remains the same. So we must have $k = 1$. So a moving particle with velocity u will have a mass

$$m(u) = \frac{m_0}{\sqrt{1-\frac{u^2}{c^2}}}. \qquad (9.16)$$

9.2 Experimental Verification of Relativistic Mass

The relativistic mass formula has been verified by many scientists, however, we shall discuss the experiment done by Guye and Lavanchy.

The Experiment Setup is as Follows

A cathode–anode system CA is arranged in a vacuum tube where cathode and anode are connected together to the high potential terminal which produces several thousand volts. The cathode rays are collimated into fine beam by the hole B in the anode and then continue a straight line path to strike a photographic plate at O (see Fig. 9.2). Another arrangement is done to produce electric by applying a potential difference between the plates P and Q and magnetic fields by using an electromagnet M indicated

Fig. 9.2 The experiment setup of Guye and Lavanchy experiment

by a circle in the path of the electron beam. Both of these fields are successively applied and adjusted in such a way that the deflection of the electron beam would be the same.

The force experienced by electron due to electric field X as Xe and that due to magnetic field H as Hev where v be the velocity of the beam of low speed electrons and e is the charge of the electron. Due to the above assumption, $Hev = Xe$, i.e.

$$v = \frac{X}{H}. \tag{9.17}$$

Since the beam follows a circular path of radius, say a due to magnetic field, we have $Hev = \frac{mv^2}{a}$ where m is mass of the beam of electrons. This implies

$$m = \frac{Hea}{v}. \tag{9.18}$$

If X^1, H^1, v^1, m^1 are corresponding quantities to produce the same deflections (here, beam of high speed electrons), then

$$v^1 = \frac{X^1}{H^1} \tag{9.19}$$

$$m^1 = \frac{H^1 ea}{v^1}. \tag{9.20}$$

Using the above equations, we get

$$\frac{m^1}{m} = \frac{H^{1^2} X}{H^2 X^1}. \tag{9.21}$$

Guye and Lavanchy made nearly 2,000 determinations of $\frac{m^1}{m}$ for electrons with velocity ranging from 26 to 48 % that of light and showed that the relativistic formula of mass is almost correct with an accuracy of 1 part in 2,000.

9.3 Lorentz Transformation of Relativistic Mass

Let us consider the relativistic mass of a moving particle with velocity u in S frame be m with rest mass m_0 and corresponding quantity be m^1 in S^1 which is moving with constant velocity v along x-direction. Here,

$$m = \frac{m_0}{\sqrt{1 - \frac{u^2}{c^2}}}. \tag{9.22}$$

Therefore, the Lorentz Transformation of relativistic mass is given by

$$m^1 = \frac{m_0}{\sqrt{1 - \frac{u^{1^2}}{c^2}}}. \tag{9.23}$$

We know from note-6 of chapter seven that

$$\sqrt{1 - \left(\frac{u^1}{c}\right)^2} = \frac{\sqrt{1 - (\frac{v}{c})^2}\sqrt{1 - (\frac{u}{c})^2}}{(1 - \frac{u_x v}{c^2})}. \tag{9.24}$$

Using the result of Eq. (9.24) in (9.23), we get

$$m^1 = \frac{m_0(1 - \frac{u_x v}{c^2})}{\sqrt{1 - (\frac{v}{c})^2}\sqrt{1 - (\frac{u}{c})^2}} = \frac{m(1 - \frac{u_x v}{c^2})}{\sqrt{1 - (\frac{v}{c})^2}}. \tag{9.25}$$

The inverse transformation is given by

$$m = \frac{m^1(1 + \frac{u_x^1 v}{c^2})}{\sqrt{1 - (\frac{v}{c})^2}} \tag{9.26}$$

Note 1: Figure 9.3 indicates that the relativistic mass increases without bound as v approaches to c.

Example 9.1 A particle of rest mass m_0 and velocity fc ($0 < f < 1$) collides with another particle of rest mass m_0 at rest and sticks. What will be the velocity and mass of the resulting particle?

Hint Let M_0 be the rest mass and V be the velocity of the resulting particle.

$$\text{Conservation of momentum}: \frac{m_0 fc}{\sqrt{1 - \frac{f^2 c^2}{. c^2}}} + m_0 \times 0 = \frac{M_0 V}{\sqrt{1 - \frac{V^2}{c^2}}}$$

Fig. 9.3 The variation of mass is shown with respect to the velocity

Conservation of mass : $m_0 + \dfrac{m_0}{\sqrt{1 - \frac{f^2 c^2}{c^2}}} = \dfrac{M_0}{\sqrt{1 - \frac{V^2}{c^2}}}.$

Solving these equations, one can get V and M_0.

Example 9.2 A particle of rest mass m_1 and velocity v collides with another particle of rest mass m_2 at rest and sticks. Show that the velocity and mass of the resulting particle are given by

$$M^2 = m_1^2 + m_2^2 + \frac{2m_1 m_2}{\sqrt{1 - \frac{v^2}{c^2}}}, \quad V = \frac{m_1 v}{m_1 + m_2 \sqrt{1 - \frac{v^2}{c^2}}}.$$

Hint M be the rest mass and V be the velocity of the resulting particle.

Conservation of momentum : $\dfrac{m_1 v}{\sqrt{1 - \frac{v^2}{c^2}}} + m_2 \times 0 = \dfrac{MV}{\sqrt{1 - \frac{V^2}{c^2}}}$

Conservation of mass : $m_2 + \dfrac{m_1}{\sqrt{1 - \frac{v^2}{c^2}}} = \dfrac{M}{\sqrt{1 - \frac{V^2}{c^2}}}.$

Solving these equations, one can get V and M.

Example 9.3 Find the velocity of the particle for which the relativistic mass of the particle exceeds its rest mass by a given fraction f.

Hint Let v be the velocity of the particle. Now,

$$f = \frac{m - m_0}{m_0} = \frac{m}{m_0} - 1 = \frac{1}{m_0} \frac{m_0}{\sqrt{1 - \frac{v^2}{c^2}}} - 1 \implies v = \text{etc.}$$

Example 9.4 Find the velocity of the particle for which the relativistic mass of the particle a times its rest mass.

Hint Let v be the velocity of the particle. Now,

$$\frac{m}{m_0} = a \implies \frac{m_0}{\sqrt{1 - \frac{v^2}{c^2}}} = am_0 \implies v = \text{etc.}$$

Example 9.5 Show that for inelastic collisions of two particles, the rest mass of the combined particles is greater than the original rest masses.

Hint In an inelastic collision, a particle of rest mass m_1 and velocity v collides with another particle of rest mass m_2 at rest and sticks. Let M be the rest mass and V be the velocity of the resulting particle.

$$\text{Conservation of momentum}: \frac{m_1 v}{\sqrt{1 - \frac{v^2}{c^2}}} + m_2 \times 0 = \frac{MV}{\sqrt{1 - \frac{V^2}{c^2}}}$$

$$\text{Conservation of mass}: m_2 + \frac{m_1}{\sqrt{1 - \frac{v^2}{c^2}}} = \frac{M}{\sqrt{1 - \frac{V^2}{c^2}}}.$$

Solving these equations, one can get M as

$$M^2 = m_1^2 + m_2^2 + \frac{2m_1 m_2}{\sqrt{1 - \frac{v^2}{c^2}}}$$

Since $\frac{1}{\sqrt{1 - \frac{v^2}{c^2}}} > 1$, we have

$$M^2 > m_1^2 + m_2^2 + 2m_1 m_2 = (m_1 + m_2)^2, \text{etc.}$$

Example 9.6 A particle of rest mass m_0 and velocity v collides with another particle of rest mass m_0 at rest and sticks (i.e. inelastic collision). Show that the rest mass of the combined particles is greater than the original rest masses by an amount $\frac{m_0 v^2}{4c^2}$ (neglect $\frac{v^4}{c^4}$ and higher order).

Hint In an inelastic collision, a particle of rest mass m_0 and velocity v collides with another particle of rest mass m_0 at rest and sticks. Let M be the rest mass and V be the velocity of the resulting particle.

$$\text{Conservation of momentum}: \frac{m_0 v}{\sqrt{1 - \frac{v^2}{c^2}}} + m_0 \times 0 = \frac{MV}{\sqrt{1 - \frac{V^2}{c^2}}}$$

$$\text{Conservation of mass}: m_0 + \frac{m_0}{\sqrt{1 - \frac{v^2}{c^2}}} = \frac{M}{\sqrt{1 - \frac{V^2}{c^2}}}.$$

Solving these equations, one can get M as

$$M^2 = 2m_0^2 + \frac{2m_0^2}{\sqrt{1 - \frac{v^2}{c^2}}} = 2m_0^2 + 2m_0^2 \left(1 - \frac{v^2}{c^2}\right)^{-\frac{1}{2}}.$$

Expanding binomially and neglecting $\frac{v^4}{c^4}$ and higher order, we get

$$M = 2m_0 \left(1 + \frac{v^2}{4c^2}\right)^{\frac{1}{2}} \implies M - 2m_0 = \frac{m_0 v^2}{4c^2}$$

[again, expanding binomially and neglecting $\frac{v^4}{c^4}$ and higher order]

Example 9.7 The average lifetime of π-mesons at rest is t s. A laboratory measurement on π-mesons yields an average lifetime of nt s. What is the speed of the π-mesons in the laboratory.

Hint Here, t be the lifetime of the rest π-mesons. Let v be its velocity. According to time dilation,

$$t^1 = nt = \frac{t}{\sqrt{1 - \frac{v^2}{c^2}}} \implies v = \text{etc.}$$

Example 9.8 Let a constant force F applied on an object with rest mass m_0 at a rest position. Prove that its velocity after a time t is $v = \frac{cFt}{\sqrt{m_0^2 c^2 + F^2 t^2}}$. Prove also that the above result is in agreement with classical result. Further find v after a very long time.

Hint We know

$$F = \frac{\mathrm{d}}{\mathrm{d}t}(mv) = \frac{\mathrm{d}}{\mathrm{d}t}\left(\frac{m_0 v}{\sqrt{1 - \frac{v^2}{c^2}}}\right)$$

Now, integrating we get

$$\int F\, dt = \int d\left(\frac{m_0 v}{\sqrt{1 - \frac{v^2}{c^2}}}\right).$$

This implies

$$Ft = \frac{m_0 v}{\sqrt{1 - \frac{v^2}{c^2}}}\ \text{etc.}.$$

If t is small, then $t^2 + \frac{m_0^2 c^2}{F^2} \approx \frac{m_0^2 c^2}{F^2}$. Thus

$$v = \frac{cFt}{\sqrt{m_0^2 c^2 + F^2 t^2}} = \frac{cFt}{F\sqrt{t^2 + \frac{m_0^2 c^2}{F^2}}} \approx \frac{cFt}{F\sqrt{\frac{m_0^2 c^2}{F^2}}} = \frac{Ftc}{m_0 c}.$$

This implies

$$F \approx \frac{m_0 v}{t} = m_0 f.$$

$[f = \frac{v}{t} = \text{acceleration}]$
If t is large enough, then

$$v = \frac{cFt}{\sqrt{m_0^2 c^2 + F^2 t^2}} = \frac{cF}{\sqrt{\frac{m_0^2 c^2}{t^2} + F^2}} \approx \frac{Fc}{\sqrt{F^2}} = c$$

[since $\frac{m_0^2 c^2}{t^2} \approx 0$ for large t]

Example 9.9 Which of the following objects should be heaviest? (i) Mass m_0 moving with velocity $0.5c$; (ii) Mass $0.5m_0$ moving with velocity $0.7c$; (iii) Mass $2m_0$, moving with velocity $0.4c$ and (iv) Mass $3m_0$, moving with velocity $0.1c$.

Hint Use relativistic mass.

Chapter 10
Relativistic Dynamics

10.1 Four Forces or Minkowski Force

We proceed to generalize Newton's equation of motion

$$F_j = \frac{d(mv_j)}{dt} \ , \ j = 1, 2, 3 \tag{10.1}$$

which is not invariant under Lorentz Transformation. Its relativistic generalization should be a four-vector equation, the spatial part of which would reduce to equation (10.1) in the limit $\beta = \frac{v}{c} \longrightarrow 0$.

We shall bring the following changes:

1. Since t is not Lorentz invariant, it should be replaced by proper time τ.
2. The rest mass m_0 can be taken as an invariant property of the particle.
3. In place of v_j, world velocity u_μ should be substituted.
4. The left-hand side force F_i should be replaced by some four vectors K_μ (called *Minkowski force*).

Thus taking into account all these changes, the relativistic generalization of Eq. (10.1) is

$$K_\mu = \frac{d(m_0 u_\mu)}{d\tau} \ \mu = 1, 2, 3, 4. \tag{10.2}$$

The spatial part of the above equation can be written as

$$\frac{ad(m_0 a v_j)}{dt} = K_j, \ a = \frac{1}{\sqrt{1 - \frac{v^2}{c^2}}}$$

or,

© The Author(s), under exclusive license to Springer Nature Singapore Pte Ltd. 2022   137
F. Rahaman, *The Special Theory of Relativity*, La Matematica per il 3+2 136,
https://doi.org/10.1007/978-981-19-0497-4_10

$$\frac{\mathrm{d}(m_0 a v_j)}{\mathrm{d}t} = \frac{K_j}{a}. \tag{10.3}$$

If we continue to use the classical definition of force and define force as being the time rate of change of momentum in all Lorentz systems, then the classical force should be defined as

$$\vec{F} = \frac{\mathrm{d}(m\,\vec{v}\,)}{\mathrm{d}t} \implies F_j = \frac{\mathrm{d}(mv_j)}{\mathrm{d}t}. \tag{10.4}$$

This is called *Relativistic force*.

Actually, it is measured from S frame of a moving particle of rest mass m_0.

Hence, the space components of the force K_μ is related to F_j as

$$K_j = a F_j \tag{10.5}$$

[these are Relativistic force, i.e. space components of the Minkowski forces]

Note 1: Minkowski force and four-velocity vector are orthogonal to each other.

Proof Here,

$$\begin{aligned}
K_\mu \cdot u_\mu &= \frac{\mathrm{d}(m_0 u_\mu)}{\mathrm{d}\tau} \cdot u_\mu \\
&= \frac{1}{2}\frac{\mathrm{d}(m_0 u_\mu \cdot u_\mu)}{\mathrm{d}\tau} \\
&= \frac{1}{2}\frac{\mathrm{d}(-m_0 c^2)}{\mathrm{d}\tau} = 0.
\end{aligned}$$

If a force K_μ does not depend on velocity u_μ, then this force is not consistent with special relativity. Actually, in this case $K_\mu \cdot u_\mu \neq 0$ and this contradicts the above requirements.

Now, we try to find the time-like part of the force vector K_μ, i.e. the fourth component of Minkowski force, with the help of the above theorem.

We write the result $K_\mu \cdot u_\mu = 0$ explicitly which yields

$$K_\mu \cdot u_\mu = K_1 u_1 + K_2 u_2 + K_3 u_3 + K_4 u_4 = 0$$

or

$$a^2 F_j v_j + K_4 a i c = 0$$

or

$$a^2 \vec{F} \cdot \vec{v} + K_4 a i c = 0$$

or

$$K_4 = \left(\frac{ia}{c}\right) \vec{F} \cdot \vec{v}.$$ (10.6)

Therefore, the Minkowski force is given by

$$K_\mu = \left[aF_1, aF_2, aF_3, \left(\frac{ia}{c}\right) \vec{F} \cdot \vec{v}\right]; \quad a = \frac{1}{\sqrt{1 - \frac{v^2}{c^2}}}.$$ (10.7)

10.2 Four Momenta

In Newtonian mechanics, the momentum vector p is obtained by multiplying the velocity vector v by a mass m which is independent of the reference frame. The four-momentum vector p_μ is obtained similarly from the four-velocity vector $u_\mu = [av_1, av_2, av_3, iac]$ by multiplication with mass factor independent of the frame of reference. This mass factor is called the rest mass, m_0. Hence we get the *four-momentum vector* as

$$p_\mu = m_0[av_1, av_2, av_3, iac] = am_0[v_1, v_2, v_3, ic].$$ (10.8)

The quantity in front of the bracket is generally called the relativistic mass. We therefore have

$$m = am_0 = \frac{m_0}{\sqrt{1 - \frac{v^2}{c^2}}}.$$

The spatial components are

$$p_j = \frac{m_0 v_j}{\sqrt{1 - \frac{v^2}{c^2}}} = mv_j, \quad j = 1, 2, 3.$$

This equation gives the relativistic definition of classical linear momentum of the particle.

The four or time component is

$$p_4 = m_0 u_4 = \frac{m_0 i c}{\sqrt{1 - \frac{v^2}{c^2}}} = icm.$$ (10.9)

The norm of the four-momentum vector is

$$p_\mu \cdot p_\mu = -m_0^2 c^2.$$ (10.10)

10.3 Relativistic Kinetic Energy

Minkowski force is defined by the rate of change of momentum and here, time t is replaced by proper time τ. Therefore,

$$K_\mu = \frac{dp_\mu}{d\tau} = \frac{d(m_0 u_\mu)}{d\tau}. \tag{10.11}$$

Thus,

$$K_j = \frac{d(am_0 v_j)}{d\tau}$$

From Eq. (10.4), we can write,

$$F_j = \frac{d(am_0 v_j)}{dt}. \tag{10.12}$$

Again, we know that Minkowski force and four-velocity vector are orthogonal to each other, then we have

$$u_j K_j + u_4 K_4 = 0$$

This implies

$$a v_j \frac{d(am_0 v_j)}{d\tau} = -a i c \frac{d(am_0 i c)}{d\tau}$$

or,

$$v_j \frac{d(am_0 v_j)}{dt} = v_j F_j \equiv \vec{v} \cdot \vec{F} = \frac{d(am_0 c^2)}{dt} \tag{10.13}$$

[using Eq. (10.12)].

We know that the rate of change of kinetic energy (T) of a particle is given by the rate at which the force does work on it which is $\vec{v} \cdot \vec{F}$. Using the above definition of the rate of change of kinetic energy, we get from Eq. (10.13)

$$\vec{v} \cdot \vec{F} = \frac{dT}{dt} = \frac{d(am_0 c^2)}{dt}. \tag{10.14}$$

This implies

$$T = am_0 c^2 + T_0, \quad T_0 \text{ is an integration constant.}$$

When, $v = 0$, then $T = 0$, therefore, $T_0 = -m_0 c^2$, hence

$$T = am_0c^2 - m_0c^2 = (m - m_0)c^2. \tag{10.15}$$

This is the *relativistic kinetic energy* of a freely moving particle.

Actually, it is measured from S frame of a moving particle of rest mass m_0, i.e. kinetic energy of a moving particle with relativistic effects.

Note 2: Relativistic kinetic energy of a moving particle is equal to c^2 times its gain in mass due to motion.

Alternative derivation of relativistic kinetic energy:

We have

$$m = \frac{m_0}{\sqrt{1 - \frac{v^2}{c^2}}}, \text{ i.e. } m^2c^2 - m^2v^2 = m_0^2c^2.$$

From this, we can get

$$2\,mc^2\mathrm{d}m - 2mv^2\mathrm{d}m - 2m^2v\mathrm{d}v = 0,$$

i.e.

$$v^2\mathrm{d}m + mv\mathrm{d}v = c^2\mathrm{d}m. \tag{10.16}$$

When a force $\overrightarrow{F}$ acts on the mass m so as to move it through a distance $\mathrm{d}\overrightarrow{x}$, the work done is $\overrightarrow{F} \cdot \mathrm{d}\overrightarrow{x}$ which is converted into kinetic energy T as

$$T = \int_{v=0}^{v} \overrightarrow{F} \cdot \mathrm{d}\overrightarrow{x}$$

$$= \int_{v=0}^{v} \frac{\mathrm{d}\overrightarrow{p}}{\mathrm{d}t} \cdot \mathrm{d}\overrightarrow{x}$$

$$= \int_{v=0}^{v} \frac{\mathrm{d}(m\overrightarrow{v})}{\mathrm{d}t} \cdot \mathrm{d}\overrightarrow{x}$$

$$= \int_{v=0}^{v} \mathrm{d}(m\overrightarrow{v}) \cdot \frac{\mathrm{d}\overrightarrow{x}}{\mathrm{d}t}$$

$$= \int_{v=0}^{v} (m\mathrm{d}\overrightarrow{v} + \overrightarrow{v}\,\mathrm{d}m) \cdot \overrightarrow{v}$$

$$= \int_{v=0}^{v} (m\overrightarrow{v} \cdot \mathrm{d}\overrightarrow{v} + v^2\mathrm{d}m).$$

Fig. 10.1 Variation of classical and relativistic kinetic energy with respect to the velocity

Since $\vec{v} \cdot \vec{dv} = v dv$, therefore, using (10.16), we get

$$T = \int_{m=m_0}^{m} c^2 dm = mc^2 - m_0 c^2 = (m - m_0)c^2 = m_0 c^2 (a - 1) \qquad (10.17)$$

[when, $v = 0$, $m = m_0$ = rest mass, $a = \frac{1}{\sqrt{1 - \frac{v^2}{c^2}}}$].

Classical result: The above relativistic expression for kinetic energy T must reduce to classical result, $\frac{1}{2} m_0 v^2$, $\frac{v}{c} \ll 1$.

To check:

Now,

$$T = m_0 c^2 (a - 1)$$

$$= m_0 c^2 \left(\frac{1}{\sqrt{1 - \frac{v^2}{c^2}}} - 1 \right)$$

$$= m_0 c^2 \left[\left(1 - \frac{v^2}{c^2} \right)^{-\frac{1}{2}} - 1 \right].$$

Expanding binomially and neglecting $\frac{v^4}{c^4}$ and higher order, we get

$$T \approx m_0 c^2 \left[\left(1 + \frac{v^2}{2c^2} \right) - 1 \right] = \frac{1}{2} m_0 v^2.$$

Hence the result.

Figure 10.1 shows how the kinetic energy of a moving body varies with its speed according to both classical and relativistic mechanics. At low speeds, i.e. $\frac{v}{c} \ll 1$, the formulae give the same results, but diverge at speeds approaching that of light. Relativity theory indicates that a body would need infinite kinetic energy to travel with speed of light, whereas in classical mechanics it would need only a kinetic energy of half its rest energy to have this speed. This again confirms that c plays the role of limiting velocity.

10.4 Mass–Energy Relation

The expression of kinetic energy $T = (m - m_0)c^2$ suggests that the kinetic energy of a moving body is equal to the increase in mass times the square of the speed of light. Therefore, m_0c^2 may be regarded as *rest energy* of a body of rest mass m_0. It is argued that the rest mass has energy m_0c^2 due the presence of an internal store of energy. Thus, total energy E of the body would be the sum of rest energy and kinetic energy (relativistic due to motion). Hence,

$$E = m_0c^2 + (m - m_0)c^2 = mc^2 = \frac{m_0c^2}{\sqrt{1 - \frac{v^2}{c^2}}}. \tag{10.18}$$

It is the famous principle of mass–energy equivalence which states a universal equivalence of mass and energy. Here, E is the relativistic energy of a particle. Einstein argued that mass can be regarded as the source of energy in other words, mass is transformed into energy and vice versa. Thus, law of conservation of mass automatically implies the so-called law of conservation of energy.

Recall the four or time component of momentum as

$$p_4 = icm = \frac{iE}{c}. \tag{10.19}$$

Hence the *four-momentum vector* takes the form as

$$p_\mu = \left[p_1, p_2, p_3, \frac{iE}{c} \right]. \tag{10.20}$$

10.5 Relation Between Momentum and Energy

Total energy and momentum are conserved quantities and the rest energy of a particle is invariant quantity. Now, we try to find how the total energy, rest energy and momentum of a particle are related. We know, total energy is

$$E = mc^2 = \frac{m_0c^2}{\sqrt{1 - \frac{v^2}{c^2}}}.$$

This implies

$$E^2 = \frac{m_0^2c^4}{1 - \frac{v^2}{c^2}}. \tag{10.21}$$

The momentum is given by

$$p = mv = \frac{m_0 v}{\sqrt{1 - \frac{v^2}{c^2}}}.$$

This gives

$$p^2 c^2 = \frac{m_0^2 v^2 c^2}{1 - \frac{v^2}{c^2}}. \tag{10.22}$$

Subtracting (10.21) from (10.20), we get

$$E^2 - p^2 c^2 = m_0^2 c^4. \tag{10.23}$$

The above relation holds for a single particle. For particles in a system that are moving with respect to other, the sum of their individual rest energies may not equal the rest energy of the system.

Alternative

We know

$$p_\mu \cdot p_\mu = [p_4 p_4 + p_1 p_1 + p_2 p_2 + p_3 p_3] = -\frac{E^2}{c^2} + p^2. \tag{10.24}$$

Also,

$$p_\mu \cdot p_\mu = m_0^2 u_\mu \cdot u_\mu = -m_0^2 c^2 \tag{10.25}$$

Equating (10.24) and (10.25), we get,

$$-\frac{E^2}{c^2} + p^2 = -m_0^2 c^2 \implies E^2 - p^2 c^2 = m_0^2 c^4$$

Note 3: Since $m_0 c^2$ is an invariant quantity, therefore, $E^2 - p^2 c^2$ is also an invariant quantity.

Note 4: Differentiating Eq. (10.23) with respect to p, we get

$$\frac{dE}{dp} = \frac{pc}{\sqrt{p^2 + m_0^2 c^2}} = \frac{pc^2}{E} = v.$$

This is another useful relation between momentum and energy.

10.6 Evidence in Support of Mass–Energy Relation

In fusion process, when two neutrons and protons make helium $_2\text{He}^4$ nucleus, enormous energy is found. The explanation is as follows:

The mass of two protons $= 2 \times$ mass of hydrogen nucleus

$$= 2 \times m(_1\text{H}^1) = 2 \times (1.00815) \text{ a.m.u.}$$

Mass of two neutrons $= 2 \times m(_0n^1) = 2 \times (1.00898) \text{ a.m.u.}$
Therefore,

$$2 \times m(_1\text{H}^1) + 2 \times m(_0n^1) = 2 \times (1.00815 + 1.00898) \text{ a.m.u.} = 4.03426 \text{ a.m.u..}$$

The difference in mass of the $_2\text{He}^4$ nucleus and total mass of the constituent nucleus in free state is

$$(4.03426 - 4.00387) = 0.03039 \text{ a.m.u.} = .03039 \times 931.1 \text{ MeV} = 28.3 \text{ MeV.}$$

This explains why during fusion process of two neutrons and two protons, tremendous amount of energy is found.

Scientists believe that this fusion process occurs in the Sun to provide solar energy.

10.7 Force in Special Theory of Relativity

A force is not in general proportional to acceleration in case of special theory of relativity.

In general, force is given by

$$\vec{F} = \frac{d\vec{p}}{dt} = \frac{d(m\vec{v})}{dt}.$$

This implies

$$\vec{F} = m\frac{d\vec{v}}{dt} + \vec{v}\frac{dm}{dt}. \tag{10.26}$$

Using mass–energy relation, $E = mc^2$, we have

$$\frac{dm}{dt} = \frac{1}{c^2}\frac{dE}{dt} = \frac{1}{c^2}\frac{d(T + m_0c^2)}{dt} = \frac{1}{c^2}\frac{dT}{dt}$$

$[T = \text{kinetic energy and } m_0c^2 = \text{rest energy}].$

Again we know

$$\frac{dT}{dt} = \frac{(\vec{F}.d\vec{x})}{dt} = \vec{F}.\frac{d\vec{x}}{dt} = \vec{F}.\vec{v}.$$

Hence,

$$\frac{dm}{dt} = \frac{1}{c^2}(\vec{F}.\vec{v}). \tag{10.27}$$

Using (10.26) and (10.27), we get

$$\vec{F} = m\frac{d\vec{v}}{dt} + \left(\frac{1}{c^2}\right)(\vec{v}\,\vec{F}.\vec{v}). \tag{10.28}$$

The acceleration $\vec{f}$ is defined by $\vec{f} = \frac{d\vec{v}}{dt}$. Therefore,

$$\vec{f} = \frac{\vec{F}}{m} - \left(\frac{1}{mc^2}\right)(\vec{v}\,\vec{F}.\vec{v}). \tag{10.29}$$

Thus force is not in general, proportional to acceleration.

Also Eq. (10.29) indicates that in general, acceleration $\vec{f}$ is not parallel to the force in relativity, since the last term is in the direction of the velocity $\vec{v}$. If the velocity is perpendicular to the force, then acceleration $\vec{f}$ is parallel to the force.

10.8   Covariant Formulation of Newton's Law

The covariant formulation of Newton's law implies that the force is parallel to the acceleration only when the velocity is either parallel or perpendicular to the acceleration.

Following Newton's law, we write the force as

$$K_\mu = \frac{dp_\mu}{d\tau} = \frac{d(m_0 u_\mu)}{d\tau}$$

The spatial components are

$$K_j = \frac{d(m_0 u_j)}{d\tau} = a\frac{d(am_0 v_j)}{dt}$$

[where $a = \frac{1}{\sqrt{1-\frac{v^2}{c^2}}}$, $v_j v_j = v^2$, $u_j = a v_j$].

Thus

$$K_j = m_0 b_j a^2 + \frac{(a^4 m_0 v_j v_k b_k)}{c^2}, \tag{10.30}$$

where $b_j = \frac{d(v_j)}{dt}$ = acceleration.

$$\left[\frac{d(v^2)}{dt} = \frac{d(v_k v_k)}{dt} = 2v_k b_k \right].$$

Note that spatial components of the force will be parallel to the acceleration if second term in (10.30) vanishes. Now, if the velocity vector is perpendicular to the acceleration vector, then $u_j b_j = 0$ and the second term vanishes. Hence,

$$K_j = m_0 b_j a^2.$$

We know the space components of the force K_μ is related to F_j(Newtonian force) as $K_j = a F_j$. Therefore,

$$\left(\frac{1}{a} \right) K_j = m_0 b_j a = \frac{m_0 b_j}{\sqrt{1 - \frac{v^2}{c^2}}}. \tag{10.31}$$

The coefficient of b_j of Eq. (10.31) is referred as the *Transverse mass* of the body.

On the other hand, if the velocity is parallel to the acceleration, then $\overrightarrow{b} = h \overrightarrow{v}$ [h = constant].

Thus from (10.30), we get

$$K_j = m_0 b_j a^2 + \frac{(a^4 m_0 v_j v_k h v_k)}{c^2}$$

$$= m_0 b_j a^2 + \frac{(a^4 m_0 h v_j v^2)}{c^2}$$

$$= m_0 b_j a^4$$

$[v_k v_k = v^2, b_j = h v_j]$.

The space components of the force K_μ are related to F_j (Newtonian force) as $K_j = a F_j$. Therefore,

$$F_j = \left(\frac{1}{a} \right) K_j = m_0 b_j a^3 = m_0 b_j \left(1 - \frac{v^2}{c^2} \right)^{-\frac{3}{2}}. \tag{10.32}$$

The coefficient of b_j of Eq. (10.32) is referred as the *Longitudinal mass* of the body.

10.9   Examples of Longitudinal Mass and Transverse Mass

The case in which the force $\overrightarrow{F}$ is parallel to velocity, then the acceleration $\overrightarrow{b}$ is parallel to both $\overrightarrow{v}$ and $\overrightarrow{F}$. The particle moves in a straight line such as a charged particle starts from rest in a uniform electric field is the example of longitudinal mass. The case in which the force $\overrightarrow{F}$ is perpendicular to the velocity $\overrightarrow{v}$, for then $\overrightarrow{F}.\overrightarrow{v} = 0$. The force on a charged particle moving with velocity $\overrightarrow{v}$ in a magnetic field is the example of transverse mass.

10.10   The Lorentz Transformation of Momentum

The four momenta can be written as

$$p_\mu = [p_1, p_2, p_3, p_4], \tag{10.33}$$

where $p_4 = (\frac{iE}{c})$. We know Lorentz Transformation of any four vectors A_μ ($\mu = 1, 2, 3, 4$) may be expressed as

$$A_\mu^1 = a_{\mu\nu} A_\nu,$$

where $a_{\mu\nu}$ is given by

$$a_{\mu\nu} = \begin{bmatrix} a & 0 & 0 & ia\beta \\ 0 & 1 & 0 & 0 \\ 0 & 0 & 1 & 0 \\ -ia\beta & 0 & 0 & a \end{bmatrix}$$

$[\beta = \frac{v}{c}$ and $a = \frac{1}{\sqrt{1-\beta^2}}]$.

Hence,

$$\begin{bmatrix} p_1^1 \\ p_2^1 \\ p_3^1 \\ p_4^1 \end{bmatrix} = \begin{bmatrix} a & 0 & 0 & ia\beta \\ 0 & 1 & 0 & 0 \\ 0 & 0 & 1 & 0 \\ -ia\beta & 0 & 0 & a \end{bmatrix} \begin{bmatrix} p_1 \\ p_2 \\ p_3 \\ p_4 \end{bmatrix}. \tag{10.34}$$

This implies

$$p_1^1 = ap_1 + ia\beta p_4 = ap_1 + ia\left(\frac{v}{c}\right)\left(\frac{iE}{c}\right),$$

i.e.

$$p_1^1 = a \left[p_1 - \left(\frac{vE}{c^2} \right) \right] \tag{10.35}$$

$$p_2^1 = p_2, \quad p_3^1 = p_3 \tag{10.36}$$

$$p_4^1 = -ia\beta p_1 + ap_4$$

or,

$$\left(\frac{iE^1}{c} \right) = -ia \left(\frac{v}{c} \right) p_1 + a \left(\frac{iE}{c} \right),$$

i.e.

$$E^1 = a \left[E - v p_1 \right]. \tag{10.37}$$

The inverse transformations are obtained by replacing v by $-v$ as

$$p_1 = a \left[p_1^1 + \left(\frac{vE^1}{c^2} \right) \right], \quad p_2 = p_2^1, \quad p_3 = p_3^1, \quad E = a \left[E^1 + v p_1^1 \right]. \tag{10.38}$$

Let a body be at rest in the S^1 frame so that its rest energy is $m_0 c^2$. This means for observer in S frame the body is moving with velocity v. So, in S^1 frame,

$$p_1^1 = p_2^1 = p_3^1 = 0, \quad E^1 = m_0 c^2.$$

Hence,

$$p_1 = \frac{m_0 v}{\sqrt{1 - \frac{v^2}{c^2}}}, \quad p_2 = 0, \quad p_3 = 0, \quad E = \frac{m_0 c^2}{\sqrt{1 - \frac{v^2}{c^2}}}. \tag{10.39}$$

Suppose a body be at rest in the S^1 frame and emits radiation, then its energy is E^1. Therefore, the components of the momentum four vectors in S frame are

$$p_1 = \frac{vE^1}{c^2 \sqrt{1 - \frac{v^2}{c^2}}}, \quad p_2 = 0, \quad p_3 = 0, \quad E = \frac{E^1}{\sqrt{1 - \frac{v^2}{c^2}}}. \tag{10.40}$$

10.11   The Expression $p^2 - \frac{E^2}{c^2}$ Is Invariant Under Lorentz Transformation

We know

$$p_1^1 = a\left[p_1 - \left(\frac{vE}{c^2}\right)\right], \quad p_2^1 = p_2, \quad p_3^1 = p_3, \quad E^1 = a\left[E - vp_1\right].$$

Now,

$$p^{1^2} - \frac{E^{1^2}}{c^2} = p_1^{1^2} + p_2^{1^2} + p_3^{1^2} - \frac{E^{1^2}}{c^2}$$

$$= a^2\left[p_1 - \left(\frac{vE}{c^2}\right)\right]^2 + p_2^2 + p_3^2 - \frac{a^2\left[E - vp_1\right]^2}{c^2}$$

$$= p^2 - \frac{E^2}{c^2} \quad \text{(using, } p^2 = p_1^2 + p_2^2 + p_3^2).$$

Alternative

The relation between momentum and energy is given by

$$E^2 = p^2 c^2 + m_0^2 c^4.$$

From this relation, we can write

$$E^2 - p^2 c^2 = m_0^2 c^4.$$

We know, the rest mass and velocity of light are invariant quantities, therefore, $E^2 - p^2 c^2$ is also an invariant quantity, i.e.

$$E^{1^2} - p^{1^2} c^2 = E^2 - p^2 c^2.$$

This gives

$$p^{1^2} - \frac{E^{1^2}}{c^2} = p^2 - \frac{E^2}{c^2}.$$

Note 5: Let two particles of respective rest masses m_1 and m_2 collide and after collision their masses be m_3 and m_4. Let p_1 and p_2 be the initial and p_3 and p_4 be the final four momenta of the particles.

Then conservation of four momenta becomes

$$p_1^\mu + p_2^\mu = p_3^\mu + p_4^\mu, \tag{i}$$

$$p_j^\mu p_j^\mu = -m_j^2 c^2, \tag{ii}$$

$$p^{\mu}_j = \left(p_{xj}, p_{yj}, p_{zj}, \frac{iE_j}{c} = icm_j \right), \qquad (iii)$$

where [$j = 1, 2, 3, 4$].
Squaring both sides of (i) we get

$$- m_1^2 c^2 - m_2^2 c^2 + 2 p_1^{\mu} \cdot p_2^{\mu} = -m_3^2 c^2 - m_4^2 c^2 + 2 p_3^{\mu} \cdot p_4^{\mu}, \qquad (iv)$$

which valid for all inertial frame.

Now separating time and space parts yields the conservation of energy and momentum,

$$E_1 + E_2 = E_3 + E_4 \qquad (v)$$

$$\vec{p}_1 + \vec{p}_2 = \vec{p}_3 + \vec{p}_4 \qquad (vi)$$

Suppose the mass m_2 initially was at rest, then

$$p_2^{\mu} = (0, 0, 0, icm_2) \qquad (vii)$$

Equation (iv) implies

$$m_1^2 c^2 + m_2^2 c^2 + 2 m_2 E_1 = m_3^2 c^2 + m_4^2 c^2 + 2 \left(\frac{E_3 E_4}{c^2} - \vec{p}_3 \cdot \vec{p}_4 \right). \qquad (viii)$$

Equations (v) and (vi) become

$$E_1 + m_2^2 c^2 = E_3 + E_4, \qquad (ix)$$

$$\vec{p}_1 = \vec{p}_3 + \vec{p}_4. \qquad (x)$$

Now eliminating p_4 and E_4 and using $\frac{E_4^2}{c^2} = p_3^2 + m_3 c^2$, one can obtain from (viii)

$$m_1^2 c^2 + m_2^2 c^2 + 2 m_2 E_1 = m_4^2 c^2 - m_3^2 c^2 + 2 E_3 \left(\frac{E_1}{c^2} + m_2 \right) - 2 \vec{p}_3 \cdot \vec{p}_1. \quad (xi)$$

Remember that

$$\vec{p}_3 \cdot \vec{p}_1 = \mid \vec{p}_3 \mid\mid \vec{p}_1 \mid \cos \theta_1,$$

where θ_1 is the scattering angle of m_3 particle. If the energy of m_1 particle is known, then energy of m_3 particle will be known (as a function of scattering angle).

Note 6: Consider the decay of moving particle $a \rightarrow 1 + 2 + 3 + \dots + n$.
Now conservation of energy and momentum gives

$$E_a = E_1 + \dots + E_n = \sum_{i=1}^{n} E_i \qquad\qquad (i)$$

$$\vec{P}_a = \vec{P}_1 + \vec{P}_2 + \dots + \vec{P}_n = \sum_{i=1}^{n} \vec{P}_i \qquad\qquad (ii)$$

Also

$$E_i^2 = (P_i c)^2 + (M_i c^2)^2, i = a, 1, 2, \dots n \qquad\qquad (iii)$$

Now,

$$(M_a c^2)^2 = (E_a)^2 - \vec{P}_a \cdot \vec{P}_a c^2$$

$$= \left(\sum E_i\right)^2 - \left(\sum \vec{P}_i\right) \cdot \left(\sum \vec{P}_i\right) c^2, \qquad i = 1, 2, \dots n.$$

For two particles

$$\vec{P}_a \cdot \vec{P}_a = (\vec{P}_1 + \vec{P}_2) \cdot (\vec{P}_1 + \vec{P}_2)$$

$$= P_1^2 + P_2^2 + 2\vec{P}_1 \cdot \vec{P}_2 = P_1^2 + P_2^2 + 2P_1 P_2 \cos\theta,$$

where θ is the angle between the directions of the vectors P_1 and P_2.
Here,

$$\vec{P} \cdot \vec{P} = P_x^2 + P_y^2 + P_z^2 = P^2.$$

Example 10.1 Show that $m_0 = \frac{p^2 c^2 - T^2}{2Tc^2}$, where, m_0 is the rest mass of a particle with momentum, p and kinetic energy T.

Hint: We know

$$E^2 = p^2 c^2 + m_0^2 c^4 \qquad\qquad (1)$$

$$T = mc^2 - m_0 c^2 = E - m_0 c^2. \qquad\qquad (2)$$

Now,

$$T^2 = E^2 - 2Em_0 c^2 + m_0^2 c^4 = p^2 c^2 + 2m_0^2 c^4 - 2Em_0 c^2$$

or,

$$p^2 c^2 - T^2 = 2T m_0 c^2. \text{ etc.}$$

Example 10.2 Show that if a particle is highly relativistic, then the fractional difference between c and v is approximately $\frac{1}{2}\left(\frac{m_0c^2}{E}\right)^2$ where E is the relativistic energy.

Hint: Since the particle is highly relativistic, then $\frac{v^2}{c^2}$ is very close to unity, therefore, $1 - \frac{v^2}{c^2}$ is a very small quantity, i.e.

$$1 - \frac{v^2}{c^2} = \epsilon, \text{ say}$$

or

$$\frac{v^2}{c^2} = 1 - \epsilon \Longrightarrow \frac{v}{c} = 1 - \frac{1}{2}\epsilon$$

[expanding binomially and neglecting higher order of ϵ].
  This implies

$$\epsilon = \frac{2(c-v)}{c}.$$

Now,

$$E = mc^2 = \frac{m_0c^2}{\sqrt{1 - \frac{v^2}{c^2}}} = \frac{m_0c^2}{\sqrt{\epsilon}}.$$

Replacing ϵ, we get

$$\frac{(c-v)}{c} = \frac{1}{2}\left(\frac{m_0c^2}{E}\right)^2.$$

Example 10.3 Can a massless particle exist?

Hint: We know total energy and relativistic momentum

$$E = mc^2 = \frac{m_0c^2}{\sqrt{1 - \frac{v^2}{c^2}}} \text{ and } p = mv = \frac{m_0v}{\sqrt{1 - \frac{v^2}{c^2}}}.$$

When $m_0 = 0$ and $v < c$, then it is evident that $E = p = 0$. Thus a massless particle with a speed less than c can have neither energy nor momentum. However, when $m_0 = 0$ and $v = c$, then, $E = \frac{0}{0}$ and $p = \frac{0}{0}$, which are indeterminate. These imply E and p can have any values. Thus, massless particles with non-zero energy and momentum must travel with the speed of light.
  Examples of that particle are photon and neutrino.

Example 10.4 Show that integral of the world force over the world line vanishes if the rest mass is assumed not to change.

Hint:

$$\int K_\mu \cdot dx_\mu = \int K_\mu \cdot \frac{dx_\mu}{d\tau} d\tau = \int K_\mu \cdot u_\mu d\tau$$

$$= \int \frac{d(m_0 u_\mu)}{d\tau} \cdot u_\mu d\tau = \int \frac{d(\frac{1}{2} m_0 u_\mu \cdot u_\mu)}{d\tau} d\tau$$

$$= -\int \frac{d(\frac{1}{2} m_0 c^2)}{d\tau} d\tau = 0.$$

Since rest mass is not changed.

Example 10.5 Find the integral of the relativistic force over 3D path.

Hint: When a relativistic force $\vec{F}$ acts on a mass m, so as to move it through a distance $d\vec{x}$, then $\int \vec{F} \cdot d\vec{x} = I$ is the work done. Thus,

$$I = \int \vec{F} \cdot d\vec{x} = \int \frac{d(m\vec{v})}{dt} \cdot d\vec{x}$$

$$= \int \frac{d(am_0 \vec{v})}{dt} \cdot d\vec{x} = \int d(am_0 \vec{v}) \cdot \frac{d\vec{x}}{dt}$$

$$= \int d(am_0 \vec{v}) \cdot \vec{v} = \int am_0 \vec{v} \cdot d\vec{v}$$

$$= \int am_0 v dv = \int \frac{m_0 v dv}{\sqrt{1 - \frac{v^2}{c^2}}}$$

[since $v^2 = \vec{v} \cdot \vec{v} \implies v dv = \vec{v} \cdot d\vec{v}$ and $a = \frac{1}{\sqrt{1 - \frac{v^2}{c^2}}}$]

After integration, we get

$$I = -c^2 m_0 \sqrt{1 - \frac{v^2}{c^2}} + D.$$

When $v = 0$, work done is zero $\implies D = m_0 c^2$.
    Thus,

$$I = m_0 c^2 - c^2 m_0 \sqrt{1 - \frac{v^2}{c^2}} = T = \text{ relativistic kinetic energy.}$$

The above relativistic expression for kinetic energy T must reduce to classical result, $\frac{1}{2} m_0 v^2$, when, $\frac{v}{c} << 1$.
    Now,

$$T = m_0 c^2 \left[1 - \left(1 - \frac{v^2}{c^2} \right)^{\frac{1}{2}} \right]$$

Expanding binomially and neglecting $\frac{v^4}{c^4}$ and higher order, we get

$$T \approx m_0 c^2 \left[\left(1 - 1 + \frac{v^2}{2c^2} \right) \right] = \frac{1}{2} m_0 v^2.$$

Hence the result.

Example 10.6 Calculate the amount gain in the mass of earth in a year, if approximately 2 cal of radiant energy are received by each square centimetre of earth surface per minute.

[Given, radius of earth $= 6.4 \times 10^3$ km.]

Hint: Total surface area of earth is $4\pi r^2$, therefore, energy received in one minute is $4\pi r^2 \times$ energy received by each square centimetre of earth surface per minute which is given by (since $2\,\text{cal} = 2 \times 4.2\,\text{J}$)

$$= 4 \times 3.14 \times (6.4 \times 10^6)^2 \times 2 \times 4.2 \times 10^4\,\text{J}$$

Energy received in 1 year

$$= 4 \times 3.14 \times (6.4 \times 10^6)^2 \times 2 \times 4.2 \times 10^4 \times 60 \times 24 \times 365\,\text{J}.$$

Hence, annual gain in the mass of earth

$$\Delta m = \frac{\Delta E}{c^2} = \frac{4 \times 3.14 \times (6.4 \times 10^6)^2 \times 2 \times 4.2 \times 10^4 \times 60 \times 24 \times 365}{(3 \times 10^8)^2}$$
$$= 2.524 \times 10^8\,\text{kg per year.}$$

Example 10.7 A particle of rest mass m_0 moving with relativistic velocity v has got a momentum p and kinetic energy T. Show that $\frac{pv}{T} = \frac{T + 2m_0 c^2}{T + m_0 c^2}$.

Hint: We know

$$E = mc^2 = \frac{pc^2}{v}.$$

Therefore,

$$p = \frac{vE}{c^2} = \frac{v}{c^2}(T + m_0 c^2). \qquad (1)$$

Again,

$$E = T + m_0 c^2 = \frac{m_0 c^2}{\sqrt{1 - \frac{v^2}{c^2}}} \implies v^2 = \frac{c^2(T^2 + 2 m_0 T c^2)}{(T + m_0 c^2)^2}. \tag{2}$$

Hence,

$$\frac{pv}{T} = \text{etc.}$$

Example 10.8 Find the velocity that one electron must be given so that its momentum is a times its rest mass times speed of light. What is the energy of this speed?

Hint: The momentum of an electron of rest mass $m_0 = 9 \times 10^{-31}$ kg moving with velocity v is given by

$$p = a \times m_0 c^2 = \frac{m_0 v}{\sqrt{1 - \frac{v^2}{c^2}}} \implies v =$$

Now,

$$E = \frac{m_0 c^2}{\sqrt{1 - \frac{v^2}{c^2}}} = \text{etc.}$$

Example 10.9 Calculate the velocity of an electron having a total energy of a MeV (rest mass of the electron is $m_0 = 9 \times 10^{-31}$ kg).

Hint: Here,

$$E = \frac{m_0 c^2}{\sqrt{1 - \frac{v^2}{c^2}}} = a \, \text{MeV} = a \times 10^6 \, \text{eV} = a \times 10^6 \times 1.6 \times 10^{-19} \, \text{J} \implies v = \text{etc.}$$

Example 10.10 A π-meson of rest mass m_π decays into μ-meson of mass m_μ and a neutrino of mass m_ν. Show that the total energy of the μ-meson is $\frac{1}{2 m_\pi}[m_\pi^2 + m_\mu^2 - m_\nu^2]c^2$.

Hint: According to principle of conservation of momentum, if the μ-meson of mass m_μ moves with momentum p, then neutrino of mass m_ν moves with momentum $-p$. Therefore, according to the momentum–energy relation, we have

$$E_\mu^2 = p^2 c^2 + m_\mu^2 c^4, \quad E_\nu^2 = p^2 c^2 + m_\nu^2 c^4 \tag{1}$$

From these, we get

$$E_\mu^2 - E_\nu^2 = (m_\mu^2 - m_\nu^2)c^4 \tag{2}$$

Total energy

$$E_\mu + E_\nu = E = m_\pi c^2 \tag{3}$$

Solving (2) and (3), one can get, total energy of the μ-meson as $E_\mu =$ etc.

A π-meson of rest mass m_π decays into μ-meson of mass m_μ and a neutrino of mass m_ν. Show that the total energy of the μ-meson is $\frac{1}{2m_\pi}[m_\pi^2 + m_\mu^2 - m_\nu^2]c^2$.

Example 10.11 A particle of rest mass M decays spontaneously into two masses m_1 and m_2. Obtain energies of the daughter masses. Show also that their kinetic energy is given by $T_i = \Delta M \left(1 - \frac{m_i}{M} - \frac{\Delta M}{2M}\right) c^2$ where $\Delta M = M - m_1 - m_2$, i.e. difference between initial and final masses.

Hint: According to principle of conservation of momentum, if the m_1 mass moves with momentum p, then other mass m_2 moves with momentum $-p$. Therefore, according to the momentum–energy relation, we have

$$E_1^2 = p^2 c^2 + m_1^2 c^4, \quad E_2^2 = p^2 c^2 + m_2^2 c^4 \tag{1}$$

(here, energies of the daughter masses are E_1 and E_2, respectively)

From these, we get

$$E_1^2 - E_2^2 = (m_1^2 - m_2^2)c^4. \tag{2}$$

Total energy

$$E_1 + E_2 = E = Mc^2. \tag{3}$$

Solving (2) and (3), one can get, energies E_1 and E_2 of the daughter masses as

$$E_1 = \frac{(M^2 + m_1^2 - m_2^2)c^2}{2M}, \quad E_2 = \frac{(M^2 + m_2^2 - m_1^2)c^2}{2M}.$$

The relativistic kinetic energies T_1 and T_2 of the daughter masses are

$$T_1 = E_1 - m_1 c^2, \quad T_2 = E_2 - m_1 c^2$$

which yields

$$T_1 + T_2 = E_1 + E_1 - m_1 c^2 - m_2 c^2 = Mc^2 - m_1 c^2 - m_2 c^2 = \Delta M c^2.$$

Now,

$$T_1 = \frac{(M^2 + m_1^2 - m_2^2)c^2}{2M} - m_1 c^2 = \frac{(M - m_1 - m_2)(M - m_1 + m_2)c^2}{2M}$$

$$= \Delta M \left(1 - \frac{m_1}{M} - \frac{\Delta M}{2M}\right) c^2, \quad \text{etc.}$$

Example 10.12 Calculate the velocity of an electron accelerated by a potential of a MV (rest mass of the electron is $m_0 = 9 \times 10^{-31}$ kg).

Hint: Here,

$$T = (m - m_0)c^2 = \frac{m_0 c^2}{\sqrt{1 - \frac{v^2}{c^2}}} - m_0 c^2 = a \times 10^6 \, \text{eV} = a \times 10^6 \times 1.6 \times 10^{-19} \, \text{J} \Rightarrow v = \text{etc.}$$

Example 10.13 By what fraction does the mass of water increase due to increase in its thermal energy, when it is heated from a^0 C to b^0 C?

Hint: Let mass of the water to be heated be M kg. Energy needed in heating the water from a^0 C to b^0 C is

$$\triangle E = M \times 1 \times (b - a) \, \text{kilo calories} = M \times 1 \times (b - a) \times 1000 \times 4.2 \, \text{J}$$

Mass–energy equivalent relation $\triangle M = \frac{\triangle E}{c^2}$ gives the required $\triangle M$.
The fractional increase in mass of water due to heating is $\frac{\triangle M}{M}$, etc.

Example 10.14 A particle of rest mass m_0 and kinetic energy $2m_0 c^2$ strikes and sticks to a stationary particle of rest mass $2m_0$. Find the rest mass M_0 of the composite particle.

Hint:

$$\odot \longrightarrow v \qquad\qquad \otimes$$

$$m_0, T = 2m_0 c^2 \qquad\qquad T = 0, 2m_0$$

Total energy of the particle of rest mass m_0 moving with velocity v is

$$E = mc^2 = \frac{m_0 c^2}{\sqrt{1 - \frac{v^2}{c^2}}} = 2m_0 c^2 + m_0 c^2. \tag{1}$$

From energy conservation, we have

$$(2m_0 c^2 + m_0 c^2) + 2m_0 c^2 = \frac{M_0 c^2}{\sqrt{1 - \frac{V^2}{c^2}}} \tag{2}$$

(Velocity of the composite particle is V)
Conservation of momentum

Fig. 10.2 The angle between two decay particles is θ

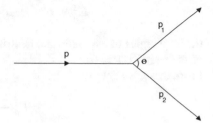

$$\frac{m_0 v}{\sqrt{1 - \frac{v^2}{c^2}}} + 0 = \frac{M_0 V}{\sqrt{1 - \frac{V^2}{c^2}}} \tag{3}$$

Solving from (1)–(3), one can get the solutions for V and M_0 as $V = \frac{2\sqrt{2}c}{5}$ and $M_0 = \sqrt{17}m_0$.

Example 10.15 An unstable particle of rest mass m_0 and momentum p decays into two particles of rest masses m_1 and m_2, momentum p_1 and p_2 and total energies E_1 and E_2, respectively. Show that

$$m_0 c^4 = (m_1 + m_2)c^4 + 2E_1 E_2 - 2m_1 m_2 c^4 - 2p_1 p_2 c^2 \cos \theta,$$

where θ is the angle between the two decay particles (Fig. 10.2).

Hint: From principle of conservation of energy $E_0 = E_1 + E_2$, we have

$$\sqrt{m_0 c^4 + p_0^2 c^2} = \sqrt{m_1 c^4 + p_1^2 c^2} + \sqrt{m_2 c^4 + p_2^2 c^2}. \tag{1}$$

From principle of conservation of momentum, we have

$$p^2 = p_1^2 + p_2^2 + 2p_1 p_2 \cos \theta \tag{2}$$

Squaring (1) and using (2), one can obtain the desired result.

Example 10.16 A meson of rest mass π comes to rest and disintegrates into a muon of rest mass μ and a neutrino of zero rest mass. Show that the kinetic energy of motion the muon is $T = \frac{(\pi - \mu)^2 c^2}{2\pi}$.

Hint:

$$\pi \longrightarrow \mu + \nu$$

According to principle of conservation of momentum, if muon moves with momentum p, then neutrino moves with momentum $-p$. Therefore, according to the momentum–energy relation, we have

$$E_\mu^2 = p^2c^2 + \mu^2c^4, \quad E_\nu^2 = p^2c^2 + \nu^2c^4 \tag{1}$$

(here, energies of the muon and neutrino are E_μ and E_ν, respectively, and rest mass of the neutrino $\nu = 0$)
From these, we get

$$E_\mu^2 - E_\nu^2 = \mu^2c^4. \tag{2}$$

Total energy

$$E_\mu + E_\nu = E = \pi c^2 \tag{3}$$

Solving (2) and (3), one can get, energies of muon and neutrino as

$$E_\mu = \frac{(\mu^2 + \pi^2)c^2}{2\pi}, \quad E_\nu = \frac{(\mu^2 - \pi^2)c^2}{2\pi}. \tag{4}$$

The kinetic energy of the meson is

$$T = E_\mu - \mu c^2 = \frac{(\mu^2 + \pi^2)c^2}{2\pi} - \mu c^2 = \text{etc.}$$

Example 10.17 A rocket propels itself rectilinearly through empty space by emitting pure radiation in the direction opposite to its motion. If v is the final velocity relative to its initial frame, prove that the ratio of the initial to the final rest mass of the rocket is given by $\frac{M_i}{M_f} = \sqrt{\frac{c+v}{c-v}}$.

Hint: Principle of conservation of energy

$$M_i c^2 = \frac{M_f c^2}{\sqrt{1 - \frac{v^2}{c^2}}} + E_\gamma \tag{1}$$

E_γ is the energy of radiation. Since it is massless, therefore, its momentum $p = \frac{E_\gamma}{c}$.
Principle of conservation of momentum

$$p = \frac{E_\gamma}{c} = \frac{M_f v}{\sqrt{1 - \frac{v^2}{c^2}}}. \tag{2}$$

Assuming rocket started from rest.
Eliminating E_γ from (1) and (2), one can get the required result.

Example 10.18 A particle of rest mass M moving at a velocity u collides with a stationary particle of rest mass m. If the particles stick together, show that the speed of the composite ball is equal to $\frac{u\gamma M}{\gamma M + m}$ where $\gamma = \frac{1}{\sqrt{1 - \frac{u^2}{c^2}}}$.

Hint: Principle of conservation of energy

$$mc^2 + M\gamma c^2 = \frac{(M+m)c^2}{\sqrt{1 - \frac{v^2}{c^2}}} \tag{1}$$

Principle of conservation of momentum

$$M\gamma u = \frac{(M+m)v}{\sqrt{1 - \frac{v^2}{c^2}}} \tag{2}$$

[v is the velocity of the combined ball]
Using (1) and (2), one can solve for v.

Example 10.19 Prove the following relations:

(i) $T = c\sqrt{m_0^2 c^2 + p^2} - m_0 c^2$ (ii) $p = \dfrac{\sqrt{T^2 + 2m_0 c^2 T}}{c}$ (iii) $m_0 = \dfrac{\sqrt{E^2 - p^2 c^2}}{c^2}$

Hint: We know

$$E^2 = p^2 c^2 + m_0^2 c^4, \quad E = T + m_0 c^2.$$

Now,

(i) $T = E - m_0 c^2 = \sqrt{p^2 c^2 + m_0^2 c^4} - m_0 c^2 = c\sqrt{p^2 + m_0^2 c^2} - m_0 c^2.$

(ii) $E^2 = (T + m_0 c^2)^2 = p^2 c^2 + m_0^2 c^4 \Longrightarrow p = \dfrac{\sqrt{T^2 + 2m_0 c^2 T}}{c}.$

(iii) $E^2 = p^2 c^2 + m_0^2 c^4 \Longrightarrow m_0 = \dfrac{\sqrt{E^2 - p^2 c^2}}{c^2}.$

Example 10.20 What is the speed of an electron whose kinetic energy equals a times its rest energy?

Hint:

$$T = (m - m_0)c^2 = \left(\frac{m_0}{\sqrt{1 - \frac{v^2}{c^2}}} - m_0 \right) c^2 = a m_0 c^2 \text{ etc.}$$

Example 10.21 Two particles of rest masses m_1 and m_2 move with velocities u_1 and u_2 collide with another and sticks (i.e. inelastic collision). Show that the rest mass and the velocity of the combined particles are

$$m_3^2 = m_1^2 + m_2^2 + 2m_1 m_2 \gamma_1 \gamma_2 \left(1 - \frac{u_1 u_2}{c^2}\right), \quad u_3 = \frac{m_1 \gamma_1 u_1 + m_2 \gamma_2 u_2}{m_1 \gamma_1 + m_2 \gamma_2}.$$

$$\gamma_1 = \frac{1}{\sqrt{1 - \frac{u_1^2}{c^2}}}, \quad \gamma_2 = \frac{1}{\sqrt{1 - \frac{u_2^2}{c^2}}}.$$

Hint: In an inelastic collision, let m_3 be the rest mass and u_3 be the velocity of the resulting particle.

Conservation of momentum : $m_1 \gamma_1 u_1 + m_2 \gamma_2 u_2 = m_3 \gamma_3 u_3$

Conservation of mass : $m_1 \gamma_1 + m_2 \gamma_2 = m_3 \gamma_3$

$$\gamma_3 = \frac{1}{\sqrt{1 - \frac{u_3^2}{c^2}}}.$$

Solving these equations, one can get m_3 and u_3.

Example 10.22 A particle with kinetic energy T_0 and rest energy E_0 strikes an identical particle at rest and gets scattered at an angle θ. Show that its kinetic energy T after scattering is given by $T = \frac{T_0 \cos^2 \theta}{1 + \frac{T_0 \sin^2 \theta}{2E_0}}$.

Hint: Consider a particle with mass m_1 and momentum p_1 strikes a particle of identical mass at rest. After collision, the particle and rest particle get scattered at angles θ, ϕ, respectively. We have

$$E^2 = (T + E_0)^2 = c^2 p^2 + m_0^2 c^4 = c^2 p^2 + E_0^2 \implies p^2 c^2 = 2T E_0 + T^2. \quad (1)$$

Principle of conservation of momentum

$$p_1 = p_3 \cos \theta + p_4 \cos \phi. \quad (2)$$

$$0 = p_3 \sin \theta - p_4 \sin \phi. \quad (3)$$

Principle of conservation of energy

$$E_1 + E_2 = E_3 + E_4 \implies E_0 + T_0 + E_0 = E_0 + T + E_0 + T_4 \implies T_4 = T_0 - T.$$
(4)

From (2) and (3), we get

$$p_4 c^2 = (p_1^2 + p_3^2 - 2p_1 p_3 \cos\theta)c^2.$$
(5)

Using the relation (1) in (5), we get

$$2T_4 E_0 + T_4^2 = 2T_0 E_0 + T_0^2 + 2T E_0 + T^2 - 2\sqrt{2T_0 E_0 + T_0^2}\sqrt{2T E_0 + T^2}\cos\theta.$$

Replacing T_4 by $T_0 - T$ and after some simplification, one will get

$$T = \frac{T_0 \cos^2\theta}{1 + \frac{T_0 \sin^2\theta}{2E_0}}.$$

Example 10.23 For Minskowski force K_μ, show that $K_\mu \cdot K_\mu = \left[\frac{1 - \frac{v^2}{c^2}\cos^2\theta}{1 - \frac{v^2}{c^2}}\right]F^2$, where velocity $\vec{v}$ makes θ angle with relativistic force $\vec{F}$.

Hint: The Minkowski force is given by

$$K_\mu = \left[aF_1, aF_2, aF_3, \left(\frac{ia}{c}\right)\vec{F}\cdot\vec{v}\right]; \quad a = \frac{1}{\sqrt{1 - \frac{v^2}{c^2}}}.$$

Now,

$$K_\mu \cdot K_\mu = K_1 K_1 + K_2 K_2 + K_3 K_3 + K_4 K_4$$

$$= a^2 F_1^2 + a^2 F_2^2 + a^2 F_3^2 + \frac{i^2 a^2}{c^2}(\vec{F}\cdot\vec{v})^2$$

$$= a^2 F^2 - \frac{a^2}{c^2}(Fv\cos\theta)^2 = \left[\frac{1 - \frac{v^2}{c^2}\cos^2\theta}{1 - \frac{v^2}{c^2}}\right]F^2$$

[here, $\vec{F}\cdot\vec{v} = vF\cos\theta$]

Example 10.24 Show that if a particle decays spontaneously from rest into two or more components, then the rest mass of the particle must be greater than the sum of the rest masses of the resulting components.

Hint: Let a particle of rest mass m decays spontaneously into a number of components of rest masses $m_1, m_2, m_3,...$ and velocities $v_1, v_2, v_3,...$
Conservation of energy yields

$$E = mc^2 = E_1 + E_2 + E_3 + \cdots = \frac{m_1 c^2}{\sqrt{1 - \frac{v_1^2}{c^2}}} + \frac{m_2 c^2}{\sqrt{1 - \frac{v_2^2}{c^2}}} + \frac{m_3 c^2}{\sqrt{1 - \frac{v_3^2}{c^2}}} = \ldots$$

Note that $E_i > m_i c^2$ for $i = 1, 2, 3, \ldots$. Hence,

$$m > m_1 + m_2 + m_3 + \cdots .$$

Example 10.25 A particle of momentum p_1, rest mass m_1 is incident upon a stationary particle of rest mass m_2. Show that the velocity of the centre of the mass system is equal to

$$v = \frac{p_1 c^2}{E_1 + E_2} = \frac{p_1 c^2}{\sqrt{p_1 c^2 + m_1^2 c^4 + m_2 c^2}}$$

Hint: In rest system S, total momentum is $p = p_1 + 0 = p_1$. Let the centre of the mass system is S^1 which is moving with velocity v along x-axis. In S^1 system, the total momentum is zero. The total energy in S system before collision is

$$E = E_1 + E_2 = \sqrt{p_1 c^2 + m_1^2 c^4} + m_2 c^2 \quad \text{(here, } p_2 = 0\text{)}.$$

In S system,

$$p_{1x} = p_1, \quad p_{1y} = p_{1z} = 0, \quad p_{2x} = p_{2y} = p_{2z} = 0.$$

In S^1 system,

$$p_{1x}^1 = \frac{p_1 - \frac{v E_1}{c^2}}{\sqrt{1 - \frac{v^2}{c^2}}}, \quad p_{1y}^1 = p_{1y} = 0, \quad p_{1z}^1 = p_{1z} = 0.$$

$$p_{2x}^1 = \frac{p_{2x} - \frac{v E_2}{c^2}}{\sqrt{1 - \frac{v^2}{c^2}}} = -\frac{\frac{v E_2}{c^2}}{\sqrt{1 - \frac{v^2}{c^2}}}$$

$$p_{2y}^1 = p_{2z}^1 = 0.$$

Now, in S^1 system,

$$p_{1x}^1 + p_{12x}^1 = 0 \implies \frac{p_1 - \frac{v(E_1 + E_2)}{c^2}}{\sqrt{1 - \frac{v^2}{c^2}}} = 0.$$

This will give

$$v = \frac{p_1 c^2}{E_1 + E_2} = \frac{p_1 c^2}{\sqrt{p_1 c^2 + m_1^2 c^4 + m_2 c^2}}.$$

Example 10.26 A particle of rest mass m_0 describes the trajectory $x = f(t)$, $y = g(t)$, $z = 0$ in an inertial frame S. Find the Minkowski force and the corresponding Newtonian force on the particle.

Hint: Minkowski four force is given by

$$K_\mu = \frac{dp_\mu}{d\tau} = \frac{d(m_0 u_\mu)}{d\tau}.$$

For spatial component

$$K_i = \frac{1}{\sqrt{1 - \frac{v^2}{c^2}}} \frac{d}{dt} \left(\frac{m_0 v_i}{\sqrt{1 - \frac{v^2}{c^2}}} \right),$$

where

$$v_1 = \dot{x} = \dot{f}(t), \ v_2 = \dot{y} = \dot{g}(t), \ v_3 = \dot{z} = 0, \ v^2 = v_1^2 + v_2^2 + v_3^2 = \dot{f}^2 + \dot{g}^2.$$

Let

$$\gamma = \frac{1}{\sqrt{1 - \frac{v^2}{c^2}}}, \quad \text{therefore,} \quad \dot{\gamma} - \frac{\gamma^3}{c^2}(\dot{f}\ddot{f} + \dot{g}\ddot{g}).$$

Now,

$$K_1 = \gamma \frac{d}{dt}(m_0 v_1 \gamma) = m_0 \gamma [\gamma \dot{v}_1 + v_1 \dot{\gamma}]$$

$$= m_0 \gamma^2 \left[\ddot{f} + \frac{\gamma^2}{c^2}(\dot{f}\ddot{f} + \dot{g}\ddot{g})\dot{f} \right]$$

$$K_2 = \gamma \frac{d}{dt}(m_0 v_2 \gamma) = m_0 \gamma [\gamma \dot{v}_2 + v_2 \dot{\gamma}]$$

$$= m_0 \gamma^2 \left[\ddot{g} + \frac{\gamma^2}{c^2}(\dot{f}\ddot{f} + \dot{g}\ddot{g})\dot{g} \right]$$

$$K_3 = 0$$

$$K_4 = \gamma \frac{d}{dt}(im_0 c\gamma) = \frac{im_0 \gamma^4}{c} \left[\dot{f}\ddot{f} + \dot{g}\ddot{g} \right].$$

Newtonian force

$$F_1 = m_0 \frac{dv_1}{dt} = m_0 \ddot{f}, \quad F_2 = m_0 \frac{dv_2}{dt} = m_0 \ddot{g}, \quad F_3 = 0.$$

Example 10.27 A particle of rest mass m_0 describes the parabolic trajectory $x = at$, $y = bt^2, z = 0$ in an inertial frame S. Find the Minkowski force and the corresponding Newtonian force on the particle.

Hint: Use $f(t) = at$ and $g(t) = bt^2$ in example (10.26).

Example 10.28 A particle of rest mass m_0 describes the circular path $x = a\cos t$, $y = a\sin t$, $z = 0$ in an inertial frame S. Find the Minkowski force and the corresponding Newtonian force on the particle.

Hint: Use $f(t) = a\cos t$ and $g(t) = a\sin t$ in example (10.26).

Example 10.29 A particle of rest mass m_0 moves on the x-axis along a world line described by the parametric equations $t(a) = \frac{1}{A}\sinh a$ and $x(a) = \frac{1}{A}\cosh a$, a is the parameter and A is a constant. Find four velocities, momentum and acceleration and their norms. Also find the Minkowski force.

Hint: The proper time is defined by

$$d\tau = dt\sqrt{1 - v^2},$$

where $v = \frac{dx}{dt}$ and we have assumed $c = 1$. Here, y and z coordinates are unimportant and we will be suppressed. We find that $v = \frac{dx}{dt} = \tanh a$. Therefore,

$$\tau = \int dt\sqrt{1 - v^2} = \int \frac{dt}{da}\frac{1}{\cosh a}da = \int \frac{da}{A} \implies \tau = \frac{a}{A}.$$

Hence,

$$t(\tau) = \frac{\sinh(A\tau)}{A}, \quad x(\tau) = \frac{\cosh(A\tau)}{A}.$$

Now, we calculate the four-velocity vector $u_\mu = \frac{dx_\mu}{d\tau}$ as

$$u_1 = \frac{dx}{d\tau} = \sinh(A\tau), \quad u_4 = \frac{d(it)}{d\tau} = i\cosh(A\tau).$$

Here, norm of four velocities is given by

$$u_\mu \cdot u_\mu = u_1 u_1 + u_4 u_4 = \sinh^2(A\tau) - \cosh^2(A\tau) = -1.$$

Note that particle three velocities is $v = \frac{dx}{dt} = \tanh(A\tau)$. This velocity never exceeds the speed of light, however, when $\tau = \pm\infty$, then $v \longrightarrow 1$, i.e. approaches to the velocity of light.

The four momenta is given by

$$p_\mu = m_0 u_\mu,$$

where m_0 is the rest mass of the particle. Here, norm of four momenta is given by

$$p_\mu \cdot p_\mu = -m_0^2.$$

The four accelerations $f_\mu = \frac{du_\mu}{d\tau}$ is given by

$$f_1 = \frac{du_1}{d\tau} = A\cosh(A\tau), \quad f_4 = \frac{u_4}{d\tau} = iA\sinh(A\tau).$$

Here, norm of four accelerations is given by

$$f_\mu \cdot f_\mu = f_1 f_1 + f_4 f_4 = A^2 \cosh^2(A\tau) - A^2 \sinh^2(A\tau) = A^2.$$

The Minkowski force is given by

$$K_\mu = m_0 f_\mu.$$

Example 10.30 A particle of rest mass m_0 moves on the x-axis along a world line described by the parametric equations $t = f(\sigma)$ and $x = g(\sigma)$, σ is the parameter. Find four velocities, momentum and acceleration and their norms. Also find the Minkowski force.

Hint: The proper time is defined by

$$d\tau = dt\sqrt{1 - v^2},$$

where $v = \frac{dx}{dt}$ and we have assumed $c = 1$. Here, $y-$ and z-coordinates are unimportant and we will be suppressed. We find that $v = \frac{dx}{dt} = \frac{\dot{g}}{\dot{f}}$ (here, $\dot{g} = \frac{dg}{d\sigma}$). Therefore,

$$\tau = \int dt\sqrt{1 - v^2} = \int \sqrt{\dot{f}^2 - \dot{g}^2}\, d\sigma.$$

The four velocities vector $u_\mu = \frac{dx_\mu}{d\tau}$ can be obtained as

$$u_1 = \frac{dx}{d\tau} = \frac{\dot{g}}{\sqrt{\dot{f}^2 - \dot{g}^2}}, \quad u_4 = \frac{d(it)}{d\tau} = \frac{i\dot{f}}{\sqrt{\dot{f}^2 - \dot{g}^2}} \quad \text{etc.}$$

Example 10.31 A particle is moving along x-axis. It is uniformly accelerated in the sense that the acceleration measured in its rest frame is always a, a constant. Find x and t in terms of proper time assuming the particle passes through x_0 at time $t = 0$ with zero velocity.

Hint: Here, y and z-coordinates are unimportant and we will be suppressed and we have assumed $c = 1$. We know four velocities u_μ satisfies

$$u_\mu \cdot u_\mu = u_1^2 + u_4^2 = -1 \tag{1}$$

Differentiating (1), we get

$$u_1 f_1 + u_4 f_4 = 0, \tag{2}$$

where $f_\mu = \frac{du_\mu}{d\tau}$ is the four accelerations.
   Given:

$$f_\mu \cdot f_\mu = f_1^2 + f_4^2 = a^2. \tag{3}$$

Solving these equations, we get

$$f_4 = \frac{du_4}{d\tau} = iau_1 \tag{4}$$

$$f_1 = \frac{du_1}{d\tau} = -iau_4. \tag{5}$$

From these equations, we obtain

$$\frac{d^2 u_1}{d\tau^2} = a^2 u_1.$$

The solution is obtained as

$$u_1 = A \sinh(a\tau). \tag{6}$$

A is one of the integration constants. The other integration constant is zero as for $\tau = 0$, $u_1 = 0$. Also we obtain the solution for u_4

$$u_4 = Ai \cosh(a\tau). \tag{7}$$

Equation (1) implies $A = 1$.

Again,

$$u_1 = \frac{dx_1}{d\tau} = \sinh(a\tau), \quad u_4 = \frac{dx_4}{d\tau} = \frac{d(it)}{d\tau} = i\cosh(a\tau).$$

Using the boundary condition, the solutions of these equations are obtained as

$$x(\tau) = x_0 + \frac{1}{a}[\cosh(a\tau) - 1], \quad t(\tau) = \frac{1}{a}\sinh(a\tau).$$

The world line of the particle is the hyperbola given by

$$t^2 = (x - x_0)^2 + \frac{2}{a}[x - x_0].$$

Example 10.32 A particle of rest mass M decays spontaneously into two particles with mass deficit ΔM. Show that the kinetic energy of the particle of rest mass m_i $(i = 1, 2)$ is given by $T_i = \Delta M \left(1 - \frac{m_i}{M} - \frac{\Delta M}{2M}\right) c^2$.

Hint: Let the decayed particles have rest masses m_1 and m_2. According to principle of conservation of momentum, if the m_1 mass moves with momentum p, then other mass m_2 moves with momentum $-p$. Follow Example 10.11.

Example 10.33 A particle of mass m_1 and momentum p collides with a particle of mass m_2 at rest. After the collision, the particles with mass m_1 and m_2 got momentum p' and q, respectively. Find q in terms of p, m_1, m_2.

Hint: We know momentum four vectors as

$$p^\mu = \left(p_x, p_y, p_z, \frac{iE_1}{c}\right), \quad \text{and} \quad q^\mu = \left(q_x, q_y, q_z, \frac{iE_2}{c}\right),$$

where E_1 and E_2 are energies of the particles. Suppose the initial particle is moving along x-axis. Therefore, before collision $p_x = p, p_y = 0, p_z = 0$ and after the collision $q_x = q, q_y = 0, q_z = 0$.
Hence, before collision

$$p^\mu = \left(p, 0, 0, \frac{iE_1}{c}\right), \quad q^\mu = \left(0, 0, 0, \frac{iE_2}{c}\right).$$

After collision, the four-momentum vectors become

$$p'^\mu = \left(p', 0, 0, \frac{iE_1}{c}\right), \quad q'^\mu = \left(q, 0, 0, \frac{iE_2}{c}\right).$$

According to the conservation of momentum and energy, we get

$$p^\mu + q^\mu = p'^\mu + q'^\mu.$$

Hence, we get

$$\left(p, 0, 0, \frac{i(E_1 + E_2)}{c}\right) = \left(p' + q, 0, 0, \frac{i(E'_1 + E'_2)}{c}\right).$$

This yields

$$p = p' + q. \tag{i}$$

$$E_1 + E_2 = E'_1 + E'_2. \tag{ii}$$

Using the relation, $\frac{E^2}{c^2} = p^2 + m_0^2 c^2$, before collision, we have

$$\frac{E_1}{c} = \sqrt{p^2 + m_0^2 c^2}, \quad \frac{E_2}{c} = \sqrt{0^2 + m_0^2 c^2}.$$

And after collision,

$$\frac{E'_1}{c} = \sqrt{p'^2 + m_0^2 c^2}, \quad \frac{E'_2}{c} = \sqrt{q^2 + m_0^2 c^2}.$$

Using equation (i) and (ii) we have

$$\sqrt{p^2 + m_1^2 c^2} + m_2 c = \sqrt{(p-q)^2 + m_1^2 c^2} + \sqrt{q^2 + m_2^2 c^2}.$$

After some mathematical manipulation, one can get

$$q = \frac{2pm_2(m_2 c \sqrt{p^2 + m_1^2 c^2})}{m_1^2 c + m_2^2 c + 2m_2 \sqrt{p^2 + m_1^2 c^2}}.$$

Example 10.34 A particle of mass m disintegrate mass $\frac{m}{2}$ and $\frac{m}{4}$. Find relativistic kinetic energies of the parts.

Hint: Let the masses and momentum of the parts after disintegration of the particles are m_1, m_2 and p_1, p_2, respectively.
So, by conservation of momentum, we have

$$0 = p_1 + p_2 \quad or \quad p_1 = -p_2 = p \ (say).$$

Also conservation of energy yields

$$mc^2 = E_1 + E_2,$$

or,

$$mc^2 = E_1 + \sqrt{p^2c^2 + m_2^2c^4}.$$

Also, we have

$$p^2c^2 = E_1^2 - m_1^2c^4.$$

Therefore, we get

$$mc^2 = E_1 + \sqrt{E_1^2 - m_1^2c^4 + m_2^2c^4},$$

or,

$$(mc^2 - E_1)^2 = \left[E_1^2 - \left(\frac{m}{2}\right)^2 c^4 + \left(\frac{m}{4}\right)^2 c^4 \right].$$

From here we can calculate E_1.

Now, Kinetic Energy,

$$T_1 = E_1 - m_1c^2.$$

Example 10.35 A stationary body decays into two parts each of mass m that move apart at speeds $0.6c$ relative to mass of the original body. Find the mass of the original body.

Hint: Use conservation of momentum and energy.

Example 10.36 A particle of rest mass m_0 and velocity $\frac{3c}{5}$ collides with another particle of rest mass m_0 at rest. If after collision the two particles coalesce. Find the velocity and mass of the resulting particles.

Hint: Use conservation of momentum and energy.

Example 10.37 Two particles of mass m_1 and m_2 are moving in opposite directions with the same relativistic momentum p. The two particles collide and form a particle of mass M at rest. Show that extra mass $(\times c^2)$ is equal to the total initial kinetic energy of the two particles.

Hint: Conservation of energy gives

$$E_1 + E_2 = Mc^2.$$

Now

$$(M - m_1 - m_2)c^2 = \left(\frac{E_1 + E_2}{c^2} - m_1 - m_2 \right) c^2$$

$$= E_1 - m_1c^2 + E_2 - m_2c^2$$

$$= KE_1 + KE_2.$$

Example 10.38 A particle of mass M at rest decays into two particles of masses m_1 and m_2. Find the common momentum of the particles.

Hint: Conservation of energy gives

$$Mc^2 = E_1 + E_2,$$

$$or, \ Mc^2 = E_1 + \sqrt{p^2c^2 + m^2c^4}.$$

Here,

$$E_1^2 = p^2c^2 + m_1^2c^4.$$

Thus,

$$Mc^2 - E_1 = \sqrt{E_1^2 - m_1^2c^4 + m_2c^4},$$

or,

$$E_1 = \frac{Mc^2}{2} + \frac{(m_1^2 - m_2^2)c^2}{2M}.$$

Now

$$p^2c^2 = E_1^2 - m_1^2c^4,$$

or,

$$\frac{p^2}{c^2} = \frac{(M^2 - m_1^2)^2 + (M^2 - m_2^2)^2 - M^4 - 2m_1^2m_2^2}{4M^2}.$$

Example 10.39 A particle of mass m and energy E collides with an identical particle a rest. After collision the two particles coalesce. Find the speed and mass of the formed particles.

Hint: Let the relativistic mass and velocity of the formed particles are M and v, respectively. Conservation of energy yields

$$mc^2 + E = Mc^2.$$

Conservation of momentum

$$p + 0 = P = MV.$$

We know,

$$E^2 - p^2c^2 = m^2c^4,$$

therefore,

$$V = \frac{P}{M} = \frac{pc^2}{(mc^2 + E)} = \frac{\sqrt{(E^2 - m^2c^4)}c}{mc^2 + E}.$$

Now

$$M = \frac{E + mc^2}{c^2} = \frac{M_0}{\sqrt{1 - \frac{v^2}{c^2}}},$$

therefore,

$$M_0 = \left(\frac{E + mc^2}{c^2}\right)\sqrt{1 - \frac{V^2}{c^2}}.$$

(M_0=rest mass of the formed particle)

Example 10.40 A body of rest mass m_0 moving at speed v collides and sticks to an identical body at rest. What is the mass and momentum of the final clump?

Hint: Same as above (Ex 10.39)

Example 10.41 A particle of rest mass m_0 is moving in xy-plane under the action of a constant force $\vec{F} = \vec{j} f$. If at $t = 0$ the particle starts to move from origin with three momenta $\vec{P}(0) = p_0 \vec{i}$, find the equation of the path.

Hint: We know the equation of the motion of the particle is

$$\vec{F} = \frac{d\vec{p}}{dt},$$

i.e.

$$\frac{dp_x}{dt} = 0, \quad \frac{dp_y}{dt} = f, \quad \frac{dp_z}{dt} = 0.$$

Solving these equation we get

$$p_x(t) = p_x(0) = p_0, \quad p_y(t) = ft + p_y(0) = ft, \quad p_z(t) = p_z(0) = 0.$$

Three-velocity components of the particle are v_x, v_y, v_z.
   Now,

$$m_0 a v_x = p_x,$$

implies

$$v_x = \frac{p_x}{m_0 a} = \frac{p_0 c^2}{E},$$

where $E^2 = p^2 c^2 + m_0^2 c^4$, $E = mc^2 = m_0 a c^2$, m = relativistic mass, $p^2 = p_x^2 + p_y^2 + p_z^2$, $a = \frac{1}{\sqrt{1 - \frac{v^2}{c^2}}}$. Therefore,

$$\frac{dx}{dt} = \frac{p_0 c^2}{c\sqrt{m_0^2 c^2 + p^2}} = \frac{p_0 c^2}{\sqrt{m_0^2 c^4 + p_0^2 c^2 + f^2 t^2 c^2}}.$$

This gives

$$x - x_0 = \int_0^t \frac{p_0 c^2}{\sqrt{E_0{}^2 + f^2 t^2 c^2}} dt.$$

[here, $E_0{}^2 = p_0{}^2 c^2 + m_0{}^2 c^4$].

At $t = 0$, $x = 0$, $x_0 = 0$, therefore,

$$x(t) = \frac{p_0 c}{f} \sinh^{-1} \frac{cft}{E_0}.$$

Similarly,

$$v_y = \frac{dy}{dt} = \frac{p_y}{m} = \frac{c^2 ft}{E},$$

or,

$$\frac{dy}{dt} = \frac{c^2 ft}{\sqrt{E_0{}^2 + f^2 t^2 c^2}}.$$

Solving this we get

$$y = \frac{E_0}{f} \left[-1 + \sqrt{1 + \left(\frac{cft}{E_0} \right)^2} \right].$$

Also $z(t) = 0$.

Eliminating t, we get

$$y = \frac{E_0}{f} \left(-1 + \cosh \frac{fx}{p_0 c} \right).$$

To get Newtonian limit, we have to expand cosh around the point $x = 0$.
Thus

$$y = \frac{E_0}{f} \left(-1 + 1 + \frac{f^2 x^2}{2 p_0^2 c^2} + 0(x^4) \right)$$

$$\approx \frac{E_0 f x^2}{2 p_0^2 c^2} = \frac{m_0 v_0 f c^2 x^2}{2 m_0^2 v_0^2 c^2 v_0^2} = \frac{f x^2}{2 m_0 v_0 v_0^2}.$$

This coincides with the trajectory in Newtonian mechanics with initial velocity $v_0 << c$ and $\nu_0 = \dfrac{1}{\sqrt{\left(1 - \frac{v_0^2}{c^2} \right)}}$.

Example 10.42 A particle of mass m and energy E collides with an identical particle at rest. The two particles scatter with momenta $\vec{p_1}$ and $\vec{p_2}$ making an angle θ_1 and θ_2 with the direction of motion of the incident particle. Show that

$$\tan^2 \theta_1 = \frac{2m(E - E_1)}{(E + m)(E_1 - m)}$$

$$\tan^2 \theta_2 = \frac{2m(E - E_2)}{(E + m)(E_2 - m)}.$$

Hint: Follow equation (xi) of note 5 of this chapter.

Example 10.43 A particle of mass m and relativistic energy amc^2 collides with a particle of moment bm at rest (a, b are constants) and sticks. What is the moment of the resulting composite particle ?

Hint: Initial momentum

$$p_0 = \sqrt{\frac{E^2}{c^2} - m^2 c^2} = \sqrt{a^2 m^2 c^2 - m^2 c^2},$$

etc.

Example 10.44 For one dimension, show that force is invariant under Lorentz Transformation.

Hint: Let a particle of mass m_0 nmoves with a velocity v along x-direction. Therefore,

$$E^2 = p^2 c^2 + m_0^2 c^4, \tag{i}$$

$$E = am_0 c^2, \tag{ii}$$

$$p = am_0 v, \tag{iii}$$

$$v = \frac{pc^2}{E}, \tag{iv}$$

$$F = \frac{dp}{dt}, \quad v = \frac{dx}{dt}. \tag{v}$$

[F = force, $\quad a = \frac{1}{\sqrt{1 - \frac{v^2}{c^2}}}$.]

Equation (i) implies

$$2E \frac{dE}{dt} = 2p \frac{dp}{dt} c^2,$$

or,

$$\frac{dE}{dt} = \frac{p}{E} F c^2 = vF. \tag{vi}$$

Also we know

$$t' = a\left(t - \frac{vx}{c^2}\right), \quad p' = a\left(p - \frac{vE}{c^2}\right).$$

Now,

$$F' = \frac{dp'}{dt'} = \frac{\left(dp - \frac{vdE}{c^2}\right)}{\left(dt - \frac{vdx}{c^2}\right)} = F.$$

Example 10.45 Let, a particle of rest mass m_0 moving with velocity v in S-frame along x direction. This particle is moving with velocity u with respect to frame S'-frame in the same direction. Find the components of the four momenta p'_μ.

Hint: We know from the formula of addition of velocities, in S' frame, the particle has velocity

$$w = \frac{v - u}{1 - \frac{uv}{c^2}}.$$

Now,

$$p'_4 = \frac{iE}{c} = imc = \frac{im_0}{\sqrt{1 - \frac{w^2}{c^2}}}.$$

Here,

$$\frac{1}{\sqrt{1 - \frac{w^2}{c^2}}} = \frac{1}{\sqrt{1 - \frac{1}{c^2}\left(\frac{v-u}{1-\frac{uv}{c^2}}\right)^2}} = \frac{1 - \frac{uv}{c^2}}{\sqrt{1 - \frac{v^2}{c^2}}\sqrt{1 - \frac{u^2}{c^2}}}.$$

Hence,

$$p'_4 = \frac{(im_0 c)(1 - \frac{uv}{c^2})}{\sqrt{1 - \frac{v^2}{c^2}}\sqrt{1 - \frac{u^2}{c^2}}}.$$

Now,

$$p'_1 = mw = \frac{m_0(v - u)\left(1 - \frac{uv}{c^2}\right)}{\left(1 - \frac{uv}{c^2}\right)\left(\sqrt{1 - \frac{v^2}{c^2}}\right)\left(\sqrt{1 - \frac{u^2}{c^2}}\right)}.$$

or,

$$p'_1 = \frac{m_0(v - u)}{\sqrt{1 - \frac{v^2}{c^2}}\sqrt{1 - \frac{u^2}{c^2}}}.$$

Also,

$$p'_2 = 0 \quad and \quad p'_3 = 0.$$

Example 10.46 Find the acceleration when an electron is moving with a speed of 2.7×10^{10} cm s^{-1} under a force of 2.64×10^{-20} dyne.

Hint: From the definition of force we have

$$F = \frac{d}{dt} mv = v \frac{dm}{dt} + m \frac{dv}{dt}$$

$$= v \frac{d}{dt} \left[\frac{m_0}{(1 - \frac{v^2}{c^2})^{\frac{1}{2}}} \right] + \frac{m_0}{\sqrt{1 - \frac{v^2}{c^2}}} \frac{dv}{dt}$$

$$= m_0 v \left(-\frac{1}{2} \right) \left(1 - \frac{v^2}{c^2} \right)^{-\frac{3}{2}} \left(-\frac{2v}{c^2} \right) \frac{dv}{dt} + \frac{m_0}{\sqrt{1 - \frac{v^2}{c^2}}} \frac{dv}{dt},$$

or

$$a = \frac{F \left(1 - \frac{v^2}{c^2} \right)^{\frac{3}{2}}}{m_0}. \quad \left(\because a = \frac{dv}{dt} \right)$$

Using the velocity of light, $c = 3 \times 10^{10}$cm and rest mass of the electron, $m_0 = 9.1 \times 10^{-28} g$, one get the desired result.

Example 10.47 Find the error in calculating the K.E. of a particle in classical physics from the viewpoint of relativity, if the particle moves with 0.6c velocity?

Hint: Let m_0 be the rest mass of the particle. Then the relativistic K E is

$$T = mc^2 - m_0 c^2 = \frac{m_0}{\sqrt{1 - \frac{v^2}{c^2}}} c^2 - m_0 c^2.$$

For $v = 0.6c$, one can obtain

$$T = m_0 c^2 \left(\frac{1}{\sqrt{1 - 0.36}} - 1 \right) = 0.25 m_0 c^2.$$

In classical physics, the kinetic energy of a moving mass m_0 is

$$T_0 = \frac{1}{2} m_0 v^2 = 0.18 m_0 c^2.$$

Hence, the difference

$$T - T_0 = 0.25 m_0 c^2 - 0.18 m_0 c^2 = 0.07 m_0 c^2.$$

Example 10.48 Prove that four components acceleration $a^\alpha = \frac{dU^\alpha}{d\tau}$ has only three independent components.

Hint: We know the magnitude of world velocity (four velocities) is $U^\mu \cdot U^\mu = -c^2$
Therefore,

$$\frac{d}{d\tau}(U^\mu \cdot U^\mu) = 0$$

or, $$2\frac{dU^\alpha}{d\tau} \cdot U^\alpha = 0,$$

$\Rightarrow$ $$a^\alpha \cdot U^\alpha = 0.$$

This is an constraint equation on a^α.
Hence four accelerations have only three independent components.

Example 10.49 A rocket starts from rest with a constant acceleration g. Find the distance of the rocket from the earth in T years. Also find how much time will be measured by the rocket.

Hint: Suppose the rocket is moving along y-direction. The four velocities U_μ and four accelerations a_μ follow the following conditions:

$$U_\mu \cdot U_\mu = -1 \quad or, \quad -U_t^2 + U_y^2 = -1$$
$$a_\mu \cdot U_\mu = 0 \quad or, \quad -a_t U_t + a_y U_y = 0$$
$$a_\mu \cdot a_\mu = g^2 \quad or, \quad -a_t^2 + a_y^2 = g^2$$

[assuming $c = 1$; x- and z-components are zero.]

Solving these equations, we get

$$a_y = gU_t$$
$$a_t = gU_y.$$

Now differentiating w.r.t proper time τ , we get

$$\frac{d^2 U_y}{d\tau^2} = \frac{da_y}{d\tau} = g\frac{dU_t}{d\tau} = ga_t = g^2 U_y.$$

This yields
$$U_y = A \sinh g\tau + B \cosh g\tau.$$

Initial conditions: $U_y = 0$, $\frac{dU_y}{d\tau} = g$ at $\tau = 0$.
Hence, we get

$$U_y = \frac{dy}{d\tau} = \sinh g\tau,$$

$$y_t = \frac{dt}{d\tau} = \cosh g\tau.$$

Integrating above two equations yield

$$y = \frac{1}{g}(\cosh g\tau - 1),$$

$$t = \frac{1}{g}(\sinh g\tau).$$

[we have used $y = t = 0$ *at* $\tau = 0$]
Thus T years in earth time, the rocket measures

$$\tau = \frac{1}{g} \sinh^{-1} gT.$$

The distance of the rocket from the earth is

$$y = \frac{1}{g}[\cosh(\sinh^{-1} gT) - 1].$$

Chapter 11
Photon in Relativity

11.1 Photon

In the beginning of twentieth century, Max Planck argued that light and other electromagnetic radiations consisted of individual packets of energy known as quanta. He proposed that the energy of each quantum is proportional to its frequency v. This hypothesis is known as Planck's hypothesis and is given by

$$E = hv$$

[h is known as Planck's constant]

In fact, after further development of Planck's hypothesis by Einstein, it is verified that light has a dual wave–particle nature. Diffraction, interference, etc. are of wave nature and photoelectric effect, interaction of light with atoms, etc. are of particle nature of light and are called as photons.

The relativistic mass of the photon is given by

$$m = \frac{m_0}{\sqrt{1 - \frac{v^2}{c^2}}}.$$

Here m_0 is the rest mass of the photon. From above equation, we can get

$$m_0 = m\sqrt{1 - \frac{v^2}{c^2}}.$$

Using the velocity of light as $v = c$, we get

$$m_0 = 0,$$

i.e. the rest mass of the photon is zero. Actually, in all inertial frame, light would never rest. Its velocity always is c.

© The Author(s), under exclusive license to Springer Nature Singapore Pte Ltd. 2022 181
F. Rahaman, *The Special Theory of Relativity*, La Matematica per il 3+2 136,
https://doi.org/10.1007/978-981-19-0497-4_11

Fig. 11.1 Scattering of a
photon by an electron

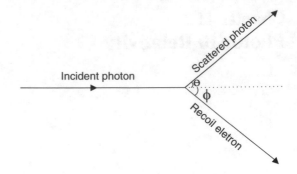

Let the direction cosines of the direction of travel of the photon are (l, m, n), then

$$\overrightarrow{p} = (lp_1 + mp_2 + np_3) = p\widehat{n} , \quad p = \sqrt{p_1^2 + p_2^2 + p_3^2}.$$

The momentum energy relation becomes

$$E^2 - p^2c^2 = 0 \Longrightarrow E = pc.$$

Using mass energy relation $E = mc^2$, we get

$$m = \frac{p}{c}.$$

Hence, we have the following important relations of the photon as

$$E = h\nu , \quad m = \frac{h\nu}{c^2} , \quad \overrightarrow{p} = \frac{h\nu}{c}\widehat{n}.$$

11.2 Compton Effect

A photon may be described as a particle of zero rest mass with momentum $\frac{h}{\lambda} = \frac{h\nu}{c}$
and energy $h\nu$. In 1921, A H Compton proposed the experiment of the scattering of
a photon by an electron that if the photon collides with an electron of rest mass m,
it will be scattered at some angle θ with a new energy $h\nu^1$. He has shown that the
change in energy is related to the scattering angle by the formula $\lambda^1 - \lambda = 2\lambda_c \sin^2 \frac{\theta}{2}$
where $\lambda_c = \frac{h}{mc}$ is the *Compton wave length*. This observed change in frequency of
wavelength of scattered radiation is known as *Compton effect*.

Let a photon of energy $h\nu$ is incident on an electron of rest mass m and be scattered
at an angle θ with energy $h\nu^1$. Here the electron which scatters the photon suffers a
momentum recoil (see Fig. 11.1).

Principle of conservation of energy

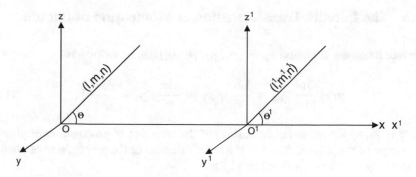

Fig. 11.2 Direction of the propagation of photon makes θ angle with x-axis

$$hv + mc^2 = hv^1 + \frac{mc^2}{\sqrt{1 - \frac{v^2}{c^2}}} \implies \frac{m^2c^4}{1 - \frac{v^2}{c^2}} = [mc^2 + h(v - v^1)]^2. \quad (11.1)$$

[v is the recoil velocity of the electron]

Principle of conservation of momentum (along and perpendicular to the direction of motion of photon)

$$\frac{hv}{c} + 0 = \frac{hv^1}{c} \cos\theta + \frac{mv}{\sqrt{1 - \frac{v^2}{c^2}}} \cos\phi \quad (11.2)$$

$$0 = \frac{hv^1}{c} \sin\theta - \frac{mv}{\sqrt{1 - \frac{v^2}{c^2}}} \sin\phi. \quad (11.3)$$

Eliminating ϕ from (11.2) and (11.3), we get

$$\frac{m^2v^2c^2}{1 - \frac{v^2}{c^2}} = h^2(v^2 - 2v^1v\cos\theta + v^{12}). \quad (11.4)$$

Subtracting (11.4) from (11.1), we get

$$\frac{v - v^1}{vv^1} = \frac{h}{mc^2}(1 - \cos\theta).$$

Using the result $v\lambda = c$, we get

$$\lambda^1 - \lambda = \frac{2h}{mc} \sin^2\frac{\theta}{2} = 2\lambda_c \sin^2\frac{\theta}{2}$$

11.3   The Lorentz Transformation of Momentum of Photon

The four momenta of photon $p_\mu = [p_1, p_2, p_3, p_4]$ can be written as

$$p_1 = \frac{h\nu l}{c}, \quad p_2 = \frac{h\nu m}{c}, \quad p_3 = \frac{h\nu n}{c}, \quad p_4 = \left(\frac{iE}{c}\right), \qquad (11.5)$$

where (l, m, n) are the direction cosines of the direction of propagation of photon and energy of the photon, $E = h\nu$. Here, the direction of the propagation of photon makes θ angle with x-axis (see Fig. 11.2).

Using the above transformation formula, we get

$$\begin{bmatrix} \frac{h\nu^1 l^1}{c} \\ \frac{h\nu^1 m^1}{c} \\ \frac{h\nu^1 n^1}{c} \\ \frac{ih\nu^1}{c} \end{bmatrix} = \begin{bmatrix} a & 0 & 0 & ia\beta \\ 0 & 1 & 0 & 0 \\ 0 & 0 & 1 & 0 \\ -ia\beta & 0 & 0 & a \end{bmatrix} \begin{bmatrix} \frac{h\nu l}{c} \\ \frac{h\nu m}{c} \\ \frac{h\nu n}{c} \\ \frac{ih\nu}{c} \end{bmatrix}. \qquad (11.6)$$

This implies

$$\frac{h\nu^1 l^1}{c} = a\left(\frac{h\nu l}{c} + i\frac{v}{c}\frac{ih\nu}{c}\right)$$

or

$$\nu^1 l^1 = \frac{\nu(l - \frac{v}{c})}{\sqrt{1 - \frac{v^2}{c^2}}} \qquad (11.7)$$

$$\frac{h\nu^1 m^1}{c} = \frac{h\nu m}{c} \implies \nu^1 m^1 = \nu m \qquad (11.8)$$

$$\frac{h\nu^1 n^1}{c} = \frac{h\nu n}{c} \implies \nu^1 n^1 = \nu n \qquad (11.9)$$

$$\frac{ih\nu^1}{c} = a\left(-i\frac{v}{c}\frac{lh\nu}{c} + \frac{ih\nu}{c}\right)$$

or

$$v^1 = \frac{v(1 - \frac{lv}{c})}{\sqrt{1 - \frac{v^2}{c^2}}}. \tag{11.10}$$

This change of the frequency is due to Doppler effects.

Substituting v^1 in Eq. (11.7), we get

$$l^1 = \frac{l - \frac{v}{c}}{1 - \frac{lv}{c}}. \tag{11.11}$$

Similarly Eqs. (11.8) and (11.9) give

$$m^1 = \frac{m\sqrt{1 - \frac{v^2}{c^2}}}{1 - \frac{lv}{c}} \tag{11.12}$$

$$n^1 = \frac{n\sqrt{1 - \frac{v^2}{c^2}}}{1 - \frac{lv}{c}}. \tag{11.13}$$

These changes in the direction are due to aberration.

If the light propagates in the xy-plane, then, $n = n^1 = 0$ and $l = \cos\theta, m = \sin\theta$ where light ray makes θ angle with x-axis. Then Eqs. (11.11) and (11.12) yield

$$\cos\theta^1 = \frac{\cos\theta - \frac{v}{c}}{1 - \frac{v}{c}\cos\theta} \tag{11.14}$$

$$\sin\theta^1 = \frac{\sin\theta\sqrt{1 - \frac{v^2}{c^2}}}{1 - \frac{v}{c}\cos\theta} \tag{11.15}$$

The last two equations give the usual formula of aberration as

$$\cot\theta^1 = \frac{\cot\theta - \frac{v}{c}\csc\theta}{\sqrt{1 - \frac{v^2}{c^2}}}. \tag{11.16}$$

11.4 Minkowski Force for Photon

Minkowski force for photon can be written in terms of four momenta of photon as

$$K_\mu = \frac{1}{h} P_\mu = \frac{1}{h}\left(p_1 = \frac{h\nu l}{c}, p_2 = \frac{h\nu m}{c}, p_3 = \frac{h\nu n}{c}, \frac{iE}{c} = \frac{ih\nu}{c}\right).$$

Therefore, the Minkowski force for photon can be written in terms of wave length λ of the photon

$$K_\mu = \left(\frac{l}{\lambda}, \frac{m}{\lambda}, \frac{n}{\lambda}, \frac{i}{\lambda}\right) \tag{11.17}$$

[using $\lambda\nu = c$]

If the photon, i.e. electromagnetic wave travels in a direction making the coordinates axes constant, angle of whose direction cosines (l, m, n) is characterized by

$$\psi = a\ exp\left(\frac{lx + my + nz}{\lambda} - \nu t\right).$$

Then, one can write

$$\psi = a\ exp(K_\mu \cdot x_\mu)\ as\ x_\mu = (x, y, z, ict).$$

We notice that $K_\mu \cdot x_\mu$ is a scalar quantity. So the phase of the electromagnetic wave ψ is also a scalar quantity and hence it is invariant under Lorentz Transformation.

Example 11.1 Show that Minkowski force for photon is a null vector.

Hint: We know

$$K_\mu = \left(\frac{l}{\lambda}, \frac{m}{\lambda}, \frac{n}{\lambda}, \frac{i}{\lambda}\right),$$

therefore,

$$\begin{aligned}
K_\mu \cdot K_\mu &= K_1 K_1 + K_2 K_2 + K_3 K_3 + K_4 K_4 \\
&= \frac{l^2}{\lambda^2} + \frac{m^2}{\lambda^2} + \frac{n^2}{\lambda^2} + \frac{i^2}{\lambda^2} \\
&= \frac{l^2 + m^2 + n^2}{\lambda^2} - \frac{1}{\lambda^2} = 0.
\end{aligned}$$

Example 11.2 Show that four momenta of photon is a null vector.

Hint: We know

$$p_\mu = \left[p_1 = \frac{h\nu l}{c}, p_2 = \frac{h\nu m}{c}, p_3 = \frac{h\nu n}{c}, p_4 = \left(\frac{iE}{c}\right) = \left(\frac{ih\nu}{c}\right)\right],$$

i.e.

Fig. 11.3 Two photons
make equal angle θ with the
direction of meson

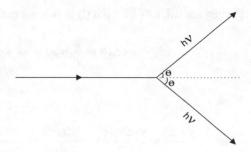

$$p_\mu = \left(\frac{lh}{\lambda}, \frac{mh}{\lambda}, \frac{nh}{\lambda}, \frac{ih}{\lambda} \right), \quad using \; \lambda \nu = c$$

therefore,

$$p_\mu \cdot p_\mu = p_1 p_1 + p_2 p_2 + p_3 p_3 + p_4 p_4$$
$$= \frac{h^2 l^2}{\lambda^2} + \frac{h^2 m^2}{\lambda^2} + \frac{h^2 n^2}{\lambda^2} + \frac{i^2}{\lambda^2}$$
$$= \frac{h^2 (l^2 + m^2 + n^2)}{\lambda^2} - \frac{h^2}{\lambda^2} = 0.$$

Example 11.3 A meson of rest mass π decays in flight into two photons. If one of
the photons is emitted at an angle θ to the direction of motion of the meson. Show
that the energy of motion of the photon is $h\nu = \frac{\pi c^2}{2\gamma \left(1 - \frac{u\cos\theta}{c}\right)}$ $\left(\text{where,} \; \gamma = \frac{1}{\sqrt{1 - \frac{u^2}{c^2}}} \right)$.

Hint: Let ν is the frequency of the decay photon. Therefore, both photons have the
energy $E = h\nu$ and momentum $p = \frac{E}{c} = \frac{h\nu}{c}$. The direction of the resultant is the
direction of motion of the meson. Here, two photons make equal angle θ with the
direction of motion of the meson, therefore, angle between two photons is 2θ (see
Fig. 11.3).

Principle of conservation of energy

$$\frac{\pi c^2}{\sqrt{1 - \frac{u^2}{c^2}}} = \gamma \pi c^2 = h\nu + h\nu = 2 h\nu. \tag{1}$$

Principle of conservation of momentum

$$\gamma \pi u = \sqrt{p^2 + p^2 + 2pp \cos 2\theta}$$

or

$$\gamma^2 \pi^2 u^2 = 2p^2 + 2p^2 \cos 2\theta = \frac{4h^2 \nu^2}{c^2} \cos^2 \theta \tag{2}$$

Putting the value of $h\nu$ from (i) in (ii), we get

$$\cos\theta = \frac{u}{c} \implies \gamma = \frac{1}{\sqrt{1 - \frac{u^2}{c^2}}} = \csc\theta.$$

Now,

$$h\nu = \frac{\gamma\pi c^2}{2} = \frac{\csc\theta\pi c^2}{2} = \frac{\pi c^2}{2\sin\theta} = \frac{\pi c^2}{2\csc\theta(1 - \cos^2\theta)} = \frac{\pi c^2}{2\gamma\left(1 - \frac{u\cos\theta}{c}\right)}$$

Example 11.4 Show that it is impossible for a photon to transfer all its energy to a free electron.

or show that an electron of finite mass cannot disintegrate into a single photon.

Hint: Suppose if possible, a photon containing energy E and momentum $p = \frac{E}{c}$ transfers all its energy to an electron of rest mass m_0 and velocity $v = fc$ where

$$0 < f < 1. \tag{1}$$

Principle of conservation of energy

$$E = \frac{m_0 c^2}{\sqrt{1 - \frac{(fc)^2}{c^2}}} - m_0 c^2. \tag{2}$$

Principle of conservation of momentum

$$p = \frac{E}{c} = \frac{m_0 fc}{\sqrt{1 - \frac{(fc)^2}{c^2}}}. \tag{3}$$

From (ii) and (iii), we get

$$\frac{m_0 fc^2}{\sqrt{1 - f^2}} = \frac{m_0 c^2}{\sqrt{1 - f^2}} - m_0 c^2 \implies f(1 - f) = 0.$$

This gives either $f = 0$ or $f = 1$ which are contradiction to condition (i). Hence the assumption was wrong, i.e. it is impossible for a photon to transfer all its energy to a free electron.

Example 11.5 An excited atom of mass m_0, initially at rest in frame S, emits a photon and recoils. The internal energy of the atom decreases by $\triangle E$ and the energy of the photon is $h\nu$. Show that $h\nu = \triangle E\left(1 - \frac{\triangle E}{2m_0 c^2}\right)$

Hint: Let M_0 be the rest mass of the atom after emission and v be the recoil velocity. Principle of conservation of energy

Fig. 11.4 Two decay γ rays
make θ_1 and θ_2 angle with
the initial direction of motion
of π meson

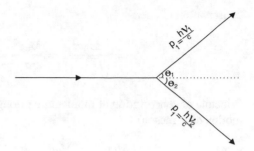

$$m_0 c^2 = \frac{M_0 c^2}{\sqrt{1 - \frac{v^2}{c^2}}} + h\nu = \frac{M_0 c^2}{a} + h\nu. \tag{1}$$

Principle of conservation of momentum

$$p = \frac{h\nu}{c} = \frac{M_0 v}{\sqrt{1 - \frac{v^2}{c^2}}} = \frac{M_0 v}{a}. \tag{2}$$

Assuming the atom is excited from rest and $a = \sqrt{1 - \frac{v^2}{c^2}}$.
Above two equations can be written as

$$\left(m_0 c - \frac{h\nu}{c}\right)^2 = \left(\frac{M_0^2 v^2}{a^2}\right) \tag{3}$$

$$\left(\frac{h\nu}{c}\right)^2 = \left(\frac{M_0^2 c^2}{a^2}\right) \tag{4}$$

Subtracting (4) from (3), we get

$$h\nu = \frac{(m_0 + M_0)(m_0 - M_0)c^2}{2m_0}.$$

Now, $\Delta E = (m_0 - M_0)c^2$, therefore we get

$$h\nu = \Delta E \left(1 - \frac{m_0 - M_0}{2m_0}\right) = \Delta E \left(1 - \frac{\Delta E}{2m_0 c^2}\right).$$

Example 11.6 A π meson of rest mass m_0 moving with velocity v disintegrates into
two γ rays. Calculate the energy distribution of γ rays from π meson.

Hint: Let ν_1 and ν_2 be the frequencies of the two decay γ rays, which make angles
θ_1 and θ_2 with initial direction of motion of π meson (see Fig. 11.4). Principle of

conservation of energy

$$\frac{M_0 c^2}{\sqrt{1 - \frac{v^2}{c^2}}} = \frac{M_0 c^2}{a} = h(v_1 + v_2). \tag{1}$$

Principle of conservation of momentum (along and perpendicular to the direction of motion of π meson)

$$\frac{M_0 v}{\sqrt{1 - \frac{v^2}{c^2}}} = \frac{M_0 v}{a} = \frac{h}{c}(v_1 \cos \theta_1 + v_2 \cos \theta_2) \tag{2}$$

$$0 = \frac{h}{c}(v_1 \sin \theta_1 - v_2 \sin \theta_2), \tag{3}$$

where $a = \sqrt{1 - \frac{v^2}{c^2}}$.

Squaring both sides of (1) and (2) and subtracting with other yields

$$M_0 c^4 = h^2(v_1 + v_2)^2 - h^2(v_1 \cos \theta_1 + v_2 \cos \theta_2)^2$$
$$= h^2[v_1^2 \sin^2 \theta_1 + v_2^2 \sin^2 \theta_2 + 2v_1 v_2(1 - \cos \theta_1 \cos \theta_2)]$$

Equation (3) implies

$$v_1 \sin \theta_1 = v_2 \sin \theta_2.$$

From these two equations, one can solve for v_1 and v_2 as

$$v_1^2 = \frac{M_0^2 c^4 \sin \theta_2}{2h^2 \sin \theta_1 [1 - \cos(\theta_1 + \theta_2)]}, \quad v_2^2 = \frac{M_0^2 c^4 \sin \theta_1}{2h^2 \sin \theta_2 [1 - \cos(\theta_1 + \theta_2)]}$$

here, $E_1 = h v_1$ and $E_2 = h v_2$. Multiplying above two values v_1 and v_2, we get

$$\sin\left(\frac{\theta_1 + \theta_2}{2}\right) = \frac{M_0 c^2}{2\sqrt{E_1 E_2}}$$

Example 11.7 A π meson of rest mass m_0 moving with velocity v disintegrates into two equal γ rays. Show that the angle between two decayed γ rays is 2θ which is given by $\sin \theta = \frac{m_0 c^2}{2hv}$ where v is the frequency of the decayed γ ray.

Hint: In example 11.6, $\theta_1 = \theta_2 = \theta$ and $v_1 = v_2 = v$.

Example 11.8 Show that a photon cannot give rise to an electron–positron pair in free space.

Fig. 11.5 Angle between electron and positron is θ

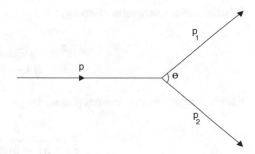

Hint: Suppose, if possible a photon of momentum $p = \frac{h\nu}{c}$ produces an electron of momentum p_1 and a positron of momentum p_2. Both have same rest mass m_0. Let the produced electron and positron makes θ angle with each other (see Fig. 11.5).

Principle of conservation of momentum gives

$$p^2 = \frac{h^2\nu^2}{c^2} = p_1^2 + p_2^2 + 2p_1p_2 \cos\theta. \tag{1}$$

Principle of conservation of energy implies $h\nu = E_1 + E_2$. Now using relation between momentum and energy $E^2 = p^2c^2 + m_0^2c^4$, we obtain

$$h\nu = \sqrt{p_1^2c^2 + m_0^2c^4} + \sqrt{p_1^2c^2 + m_0^2c^4}. \tag{2}$$

Eliminating $h\nu$ from these equations, we obtain after some simplification

$$-p_1^2 p_2^2 \sin^2\theta = m_0^2 c^2 (p_1^2 + p_2^2 + 2p_1p_2 \cos\theta). \tag{3}$$

Since minimum value of $\cos\theta$ is -1. Therefore, right-hand side is always positive. Since left-hand side is either zero or negative, therefore, we conclude that both sides must be zero. Left-hand side will be zero when $\theta = 0$ or π. For, $\theta = 0$, right-hand side is never zero. For, $\theta = \pi$, right-hand side gives $p_1 = p_2$. But, for $p_1 = p_2$, Eq. (11.1) gives $\nu = 0$, i.e. the photon is not existed. Therefore, the above process is not possible.

Example 11.9 Show that the outcome of the collision between two particles cannot be a single photon.

Hint: Let a particle of rest mass m_2 moving with velocity v_2 strikes a particle of rest mass m_1 at rest. If possible let a photon of energy $h\nu$ is produced as an outcome of this collision.

Principle of conservation of momentum:

$$\frac{m_2 v_2}{\sqrt{1 - \frac{v_2^2}{c^2}}} = \frac{h\nu}{c}.$$

Principle of conservation of energy:

$$m_1c^2 + \frac{m_2c^2}{\sqrt{1 - \frac{v_2^2}{c^2}}} = h\nu.$$

Eliminating m_2 from these equations, we get

$$\frac{v_2}{c} = \frac{h\nu}{h\nu - m_1c^2} > 1.$$

Definitely this is impossible. Hence the process cannot be possible.

Example 11.10 Show that the de Broglie wave length for a material particle rest mass m_0 and a charge q accelerated from rest through a potential difference V volt relativistically is given by $\lambda = \frac{h}{\sqrt{2m_0qV(1+\frac{qV}{2m_0c^2})}}$. Calculate the wavelength of an electron having a kinetic energy of 1 MeV.

Hint: Here, the kinetic energy of the electron is $T = qV$. Therefore, the total energy is $E = T + m_0c^2 = qV + m_0c^2$. Now, using relation between momentum and energy $E^2 = p^2c^2 + m_0^2c^4$, we obtain

$$(qV + m_0c^2)^2 = p^2c^2 + m_0^2c^4 \implies p = \sqrt{2m_0qV\left(1 + \frac{qV}{2m_0c^2}\right)}.$$

Therefore, the de Broglie wave length is given by

$$\lambda = \frac{h}{p} = \frac{h}{\sqrt{2m_0qV\left(1 + \frac{qV}{2m_0c^2}\right)}}.$$

For an electron, $q = e = 1.6 \times 10^{-19}$ coulomb, $m_0 = 9.1 \times 10^{-31}$ kg, $c = 3 \times 10^5$ km and $h = 6.62 \times 10^{-34}$. Here, eV = 1 MeV = 10^6 volt, $m_0c^2 = 0.51$ MeV. Putting all these, one can get $\lambda = 8.6 \times 10^{-3}$ Angstrom.

Example 11.11 A photon of energy E_0 collides at an angle θ with another photon of energy E. Prove that the minimum value of E_0 permitting formation of a pair of particles of mass m is $E_{th} = \frac{2m^2c^4}{E(1-\cos\theta)}$.

Hint: Here $\frac{E}{c}$ and $\frac{E_0}{c}$ are momenta of the photons with energies E and E_0, respectively. Before collision takes place, the total momenta was

$$\sqrt{\frac{E^2}{c^2} + \frac{E_0^2}{c^2} + \frac{EE_0\cos\theta}{c^2}}$$

[Here, θ be the angle between two photons (see Fig. 11.6)].

Fig. 11.6 θ be the angle
between two photons

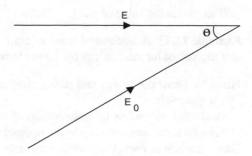

Since E_{th} is the threshold energy, therefore, the momentum of the produced particles will be zero, i.e.

$$\sqrt{\frac{E^2}{c^2} + \frac{E_{th}^2}{c^2} + \frac{E\,E_{th}\cos\theta}{c^2}} = 0. \tag{1}$$

Principle of conservation of energy

$$E + E_{th} = 2mc^2. \tag{2}$$

These two equations yield

$$E_{th} = \frac{2m^2c^4}{E(1-\cos\theta)}.$$

Example 11.12 A particle of mass M decays from the rest into two particles. One particle has mass m and the other is massless. Find the momentum of the massless particle.

Hint: Let p_1 and p be the momentum of the massive and massless particle, respectively. By conservation of momentum, we have

$$p_1 + p = 0 \; or \; p_1 = -p.$$

Conservation of energy yields

$$mc^2 = pc + E_1,$$

or,

$$mc^2 = pc + \sqrt{p_1^2c^2 + m^2c^4},$$

or,

$$mc^2 = pc + \sqrt{p^2c^2 + m^2c^4}.$$

This yields the value of p.

Example 11.13 A nucleus of mass m emits a gamma-ray of frequency γ_0. Show that the loss of internal energy by the nucleus is $h\gamma_0[1 + \frac{h\gamma_0}{2mc^2}]$.

Hint: We know the energy and momentum of γ-ray photon are $E_1 = h\gamma_0$ and $p = \frac{h\gamma_0}{c}$, respectively.

According to the law of conservation of momentum, the nucleus of mass m will recoil with same momentum p in the opposite direction.

Hence the loss of energy of the nucleus in its recoiling is

$$E_2 = \frac{1}{2}mv^2 = \frac{p^2}{2m} = \frac{1}{2m}\left(\frac{h^2\gamma_0^2}{c^2}\right).$$

Thus total loss of energy of the nucleus is

$$E_1 + E_2 = h\gamma_0 + \frac{1}{2m}\frac{h^2\gamma_0^2}{c^2}.$$

Example 11.14 A photon strikes an electron of mass m at rest mass and creates an electron–positron pair. The photon is destroyed and the positron and two electrons move off at equal speeds along the initial direction of the photon. Find energy of the photon.

Hint: We know $E^2 - p^2c^2$ invariant. We have two frames. One initial frame before collision and one final frame as after collision.

Let E, p are the total energy and momentum of the photon. In the initial frame electron of mass m is at rest.

The total energy is $E + mc^2$. So

$$(E + mc^2)^2 - p^2c^2 = constant. \tag{i}$$

In the final state of the particles,

$$2p_{e^-} + p_{e^+} = p,$$

$$E^2 - (2p_{e^-} + p_{e^+})^2c^2 = constant = (2m_{e^-} + m_{e^+})^2c^4 = (3m)^2c^4. \tag{ii}$$

So equating two frames, (L.H.S of (i) and R.H.S of (ii)) we have

$$(E + mc^2)^2 - p^2c^2 = 9m^2c^4. \tag{iii}$$

Since photon is massless,

$$E = pc.$$

Putting this in (iii), we get

$$(pc + mc^2)^2 - p^2c^2 = 9m^2c^4,$$

$$or, \ m^2c^4 + 2mpc^3 = 9m^2c^4,$$

$$or, \ p = 4mc.$$

The energy of the photon is $E = pc = 4mc^2$.

Example 11.15 A photon having frequency v strikes a nucleus of mass m_0 at rest and is absorbed. Calculate the recoil velocity of the resulting nucleus.

Hint: Here, the total energy before impact is

$$E = m_0c^2 + hv.$$

Let m' be the mass of nucleus after impact, then $E = m'c^2$. So, the law of conversation of the energy implies that
$$m_0c^2 + hv = m'c^2.$$

Also, the law of conversation of momentum gives,

$$0 + \frac{hv}{c} = m'v.$$

where v is the recoil velocity. Solving the above equations one can get the desired result.

Example 11.16 A photon of energy E is absorbed by a stationary nucleus of mass M. The collision results in an excitation of nucleus. Find the mass and speed of the excited nucleus.

Hint: Same as above.

Example 11.17 Show that a photon cannot break up into an electron and a positron.

Hint: Let p_1, p_2 and E_1, E_2 are momentum and energy of the electrons and positrons, respectively. Energy of the photon is $E = pc$, p = momentum of the electron. Conservation of energy and momentum yield,

$$E = pc = E_1 + E_2, \ \ p = p_1 + p_2.$$

Now

$$p^2 = (p_1 + p_2)^2 = \left(\frac{E_1 + E_2}{c}\right)^2 = \frac{E_1^2}{c^2} + \frac{E_2^2}{c^2} + \frac{2E_1E_2}{c^2}. \tag{i}$$

We know,

$$E_1^2 = p_1^2 c^2 + m_0^2 c^4, \quad E_2^2 = P_2^2 c^2 + m_0^2 c^4.$$

Using these, (i) implies

$$p_1 p_2 = m_0 c^2 + \frac{E_1 E_2}{c^2}. \qquad \text{`} \qquad (ii)$$

But we know $p_1 < \frac{E_1}{c}$, $p_2 < \frac{E_2}{c}$, therefore (ii) is not possible. Hence the claim.

Example 11.18 A photos of frequency *nu* is scattered through an angle θ by an electron of rest mass m_0, at rest. If the direction of recoil of the electron makes an angle ϕ with the direction of the incidence of the photon, then show that

$$tan\phi = \frac{cot\left(\frac{\theta}{2}\right)}{1 + \frac{hv}{m_0 c^2}}.$$

Hint: Conservation of Energy and Momentum yields

$$hv + m_0 c^2 = hv' + E, \qquad (i)$$

$$\frac{hv}{c} = \frac{hv'}{c} \cos\theta + p \cos\phi, \qquad (ii)$$

$$\frac{hv'}{c} \sin\theta = p \sin\phi. \qquad (iii)$$

Equations (i)–(iii) imply that

$$\frac{hv'}{c} = \frac{hv}{c} \frac{sin\phi}{sin(\theta + \phi)},$$

$$p = \frac{hv}{c} \frac{sin\phi}{sin(\theta + \phi)}.$$

Hence equation (i) provides us

$$E = hv + m_0 c^2 - \frac{hvsin\phi}{sin(\theta + \phi)}.$$

Now, putting the values of p and E in the equation $E^2 = p^2 c^2 + m_0^2 c^4$, one can get the desired results.

Example 11.19 Two identical particles of mass m_0 approaching each other with velocity v collide. After the collision a particle of mass M and a photon are produced. Can M remains stationary? Find the momenta of M and photon.

Hint: Here the particles are coming to each other with same velocity v, so their initial momentum is zero. Since photon cannot be motionless, therefore, the mass M must have some momentum to cancel the momentum of the photon. Hence M cannot be stationary.

Now conservation of momentum and energy yield,

$$0 = P + p, \quad 2m_0 ac^2 = E + e, \tag{i}$$

where P, E are the momentum and energy of M and p, e are the momentum and energy of the photon and $a = \dfrac{1}{\sqrt{1-\frac{v^2}{c^2}}}$.

Now,

$$e^2 = p^2 c^2 = P^2 c^2 = E^2 - M^2 c^4 \tag{ii}$$

Therefore from (i), one can get

$$2m_0 c^2 a = E + \sqrt{E^2 - M^2 c^4} \tag{iii}$$

Equation (iii) gives

$$E = m_0 c^2 a + \frac{M^2 c^2}{4m_0 a}.$$

Also we get

$$e = m_0 c^2 a - \frac{M^2 c^2}{4m_0 a}.$$

Hence, $p = \frac{e}{c}$.

Example 11.20 An electron of energy E and momentum p strikes a positron of mass m at rest. After collision two photons of energy E_1, E_2 are produced. Find the photons scattering angles. Show also that when the photons have same scattering angle θ then $\cos\theta = \sqrt{\dfrac{E-mc}{E+mc}}$.

Hint: Follow equation (xi) of note 5 of Chap. 10.

Example 11.21 Show that de Broglie wavelength for a material particle of rest mass m_0 and charge e accelerated from rest through a potential difference V volts relativistically given by

$$\lambda = \frac{h}{\sqrt{\left[2m_0 eV\left(1 + \frac{eV}{2m_0 c^2}\right)\right]}}.$$

Hint: We know $E^2 = p^2 c^2 + m_0^2 c^4$ and $T = eV$ (given).
Hence,

$$E = T + m_0c^2 = eV + m_0c^2.$$

$$\therefore p^2c^2 = E^2 - m_0^2c^4 \Rightarrow p = \ldots$$

Hence, de Broglie wavelength, $\lambda = \frac{h}{p}$.

Example 11.22 A particle with mass m and rest mass m_0 decays into a photon and losses its rest mass by δ amount. Find the energy of the photon.

Hint: Given rest mass of the particle before decays is m_0. Therefore its energy is $E = m_0c^2$ and it's momentum is zero.
After the decay, let the momentum of the particle is p and energy E'.

Using the conservation of Energy and momentum, we have

$$m_0c^2 = E + E' = h\nu + E'.$$

$$0 = \frac{h\nu}{c} + p.$$

We know that

$$E'^2 = p^2c^2 + (m_0 - \delta)^2c^4.$$

Therefore, combining above three equations we get

$$m_0c^2 - h\nu = \sqrt{\frac{h^2\nu^2}{c^2}c^2 + (m_0 - \delta)^2c^4}.$$

Squaring both sides we get

$$m_0^2c^4 + h^2\nu^2 - 2m_0c^2\,h\nu = h^2\nu^2 + (m_0 - \delta)^2c^4.$$

Hence,

$$\nu = \frac{c^2\delta}{h}\left(1 - \frac{\delta}{2m_0}\right).$$

etc.

Example 11.23 A particle of rest mass m is moving in the positive x-direction. During motion it decays into two photons. Two emitted photons move along x-axis but in opposite directions. If the first photon has a times the energy of the second, find the particles original speed.

Hint: Suppose the second photon is moving in the negative x-direction and its frequency be f, then its energy is hf and momentum is $-\frac{hf}{c}$. According to the problem, the energy of the first photon (moving in the positive x-direction) is ahf and hence its frequency is af and its momentum is $\frac{ahf}{c}$. Let the particle's velocity be v, then the Energy conservation and Momentum conservation yield:

Fig. 11.7 Angle of incidence is θ_1 and angle of deflection is θ_2

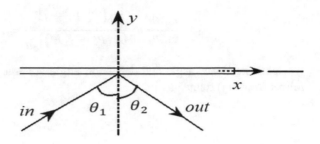

$$\frac{m}{\sqrt{1 - \frac{v^2}{c^2}}} c^2 = ahf + hf = (a+1)hf,$$

$$\frac{m}{\sqrt{1 - \frac{v^2}{c^2}}} v = \frac{ahf}{c} - \frac{hf}{c} = (a-1)\frac{hf}{c}.$$

Eliminating f and solving for v will yield the desired result.

Example 11.24 A mirror moves with velocity v parallel to its plane. Prove that the angle of incidence of light ray and angle of deflection are equal.

Hint: Suppose the energy and momentum of the light ray are E and $\vec{P}$, respectively. In the mirror frame (moving frame), let these quantities are

$$E'_{in} = E'_{out}; \quad P'_{x\ in} = P'_{x\ out}; \quad P'_{y\ in} = -P'_{y\ out}.$$

Let angle of incidence is θ_1 and angle of deflection is θ_2 (see Fig. 11.7).
Here,

$$P_{x\ in} = \frac{E}{c}\sin\theta_1; \quad P_{y\ in} = \frac{E}{c}\cos\theta_1.$$

$$P_{x\ out} = \frac{E}{c}\sin\theta_2; \quad P_{y\ out} = -\frac{E}{c}\cos\theta_2.$$

Now,

$$\frac{\tan\theta_1}{\tan\theta_2} = \frac{\left(\frac{P_x}{P_y}\right)_{in}}{\left[\frac{P_x}{(-P_y)}\right]_{out}} = \frac{(P_x)_{in}}{(P_x)_{out}} \times \frac{(P'_y)_{in}}{-(P'_y)_{out}}$$

$$= \frac{(P_x)_{in}}{(P_x)_{out}}$$

[we have assumed $P'_y = P_y$].
Using inverse Lorentz Transformation of P_x, we have

$$\frac{\tan \theta_1}{\tan \theta_2} = \frac{\left[a \left(P'_x + \frac{vE'}{c^2} \right) \right]_{in}}{\left[a \left(P'_x + \frac{v}{c^2} E' \right) \right]_{out}} = 1.$$

Hence incidence angle and deflection angle are equal in rest frame as well as mirror (moving) frame.

Chapter 12
Relativistic Lagrangian and Hamiltonian

12.1 Relativistic Lagrangian

According to classical mechanics, the Lagrangian is given by

$$L = T - V, \tag{12.1}$$

where T is the kinetic energy of the system and is a function of generalized momenta p_r (or of the generalized velocity $\dot{q}_r$); while the potential energy V is the function of generalized coordinates q_r only.

Again from classical mechanics, we get the canonical momenta as

$$p_r = \frac{\partial L}{\partial \dot{q}_r} = \frac{\partial}{\partial \dot{q}_r}(T - V) = \frac{\partial T}{\partial \dot{q}_r}. \tag{12.2}$$

Now we are to find relativistic kinetic energy T^* such that the relativistic Lagrangian function be $L^* = T^* - V$.

We know the relativistic momentum

$$p_x = m_0 \dot{x}[1 - \beta^2]^{-\frac{1}{2}} = \frac{\partial T^*}{\partial \dot{x}}$$

$$p_y = m_0 \dot{y}[1 - \beta^2]^{-\frac{1}{2}} = \frac{\partial T^*}{\partial \dot{y}} \tag{12.3}$$

$$p_z = m_0 \dot{z}[1 - \beta^2]^{-\frac{1}{2}} = \frac{\partial T^*}{\partial \dot{z}}$$

where $\beta = \frac{v}{c}$ and $v^2 = \dot{x}^2 + \dot{y}^2 + \dot{z}^2$.

Since T^* is a function of $\dot{x}$, $\dot{y}$, $\dot{z}$, so

© The Author(s), under exclusive license to Springer Nature Singapore Pte Ltd. 2022
F. Rahaman, *The Special Theory of Relativity*, La Matematica per il 3+2 136,
https://doi.org/10.1007/978-981-19-0497-4_12

$$dT^* = \frac{\partial T^*}{\partial \dot{x}} d\dot{x} + \frac{\partial T^*}{\partial \dot{y}} d\dot{y} + \frac{\partial T^*}{\partial \dot{z}} d\dot{z}$$

Using the relations given in (12.3) we obtain

$$dT^* = m_0 [1 - \beta^2]^{-\frac{1}{2}} [\dot{x} d\dot{x} + \dot{y} d\dot{y} + \dot{z} d\dot{z}]. \tag{12.4}$$

Also $v^2 = \dot{x}^2 + \dot{y}^2 + \dot{z}^2$, so

$$v dv = \dot{x} d\dot{x} + \dot{y} d\dot{y} + \dot{z} d\dot{z}$$

Therefore from (12.4), we get

$$dT^* = m_0 \left[1 - \left(\frac{v}{c} \right)^2 \right]^{-\frac{1}{2}} v dv$$

After integrating, this yields

$$T^* = -m_0 c^2 \left[1 - \left(\frac{v}{c} \right)^2 \right]^{\frac{1}{2}} + A. \tag{12.5}$$

(A = integration constant).
    We know that when $v \ll c$, $T^* = \frac{1}{2} m_0 v^2$ (classical kinetic energy).
Hence, from (12.5), one gets

$$\frac{1}{2} m_0 v^2 = -m_0 c^2 + A$$

$$\Rightarrow A = \frac{1}{2} m_0 v^2 + m_0 c^2 = m_0 c^2 \quad (\text{since } v \ll c).$$

Therefore

$$T^* = -m_0 c^2 \left[1 - \left(\frac{v}{c} \right)^2 \right]^{\frac{1}{2}} + m_0 c^2$$

$$= m_0 c^2 \left[1 - (1 - \beta^2)^{\frac{1}{2}} \right]. \tag{12.6}$$

Here V purely depends upon position, i.e. it is independent of velocity. Therefore, it remains unchanged relativistically. So the relativistic Lagrangian function is given by

$$L^* = T^* - V = m_0 c^2 \left[1 - (1 - \beta^2)^{\frac{1}{2}} \right] - V. \tag{12.7}$$

To check whether relativistic Lagrangian function gives relativistic force through Lagrangian's equation analogue to classical mechanics.

Assume Lagrangian's equation holds similar to classical mechanics in the following form:

$$\frac{d}{dt}\left[\frac{\partial L^*}{\partial v_i}\right] - \frac{\partial L^*}{\partial x_i} = 0. \qquad (12.8)$$

Here,

$$\frac{\partial L^*}{\partial v_i} = m_0 c^2 \left[1 - \left(\frac{v}{c}\right)^2\right]^{-\frac{1}{2}} \left(\frac{v_i}{c^2}\right) = m_0 v_i (1 - \beta^2)^{-\frac{1}{2}}.$$

Substituting this in Eq. (12.8) gives

$$\frac{d}{dt}\left[m_0 v_i \left(1 - \beta^2\right)^{-\frac{1}{2}}\right] = \frac{\partial V}{\partial x_i} = F_i$$

$$\Rightarrow (1 - \beta^2)^{-\frac{1}{2}} \frac{d}{dt}\left[m_0 v_i \left(1 - \beta^2\right)^{-\frac{1}{2}}\right] = (1 - \beta^2)^{-\frac{1}{2}} F_i$$

$$\Rightarrow \frac{d}{d\tau}\left[m_0 v_i \left(1 - \beta^2\right)^{-\frac{1}{2}}\right] = K_i$$

[τ = proper time and K_i = space components of Minkowski force]

This is the relativistic equation of motion.

Thus we have shown that the Lagrangian (12.7) is correct.

Now,

$$\frac{\partial L^*}{\partial v_i} = m_0 v_i (1 - \beta^2)^{-\frac{1}{2}} = p_i,$$

i.e. partial derivative of L^* with respect to the velocity is still the momentum (relativistic).

12.2 Relativistic Hamiltonian Function

Analogue to the classical mechanics, the relativistic Hamiltonian function is defined as

$$H^* = \sum p_x \dot{x} - L^*$$

where the canonical momenta are given by

$$p_r = \frac{\partial L^*}{\partial \dot{q}_r} = \frac{\partial T^*}{\partial \dot{q}_r}$$

Now,

$$H^* = \sum \frac{\partial T^*}{\partial \dot{x}} \dot{x} - m_0 c^2 \left[1 - \left(1 - \beta^2 \right)^{\frac{1}{2}} \right] + V$$

$$= \sum \frac{\partial}{\partial \dot{x}} \left[m_0 c^2 \left[1 - \left(1 - \beta^2 \right)^{\frac{1}{2}} \right] \right] \dot{x} - m_0 c^2 \left[1 - \left(1 - \beta^2 \right)^{\frac{1}{2}} \right] + V$$

$$= \sum \left[m_0 c^2 \left[1 - \left(\frac{v}{c} \right)^2 \right]^{-\frac{1}{2}} \frac{v}{c^2} \frac{\partial v}{\partial \dot{x}} \right] \dot{x} - m_0 c^2 \left[1 - \left(1 - \beta^2 \right)^{\frac{1}{2}} \right] + V.$$

Since $v^2 = \dot{x}^2 + \dot{y}^2 + \dot{z}^2$, therefore, $v \frac{\partial v}{\partial \dot{x}} = \dot{x}$. It follows that

$$H^* = \sum \left[m_0 \dot{x}^2 \left[1 - \left(\frac{v}{c} \right)^2 \right]^{-\frac{1}{2}} \right] - m_0 c^2 \left[1 - \left(1 - \beta^2 \right)^{\frac{1}{2}} \right] + V$$

$$= m_0 \left[\dot{x}^2 + \dot{y}^2 + \dot{z}^2 \right] \left[1 - \left(\frac{v}{c} \right)^2 \right]^{-\frac{1}{2}} - m_0 c^2 \left[1 - \left(1 - \beta^2 \right)^{\frac{1}{2}} \right] + V$$

$$= m_0 v^2 \left[1 - \left(\frac{v}{c} \right)^2 \right]^{-\frac{1}{2}} - m_0 c^2 \left[1 - \left(1 - \beta^2 \right)^{\frac{1}{2}} \right] + V$$

$$= m_0 c^2 \left[\left(\frac{v}{c} \right)^2 + 1 - \left(\frac{v}{c} \right)^2 \right] \left[1 - \left(\frac{v}{c} \right)^2 \right]^{-\frac{1}{2}} - m_0 c^2 + V.$$

Thus we obtain the relativistic Hamiltonian as

$$H^* = m_0 c^2 \left[1 - \left(\frac{v}{c} \right)^2 \right]^{-\frac{1}{2}} - m_0 c^2 + V. \tag{12.9}$$

Using the definition of relativistic mass, $m = m_0 \left[1 - \left(\frac{v}{c} \right)^2 \right]^{-\frac{1}{2}}$, one can write

$$H^* = m c^2 - m_0 c^2 + V$$
$$= T^* + V,$$

where T^* is the relativistic kinetic energy.

Interestingly, we note that analogous to classical Hamiltonian, the relativistic Hamiltonian is the total relativistic energy of the system.

Also,

$$p_x^2 + p_y^2 + p_z^2 = \frac{m_0^2}{1 - \frac{v^2}{c^2}} \left(\dot{x}^2 + \dot{y}^2 + \dot{z}^2 \right)$$

or,

$$p_x^2 + p_y^2 + p_z^2 = \frac{m_0^2 v^2}{1 - \frac{v^2}{c^2}} = -m_0^2 c^2 + \frac{m_0^2 c^2}{1 - \frac{v^2}{c^2}}.$$

This implies

$$\frac{1}{\sqrt{1 - \frac{v^2}{c^2}}} = \sqrt{\frac{p_x^2 + p_y^2 + p_z^2 + m_0^2 c^2}{m_0^2 c^2}}. \tag{12.10}$$

Substituting this result into Eq. (12.9), we get another expression for relativistic Hamiltonian as

$$H^* = c\sqrt{p_x^2 + p_y^2 + p_z^2 + m_0^2 c^2} - m_0^2 c^2 + V. \tag{12.11}$$

Hamiltonian canonical equations are

$$\dot{x} = \frac{\partial H^*}{\partial p_x} \quad \text{and} \quad \dot{p}_x = -\frac{\partial H^*}{\partial x}.$$

From (12.11), we obtain

$$\dot{x} = \frac{c p_x}{\sqrt{p_x^2 + p_y^2 + p_z^2 + m_0^2 c^2}} = \frac{c p_x \sqrt{1 - \frac{v^2}{c^2}}}{m_0 c}.$$

This implies

$$p_x = \frac{m_0 \dot{x}}{\sqrt{1 - \frac{v^2}{c^2}}}.$$

It is relativistic momentum.

Again,

$$\dot{p}_x = \frac{d}{dt} \left[\frac{m_0 \dot{x}}{\sqrt{1 - \frac{v^2}{c^2}}} \right] = -\frac{\partial H^*}{\partial x} = -\frac{\partial V}{\partial x} = F_x.$$

This is the relativistic force.

12.3 Covariant Lagrangian and Hamiltonian Formulation

The above expressions for Lagrangian and Hamiltonian predict the correct relativistic equations of motion in a particular Lorentz frame. Here, time coordinate t has been used as a parameter completely distinct from the spatial coordinates. For a covariant formulation, all the four coordinates have the similar meaning in world space. Here one should replace t by the invariant parameter, the proper time τ. The covariant form of the Lagrangian should be a function of the coordinates x_μ in Minkowski space and their derivatives with respect to τ. Here,

$$dt = \frac{dt}{d\tau}d\tau = t^1 d\tau \qquad (12.12)$$

[^{1} means derivative with respect to τ]

To find the connection between old Lagrangian L and new Lagrangian L^*, we use Hamilton's variational principle as

$$0 = \delta I = \delta \int_{\tau_1}^{\tau_2} L^*(x_j, x_j^1, x_4, x_4^1) d\tau = \delta \int_{t_1}^{t_2} L dt = \delta \int_{\tau_1}^{\tau_2} L t^1 d\tau, \qquad (12.13)$$

which yields

$$L^*(x_\mu, x_\mu^1) = t^1 L(x_j, \dot{x}_j, t) = t^1 L\left(x_j, \frac{x_4}{ic}, \frac{icx_j^1}{x_4^1}\right) = \frac{x_4^1}{ic} L\left(x_j, \frac{x_4}{ic}, \frac{icx_j^1}{x_4^1}\right) \qquad (12.14)$$

[over dot means derivative with respect to t, $x_4 = ict$, $\dot{x}_j = \frac{dx_j}{dt} = \frac{icx_j^1}{x_4^1}$ and $\mu = 1$, 2, 3, 4; $j = 1, 2, 3$.]

Here, one can note that the new Lagrangian $L^*(x_\mu, x_\mu^1)$ is a homogenous function of generalized velocities in the first degree whatever the functional form of old Lagrangian $L(x_j, \dot{x}_j, t)$. Then by Euler's theorem

$$L^* = x_\mu^1 \frac{\partial L^*}{\partial x_\mu^1}. \qquad (12.15)$$

Therefore,

$$\frac{dL^*}{d\tau} = x_\mu^{11} \frac{\partial L^*}{\partial x_\mu^1} + x_\mu^1 \frac{d}{d\tau}\left(\frac{\partial L^*}{\partial x_\mu^1}\right). \qquad (12.16)$$

Again since $L^* = L^*(x_\mu, x_\mu^1)$, we get

$$\frac{dL^*}{d\tau} = x_\mu^{11}\frac{\partial L^*}{\partial x_\mu^1} + x_\mu^1\frac{\partial L^*}{\partial x_\mu}. \tag{12.17}$$

Subtracting (12.17) from (12.16), we get

$$\left[\frac{d}{d\tau}\left(\frac{\partial L^*}{\partial x_\mu^1}\right) - \frac{\partial L^*}{\partial x_\mu}\right]x_\mu^1 = 0. \tag{12.18}$$

Thus the homogenous property of new Lagrangian $L^*(x_\mu, x_\mu^1)$ in generalized velocities gives covariant Lagrange equations.

For spatial part, we have

$$\frac{\partial L^*}{\partial x_j} = \frac{\partial}{\partial x_j}\left[t^1 L\left(x_j, \frac{x_4}{ic}, \frac{icx_j^1}{x_4^1}\right)\right] = \frac{x_4^1}{ic}\frac{\partial}{\partial x_j}\left[L\left(x_j, \frac{x_4}{ic}, \frac{icx_j^1}{x_4^1}\right)\right] = \frac{x_4^1}{ic}\frac{\partial L}{\partial x_j}. \tag{12.19}$$

$$\frac{\partial L^*}{\partial x_j^1} = \frac{\partial}{\partial x_j^1}\left[t^1 L\left(x_j, \frac{x_4}{ic}, \frac{icx_j^1}{x_4^1}\right)\right] = \frac{x_4^1}{ic}\frac{\partial}{\partial x_j^1}\left[L\left(x_j, \frac{x_4}{ic}, \frac{icx_j^1}{x_4^1}\right)\right]$$

$$= \frac{x_4^1}{ic}\frac{\partial}{\partial\left(\frac{icx_j^1}{x_4^1}\right)}\left[L\left(x_j, \frac{x_4}{ic}, \frac{icx_j^1}{x_4^1}\right)\right]\frac{\partial\left(\frac{icx_j^1}{x_4^1}\right)}{\partial x_j^1} = \frac{x_4^1}{ic}\frac{\partial L}{\partial \dot{x}_j}\frac{ic}{x_4^1} = \frac{\partial L}{\partial \dot{x}_j} = p_j. \tag{12.20}$$

Thus the definition of momentum remains the same.

For the fourth component, we have

$$\frac{\partial L^*}{\partial x_4} = \frac{\partial}{\partial x_4}\left[t^1 L\left(x_j, \frac{x_4}{ic}, \frac{icx_j^1}{x_4^1}\right)\right] = \frac{x_4^1}{ic}\frac{\partial}{\partial x_4}\left[L\left(x_j, \frac{x_4}{ic}, \frac{icx_j^1}{x_4^1}\right)\right]$$

$$= \frac{x_4^1}{ic}\frac{\partial}{\partial\left(\frac{x_4}{ic}\right)}\left[L\left(x_j, \frac{x_4}{ic}, \frac{icx_j^1}{x_4^1}\right)\right]\frac{\partial\left(\frac{x_4}{ic}\right)}{\partial x_4} = \frac{x_4^1}{ic}\frac{\partial L}{\partial t}\frac{1}{ic} = -\frac{x_4^1}{c^2}\frac{\partial L}{\partial t} \tag{12.21}$$

$$\frac{\partial L^*}{\partial x_4^1} = \frac{\partial}{\partial x_4^1}\left[t^1 L\left(x_j, \frac{x_4}{ic}, \frac{icx_j^1}{x_4^1}\right)\right] = \frac{\partial}{\partial x_4^1}\left[\frac{x_4^1}{ic}L\left(x_j, \frac{x_4}{ic}, \frac{icx_j^1}{x_4^1}\right)\right]$$

$$= \frac{L}{ic} + \frac{x_4^1}{ic}\frac{\partial}{\partial\left(\frac{icx_j^1}{x_4^1}\right)}\left[L\left(x_j, \frac{x_4}{ic}, \frac{icx_j^1}{x_4^1}\right)\right]\frac{\partial\left(\frac{icx_j^1}{x_4^1}\right)}{\partial x_4^1}$$

$$= \frac{L}{ic} + \frac{x_4^1}{ic}\left(-\frac{icx_j^1}{x_4^{12}}\right)\frac{\partial L}{\partial \dot{x}_j} = \frac{L}{ic} - \frac{\dot{x}_j}{ic}\frac{\partial L}{\partial \dot{x}_j} = \frac{i}{c}\left(\dot{x}_j\frac{\partial L}{\partial \dot{x}_j} - L\right) = \frac{iH}{c} = p_4. \tag{12.22}$$

Thus we have obtained the fourth component of momentum p_4. But we know $p_4 = \frac{iE}{c}$, therefore, relativistic Hamiltonian is the total relativistic energy of the system.

Using the results given in (12.19) and (12.20), we obtain from (12.18)

$$\frac{x_4^1}{ic}\frac{\partial L}{\partial \dot{x}_j} - \frac{x_4^1}{ic}\frac{d}{dt}\left(\frac{\partial L}{\partial \dot{x}_j}\right) = 0$$

[here, $\frac{d}{d\tau} = \frac{dt}{d\tau}\frac{d}{dt} = t^1\frac{d}{dt} = \frac{x_4^1}{ic}\frac{d}{dt}$]

This yields the usual Lagrange's equation

$$\frac{\partial L}{\partial \dot{x}_j} - \frac{d}{dt}\left(\frac{\partial L}{\partial \dot{x}_j}\right) = 0.$$

Using the result given in (12.21) and (12.22) the fourth component, we get as above

$$-\frac{x_4^1}{c^2}\frac{\partial L}{\partial t} - \frac{x_4^1}{ic}\frac{d}{dt}\left(\frac{iH}{c}\right) = 0 \Longrightarrow \frac{\partial L}{\partial t} + \frac{dH}{dt} = 0.$$

This is a very known result in classical mechanics.

The covariant Hamiltonian function is given by

$$H^* = p_j x_j^1 + p_4 x_4^1 - L^*.$$

The result

$$\dot{x}_j = \frac{dx_j}{dt} = \frac{icx_j^1}{x_4^1}$$

implies

$$x_j^1 = \frac{\dot{x}_j x_4^1}{ic}.$$

Thus we get

$$H^* = \frac{x_4^1}{ic}\left(p_j\dot{x}_j + p_4 ic - L\right) = \frac{x_4^1}{ic}(H - H) = 0.$$

Hence covariant Hamiltonian function identically vanishes. However, its derivatives are not, in general, zero as $H^* = p_\mu x_\mu^1 - L^*$ gives

$$x_\mu^1 = \frac{\partial H^*}{\partial p_\mu} \quad \text{and} \quad p_\mu^1 = \frac{\partial H^*}{\partial x_\mu}.$$

12.4 Lorentz Transformation of Force

Let a frame S^1 is moving with velocity v along x-axis relative to a rest frame S. Let a particle of mass m is moving with velocity $\vec{u}$ with respect to S frame. In S^1 frame let its mass and velocity be m^1 and $\vec{u}^{\,1}$, respectively. Let us also suppose that particle is moving under the action of the force $\vec{F}$ and $\vec{F}^{\,1}$ with respect to S and S^1 frames, respectively. Then

$$\vec{F} = \frac{d}{dt}(m\vec{u}) \;\; ; \;\; \vec{F}^{\,1} = \frac{d}{dt^1}(m^1\vec{u}^{\,1}).$$

Here,

$$\vec{F} = (F_x, F_y, F_z) \;, \;\; \vec{F}^{\,1} = (F_x^1, F_y^1, F_z^1) \;\; \text{and} \;\; \vec{u} = (u_x, u_y, u_z) \;\; \vec{u}^{\,1} = (u_x^1, u_y^1, u_z^1).$$

We know the following results:

$$\frac{dt^1}{dt} = \gamma\left(1 - \frac{v}{c^2}u_x\right) \;, \;\; m = \frac{m_0}{\sqrt{1 - \frac{u^2}{c^2}}} \;, \;\; u_x^1 = \frac{u_x - v}{1 - \frac{v}{c^2}u_x} \;, \;\; m^1 = \gamma m\left(1 - \frac{v}{c^2}u_x\right),$$

where $\gamma = \frac{1}{\sqrt{1 - \frac{v^2}{c^2}}}$, $u^2 = u_x^2 + u_y^2 + u_z^2$ and m_0 is the rest mass.

Now,

$$F_x^1 = \frac{d}{dt^1}(m^1 u_x^1) = \frac{d}{dt}(m^1 u_x^1)\frac{dt}{dt^1} = \frac{d}{dt}\left[\left(\frac{u_x - v}{1 - \frac{v}{c^2}u_x}\right)\gamma m\left(1 - \frac{v}{c^2}u_x\right)\right]\frac{1}{\gamma\left(1 - \frac{v}{c^2}u_x\right)}$$

$$= \frac{1}{\left(1 - \frac{v}{c^2}u_x\right)}\frac{d}{dt}[m(u_x - v)] = \frac{1}{\left(1 - \frac{v}{c^2}u_x\right)}\left[m\frac{du_x}{dt} + u_x\frac{dm}{dt} - vm_0\frac{d}{dt}\left\{\left(1 - \frac{u^2}{c^2}\right)^{-\frac{1}{2}}\right\}\right]$$

$$= \frac{1}{\left(1 - \frac{v}{c^2}u_x\right)}\left[m\frac{du_x}{dt} + u_x\frac{dm}{dt} - \frac{vm_0}{c^2}\left(1 - \frac{u^2}{c^2}\right)^{-\frac{3}{2}}u\frac{du}{dt}\right]$$

$$= \frac{1}{\left(1 - \frac{v}{c^2}u_x\right)}\left[m\frac{du_x}{dt} + u_x\frac{dm}{dt} - \frac{vm}{c^2}\left(1 - \frac{u^2}{c^2}\right)^{-1}\left(u_x\frac{du_x}{dt} + u_y\frac{du_y}{dt} + u_z\frac{du_z}{dt}\right)\right]$$

Taking logarithm of both sides of $m = \frac{m_0}{\sqrt{1 - \frac{u^2}{c^2}}}$ and differentiating with respect to t, we get

$$\frac{1}{m}\frac{dm}{dt} = \frac{1}{c^2}\left(1 - \frac{u^2}{c^2}\right)^{-1}u\frac{du}{dt}.$$

Using this result, the above expression takes the form

$$F_x^1 = \frac{1}{\left(1 - \frac{v}{c^2} u_x\right)} \left[m \frac{du_x}{dt} + u_x \frac{dm}{dt} - \frac{vm}{c^2} \left(u_x \frac{du_x}{dt} + u_y \frac{du_y}{dt} + u_z \frac{du_z}{dt} \right) \right]$$

$$- \frac{1}{\left(1 - \frac{v}{c^2} u_x\right)} \left[\frac{v}{c^2} \frac{dm}{dt} (u_x^2 + u_y^2 + u_z^2) \right]$$

$$= \frac{1}{\left(1 - \frac{v}{c^2} u_x\right)} \left[m \frac{du_x}{dt} \left(1 - \frac{v}{c^2} u_x \right) + u_x \frac{dm}{dt} \left(1 - \frac{v}{c^2} u_x \right) \right]$$

$$- \frac{1}{\left(1 - \frac{v}{c^2} u_x\right)} \left[mu_y \frac{du_y}{dt} \frac{v}{c^2} + mu_z \frac{du_z}{dt} \frac{v}{c^2} + u_y \frac{dm}{dt} \frac{vu_y}{c^2} + u_z \frac{dm}{dt} \frac{vu_z}{c^2} \right]$$

$$= \frac{d(u_x m)}{dt} - \frac{1}{\left(1 - \frac{v}{c^2} u_x\right)} \left[\frac{vu_y}{c^2} \frac{d}{dt} (mu_y) + \frac{vu_z}{c^2} \frac{d}{dt} (mu_z) \right].$$

Hence,

$$F_x^1 = F_x - \frac{v}{\left(c^2 - vu_x\right)} \left[u_y F_y + u_z F_z \right] = \frac{F_x - \frac{v}{c^2} \left[\vec{u} \cdot \vec{F} \right]}{\left(1 - \frac{v}{c^2} u_x \right)}. \tag{12.23}$$

Now,

$$F_y^1 = \frac{d}{dt^1} (m^1 u_y^1) = \frac{d}{dt} (m^1 u_y^1) \frac{dt}{dt^1}$$

$$= \frac{d}{dt} \left[\left(\frac{u_y}{\gamma (1 - \frac{v}{c^2} u_x)} \right) \gamma m \left(1 - \frac{v}{c^2} u_x \right) \right] \frac{1}{\gamma \left(1 - \frac{v}{c^2} u_x \right)}$$

$$= \frac{1}{\gamma (1 - \frac{v}{c^2} u_x)} \frac{d}{dt} (mu_y).$$

Hence,

$$F_y^1 = \frac{\sqrt{1 - \frac{v^2}{c^2}}}{\left(1 - \frac{v}{c^2} u_x \right)} F_y. \tag{12.24}$$

Similarly,

$$F_z^1 = \frac{\sqrt{1 - \frac{v^2}{c^2}}}{\left(1 - \frac{v}{c^2} u_x \right)} F_z. \tag{12.25}$$

The inverse transformation can be found as

$$F_x = \frac{F_x^1 + \frac{v}{c^2} \left[\vec{u^1} \cdot \vec{F^1} \right]}{\left(1 + \frac{v}{c^2} u_x^1 \right)}, \quad F_y = \frac{\sqrt{1 - \frac{v^2}{c^2}}}{\left(1 + \frac{v}{c^2} u_x^1 \right)} F_y^1, \quad F_z = \frac{\sqrt{1 - \frac{v^2}{c^2}}}{\left(1 + \frac{v}{c^2} u_x \right)} F_z^1. \tag{12.26}$$

Note that in the Newtonian limit, $v << c$, $\vec{F} = \vec{F}^1$.

   Transformation formula of force has an interesting consequence: the force, F_x in one frame is related to the power developed by the force in another frame, $\vec{u^1} \cdot \vec{F^1}$. Thus in special theory of relativity power and force are related. The transformations (12.26) yield the transformation of power as

$$\vec{u} \cdot \vec{F} = \frac{\vec{u^1} \cdot \vec{F^1} + vF_x^1}{(1 + \frac{v}{c^2}u_x^1)}$$

or inverse relation

$$\vec{u^1} \cdot \vec{F^1} = \frac{\vec{u} \cdot \vec{F} - vF_x}{(1 - \frac{v}{c^2}u_x)}$$

[(12.26) implies

$$u_x F_x + u_y F_y + u_z F_z = u_x F_x^1 + \frac{v}{c^2} u_x (\vec{u^1} \cdot \vec{F^1})$$

$$+ u_y F_y^1 \sqrt{1 - \frac{v^2}{c^2}} + u_z F_z^1 \sqrt{1 - \frac{v^2}{c^2}}.$$

Using

$$u_x = \frac{u_x^1 + v}{1 + \frac{v}{c^2}u_x^1}, \quad u_y = \frac{u_y^1 \sqrt{1 - \frac{v^2}{c^2}}}{1 + \frac{v}{c^2}u_x^1}, \quad u_z = \frac{u_z^1 \sqrt{1 - \frac{v^2}{c^2}}}{1 + \frac{v}{c^2}u_x^1},$$

and after some simplification, one can get

$$\vec{u} \cdot \vec{F} = \frac{\vec{u^1} \cdot \vec{F^1} + vF_x^1}{(1 + \frac{v}{c^2}u_x^1)}]$$

Particular case: If the particle is at rest in the S^1 frame, then $u^1 = 0$ and $u_x = v, u_y = u_z = 0$. The above transformations assume the following form

$$F_x^1 = F_x, \quad F_y^1 = \frac{F_y}{\sqrt{1 - \frac{v^2}{c^2}}}, \quad F_z^1 = \frac{F_z}{\sqrt{1 - \frac{v^2}{c^2}}} \qquad (12.27)$$

Alternative Proof of the Particular Case

Let a frame S^1 is moving with velocity v relative to S along (positive) x-axis and assume a particle of rest mass m_0 is at rest in S^1 frame. This means an observer in S frame observes the particle is moving with velocity v.

In S frame, the force is defined as

$$F = \frac{dp}{dt} = \frac{d}{dt}(mv) = \frac{d}{dt}\left(\frac{m_0 v}{\sqrt{1-\beta^2}}\right) = m_0 \frac{d}{dt}\left(\frac{v}{\sqrt{1-\beta^2}}\right), \qquad (12.28)$$

where $\beta = \frac{v}{c}$, m is the relativistic mass of the particle. In S^1 frame, force is defined as

$$F^1 = \frac{dp^1}{dt^1}.$$

Since the particle be at rest in S^1 frame, the inverse Lorentz Transformations for momentum and energy give

$$p_x = \frac{p_x^1 + \frac{v}{c^2}E^1}{\sqrt{1-\beta^2}} \; ; \; p_y = p_y^1 \; ; \; p_z = p_z^1. \qquad (12.29)$$

From the definition of time dilation, we have

$$\partial t = \frac{\partial t^1}{\sqrt{1-\beta^2}} \quad \text{or} \quad \partial t^1 = \partial t\sqrt{1-\beta^2}. \qquad (12.30)$$

From Eq. (12.29), we have

$$\partial p_x = \frac{\partial p_x^1 + \frac{v}{c^2}\partial E^1}{\sqrt{1-\beta^2}} \; ; \; \partial p_y = \partial p_y^1 \; ; \; \partial p_z = \partial p_z^1 \; . \qquad (12.31)$$

From energy and momentum relation $E^2 = p^2 c^2 + m_0^2 c^4$, we obtain

$$\partial E = \frac{1}{2}(pc^2 + m_0^2 c^4)^{-\frac{1}{2}}(2pc^2 \partial p) = \frac{pc^2 \partial p}{\sqrt{p^2 c^2 + m_0^2 c^4}}.$$

The inverse relation yields

$$\partial E^1 = \frac{p^1 c^2 \partial p^1}{\sqrt{p^{1^2}c^2 + m_0^2 c^4}}. \qquad (12.32)$$

We note that the body is at rest in frame S^1, therefore $p^1 = 0$ and hence from (12.32) $\partial E^1 = 0$.

Consequently, Eq. (12.31) yields

$$\partial p_x = \frac{\partial p_x^1}{\sqrt{1 - \beta^2}}. \tag{12.33}$$

From (12.30) and (12.33), we have

$$\frac{\partial p_x}{\partial t} = \frac{\partial p_x^1}{\partial t^1}.$$

Since $\partial t \to 0$ implies $\partial t^1 \to 0$, therefore taking these limits, we have

$$\frac{\mathrm{d} p_x}{\mathrm{d} t} = \frac{\mathrm{d} p_x^1}{\mathrm{d} t^1}. \tag{12.34}$$

Again from (12.30) and (12.31), we have

$$\frac{\partial p_y}{\partial t} = \frac{\partial p_y^1}{\partial t^1}\sqrt{1 - \beta^2}.$$

In the limit $\partial t \to 0$, we have

$$\frac{\mathrm{d} p_y}{\mathrm{d} t} = \sqrt{1 - \beta^2}\,\frac{\mathrm{d} p_y^1}{\mathrm{d} t^1}. \tag{12.35}$$

Similarly,

$$\frac{\mathrm{d} p_z}{\mathrm{d} t} = \sqrt{1 - \beta^2}\,\frac{\mathrm{d} p_z^1}{\mathrm{d} t^1}. \tag{12.36}$$

Hence from (12.34), (12.35) and (12.36), we have

$$\frac{\mathrm{d} p_x}{\mathrm{d} t} = \frac{\mathrm{d} p_x^1}{\mathrm{d} t^1} \qquad \text{or} \qquad F_x = F_x^1$$

$$\frac{\mathrm{d} p_y}{\mathrm{d} t} = \sqrt{1 - \beta^2}\,\frac{\mathrm{d} p_y^1}{\mathrm{d} t^1} \qquad \text{or} \qquad F_y = F_y^1\sqrt{1 - \beta^2}$$

$$\frac{\mathrm{d} p_z}{\mathrm{d} t} = \sqrt{1 - \beta^2}\,\frac{\mathrm{d} p_z^1}{\mathrm{d} t^1} \qquad \text{or} \qquad F_z = F_z^1\sqrt{1 - \beta^2}.$$

12.5   Relativistic Transformation Formula for Density

Let us consider two frames S and S^1, S^1 is moving with velocity v relative to S along x-direction. Let a parallelepiped of mass m with volume V and density ρ is moving with velocity $\vec{u}$ with respect to S frame. In S^1 frame its mass, volume, density and velocity be m^1, V^1, ρ^1 and $\vec{u}^1$, respectively. Let also m_0 and V_0 be the rest volume and mass respectively and l_x, l_y, l_z be the lengths of the edge of the body when it is at rest in S frame, then $V_0 = l_x l_y l_z$. Since the body is in motion, therefore, lengths of the edges of the body according to length contraction are given by

$$l_x\sqrt{1 - \frac{u_x^2}{c^2}},\; l_y\sqrt{1 - \frac{u_y^2}{c^2}},\; l_z\sqrt{1 - \frac{u_z^2}{c^2}}.$$

Then,

$$V = l_x l_y l_z \sqrt{1 - \frac{u_x^2}{c^2}}\sqrt{1 - \frac{u_y^2}{c^2}}\sqrt{1 - \frac{u_z^2}{c^2}} = A V_0, \qquad (12.37)$$

where

$$A = \sqrt{1 - \frac{u_x^2}{c^2}}\sqrt{1 - \frac{u_y^2}{c^2}}\sqrt{1 - \frac{u_z^2}{c^2}}. \qquad (12.38)$$

Now, the expression for the density in S frame is

$$\rho = \frac{m}{V} = \frac{m_0}{\sqrt{1 - \frac{u^2}{c^2}}}\frac{1}{V_0 A} = \frac{\rho_0}{A\sqrt{1 - \frac{u^2}{c^2}}}. \qquad (12.39)$$

[here, ρ_0 is the rest density and $u^2 = u_x^2 + u_y^2 + u_z^2$]
   The above quantities in S^1 frame

$$u_x^1 = \frac{u_x - v}{1 - \frac{vu_x}{c^2}},\quad u_y^1 = \frac{u_y\sqrt{1 - \frac{v^2}{c^2}}}{1 - \frac{vu_x}{c^2}},\quad u_z^1 = \frac{u_z\sqrt{1 - \frac{v^2}{c^2}}}{1 - \frac{vu_x}{c^2}},\quad m^1 = \frac{m_0}{\sqrt{1 - \frac{u^{1^2}}{c^2}}}.$$

Lengths of the edges in S^1 frame

$$l_x\sqrt{1 - \frac{u_x^{1^2}}{c^2}},\; l_y\sqrt{1 - \frac{u_y^{1^2}}{c^2}},\; l_z\sqrt{1 - \frac{u_z^{1^2}}{c^2}}.$$

Thus the volume in S^1 frame

$$V^1 = l_x l_y l_z \sqrt{1 - \frac{u_x^{1\,2}}{c^2}} \sqrt{1 - \frac{u_y^{1\,2}}{c^2}} \sqrt{1 - \frac{u_z^{1\,2}}{c^2}} = A^1 V_0, \qquad (12.40)$$

where

$$A^1 = \sqrt{1 - \frac{u_x^{1\,2}}{c^2}} \sqrt{1 - \frac{u_y^{1\,2}}{c^2}} \sqrt{1 - \frac{u_z^{1\,2}}{c^2}}. \qquad (12.41)$$

Therefore, the density of the body as observed from S^1 frame

$$\rho^1 = \frac{m^1}{V^1} = \frac{m_0}{\sqrt{1 - \frac{u^{1\,2}}{c^2}}} \frac{1}{V_0 A^1} = \frac{\rho_0}{A^1 \sqrt{1 - \frac{u^{1\,2}}{c^2}}} \qquad (12.42)$$

[here, $u^{1\,2} = u_x^{1\,2} + u_y^{1\,2} + u_z^{1\,2}$.]

We know the transformation of Lorentz Transformation factor as (see Chap. 7, note—6)

$$\sqrt{1 - \left(\frac{u^1}{c}\right)^2} = \frac{\sqrt{1 - (\frac{v}{c})^2} \sqrt{1 - (\frac{u}{c})^2}}{(1 - \frac{u_x v}{c^2})}$$

Using this value, ρ^1 takes the form

$$\rho^1 = \frac{\rho_0}{A^1} \frac{(1 - \frac{u_x v}{c^2})}{\sqrt{1 - (\frac{v}{c})^2} \sqrt{1 - (\frac{u}{c})^2}} = \frac{\rho A (1 - \frac{u_x v}{c^2})}{A^1 \sqrt{1 - (\frac{v}{c})^2}}.$$

Now substituting the values of A and A^1, we obtain

$$\rho^1 = \frac{\rho \sqrt{1 - \frac{u_x^2}{c^2}} \sqrt{1 - \frac{u_y^2}{c^2}} \sqrt{1 - \frac{u_z^2}{c^2}} (1 - \frac{u_x v}{c^2})}{\sqrt{1 - \frac{u_x^{1\,2}}{c^2}} \sqrt{1 - \frac{u_y^{1\,2}}{c^2}} \sqrt{1 - \frac{u_z^{1\,2}}{c^2}} \sqrt{1 - (\frac{v}{c})^2}} \qquad (12.43)$$

Particular case: If the body is at rest in S frame, then

$$u_x = u_y = u_z = 0, \quad u_x^1 = -v, \quad u_y^1 = u_z^1 = 0, \quad \rho = \rho_0$$

Then Eq. (12.42) becomes

$$\rho^1 = \frac{\rho_0}{1 - (\frac{v}{c})^2}. \qquad (12.44)$$

Thus observer sitting in S^1 frame measures the density ρ_0 of the body rest in S frame as ρ^1 given in Eq. (12.44).

Alternative Proof of the Particular Case

Let us consider two frames S and S^1, S^1 is moving with velocity v relative to S along x-direction. Let a parallelepiped of mass m_0 with volume V_0 and density ρ_0 be at rest in S frame. Then,

$$V_0 = l_x l_y l_z, \quad \rho_0 = \frac{m_0}{V_0}$$

[l_x, l_y, l_z be the lengths of the edge of the body in S frame]

The volume of parallelepiped measured in S^1 frame will be

$$V^1 = l_x \sqrt{1 - \frac{v^2}{c^2}} l_y l_z = V_0 \sqrt{1 - \frac{v^2}{c^2}} \tag{12.45}$$

[length contraction occurs only along x-direction]

Also the mass of parallelepiped measured in S^1 frame will be

$$m^1 = \frac{m_0}{\sqrt{1 - \frac{v^2}{c^2}}}. \tag{12.46}$$

Hence density in S^1 frame will be

$$\rho^1 = \frac{m^1}{V^1} = \frac{m_0}{\sqrt{1 - \frac{v^2}{c^2}}} \frac{1}{V_0 \sqrt{1 - \frac{v^2}{c^2}}} = \frac{m_0}{V_0} \frac{1}{(1 - \frac{v^2}{c^2})} = \frac{\rho_0}{(1 - \frac{v^2}{c^2})}. \tag{12.47}$$

Example 12.1 A particle of rest mass m_0 moves on the x-axis and is attached to the origin by a force $m_0 k^2 x$. If it performs oscillation of amplitude a, then show that the periodic time of this relativistic harmonic oscillation is

$$T = \frac{4}{c} \int_0^a \frac{f \, dx}{\sqrt{f^2 - m_0^2 c^4}} \quad \text{where} \quad f = m_0 c^2 + \frac{1}{2} k^2 (a^2 - x^2) m_0.$$

Also verify that as $c \to \infty$; $T = \frac{2\pi}{k}$ and show that if $\frac{ka}{c}$ is small then $T = \frac{2\pi}{k}(1 + \frac{3}{16} \frac{k^2 a^2}{c^2})$.

Hint: Let v be the velocity of Harmonic oscillator along x-axis.

Therefore, the relativistic kinetic energy of harmonic oscillator is $= (m - m_0)c^2$ where m_0 = rest mass of harmonic oscillator.

The rest energy of Harmonic Oscillator (H.O.) = $m_0 c^2$

The potential energy of H.O. = $\frac{1}{2} k^2 m_0 x^2$

$$\left[\text{we know, } -\frac{dV}{dx} = F(x); \text{ here, } F(x) = -m_0k^2x, \text{ therefore, } V = \frac{1}{2}m_0k^2x^2\right]$$

Total energy of the oscillator is given by

$$E = (m - m_0)c^2 + m_0c^2 + \frac{1}{2}kx^2m_0 = mc^2 + \frac{1}{2}k^2m_0x^2.$$

Therefore,

$$E = \frac{m_0c^2}{\sqrt{1 - \frac{v^2}{c^2}}} + \frac{1}{2}k^2m_0x^2, \tag{1}$$

where v is the velocity of the particle.

If the particle comes to rest at $x = a$, i.e. $v = 0$ at $x = a$, then from (1) we have

$$E = m_0c^2 + \frac{1}{2}k^2m_0a^2. \tag{2}$$

Equation (1) gives

$$1 - \frac{v^2}{c^2} = \frac{m_0^2c^4}{(E - \frac{1}{2}m_0k^2x^2)^2}$$

or,

$$\frac{v}{c} = \sqrt{1 - \frac{m_0^2c^4}{(E - \frac{1}{2}m_0k^2x^2)^2}}. \tag{3}$$

Therefore the period of Harmonic Oscillation is given by

$$T = 4\int_0^a dt = 4\int_0^a \frac{dx}{v}$$

$$= \frac{4}{c}\int_0^a \frac{dx}{\sqrt{1 - \frac{m_0^2c^4}{(E - \frac{1}{2}m_0k^2x^2)^2}}}$$

$$= \frac{4}{c}\int_0^a \frac{E - \frac{1}{2}m_0k^2x^2}{\sqrt{[(E - \frac{1}{2}m_0k^2x^2)^2 - m_0^2c^4]}}dx.$$

Now putting the value of E from (2), we obtain

$$T = \frac{4}{c} \int_0^a \frac{f\,dx}{\sqrt{(f^2 - m_0^2 c^4)}}, \tag{4}$$

where

$$f = m_0 c^2 + \frac{1}{2} m_0 k^2 (a^2 - x^2) \qquad \text{(proved)}$$

Now substituting $x = a \sin \phi$; $dx = a \cos \phi d\phi$ in (4), we get

$$T = \frac{4}{c} \int_0^{\frac{\pi}{2}} \frac{m_0 c^2 + \frac{1}{2} m_0 k^2 (a^2 - a^2 \sin^2 \phi) a \cos \phi}{\sqrt{\left[m_0 c^2 + \frac{1}{2} m_0 k^2 (a^2 - a^2 \sin^2 \phi) \right]^2 - m_0^2 c^4}} d\phi$$

$$= \frac{4}{c} \int_0^{\frac{\pi}{2}} \frac{(m_0 c^2 + \frac{1}{2} k^2 m_0 a^2 \cos^2 \phi) a \cos \phi}{\sqrt{\frac{1}{4} m_0^2 k^4 a^4 \cos^4 \phi + m_0^2 c^2 k^2 a^2 \cos^2 \phi}} d\phi$$

$$= \frac{4a}{c} \int_0^{\frac{\pi}{2}} \frac{m_0 c^2 (1 + \frac{1}{2} \frac{a^2 k^2}{c^2} \cos^2 \phi)}{k a m_0 c \sqrt{1 + \frac{1}{4} \frac{k^2 a^2 \cos^2 \phi}{c^2}}} d\phi$$

For $\lambda = \frac{ak}{2c}$, the period of Harmonic Oscillation is given by

$$T = \frac{4}{k} \int_0^{\frac{\pi}{2}} \frac{1 + 2\lambda^2 \cos^2 \phi}{\sqrt{1 + \lambda^2 \cos^2 \phi}} d\phi. \tag{5}$$

Now, as $c \to \infty$, $\lambda \to 0$, then from (5) we get $T = \frac{2\pi}{k}$ (verified).

Again, $\frac{ka}{c}$ is small, then from (5), we have (i.e. λ is small)

$$T = \frac{4}{k} \int_0^{\frac{\pi}{2}} (1 + 2\lambda^2 \cos^2 \phi) \left(1 - \frac{\lambda^2}{2} \cos^2 \phi + \cdots \right) d\phi$$

$$\approx \frac{4}{k} \int_0^{\frac{\pi}{2}} \left(1 + \frac{3}{2} \lambda^2 \cos^2 \phi \right) d\phi$$

$$= \frac{4}{k} \left[\frac{\pi}{2} + \frac{3}{2} \lambda^2 \frac{\pi}{4} \right]$$

$$= \frac{2\pi}{k} \left[1 + \frac{3}{4} \lambda^2 \right]$$

$$= \frac{2\pi}{k} \left[1 + \frac{3}{16} \frac{a^2 k^2}{c^2} \right].$$

Example 12.2 Show that $\frac{\mathrm{d}p_x \mathrm{d}_y \mathrm{d}p_z}{E}$ is invariant under Lorentz Transformation.

Hint: We are to show

$$\frac{\mathrm{d}p_x^1 \mathrm{d}_y^1 \mathrm{d}p_z^1}{E^1} = \frac{\mathrm{d}p_x \mathrm{d}_y \mathrm{d}p_z}{E}.$$

We know

$$p_x^1 = a\left[p_x - \left(\frac{vE}{c^2}\right)\right], \ p_y = p_y^1, \ p_z = p_z^1, \ E^1 = a\left[E - vp_x\right], \ E^2 = p^2c^2 + m_0^2c^4$$

$$[\beta = \frac{v}{c}, p^2 = p_x^2 + p_y^2 + p_z^2 \ and \ a = \frac{1}{\sqrt{1-\beta^2}}].$$

Now,

$$\frac{\mathrm{d}p_x^1 \mathrm{d}_y^1 \mathrm{d}p_z^1}{E^1} = \frac{a\left[\mathrm{d}p_x - \left(\frac{v}{c^2}\mathrm{d}E\right)\right]\mathrm{d}_y \mathrm{d}p_z}{a\left[E - vp_x\right]}$$

$$E^2 = p^2c^2 + m_0^2c^4 = (p_x^2 + p_y^2 + p_z^2)c^2 + m_0^2c^4 \implies 2E\frac{\mathrm{d}E}{\mathrm{d}p_x} = c^2 p_x.$$

Hence

$$\frac{\mathrm{d}p_x^1 \mathrm{d}_y^1 \mathrm{d}p_z^1}{E^1} = \frac{a\left[1 - \left(\frac{v}{c^2}\frac{\mathrm{d}E}{\mathrm{d}p_x}\right)\right]\mathrm{d}p_x \mathrm{d}_y \mathrm{d}p_z}{a\left[E - vp_x\right]} = \frac{\mathrm{d}p_x \mathrm{d}_y \mathrm{d}p_z}{E}.$$

Chapter 13
Electrodynamics in Relativity

13.1 Relativistic Electrodynamics

When charges are in motion, the electric and magnetic fields are associated with this motion, which have space and time variation. This phenomenon is called electromagnetism. The study involves time-dependent electromagnetic fields and the behaviour of which is described by a set of equations called Maxwell's equations.

13.2 Equation of Continuity

According to the principle of conservation of charge, the net amount of charge in an isolated system remains constant, i.e. the time rate of increase of charge within the volume, equal to the net rate of flow of charge into the volume. This statement of conservation of charge can be expressed by the equation of continuity, which can be written as

$$\nabla \cdot \overrightarrow{J} + \frac{\partial \rho}{\partial t} = 0, \tag{13.1}$$

where $\overrightarrow{J}$ = current density, ρ = charge density.

This states that 'total current flowing out of some volume must be equal to the rate of decrease of charge within the volume'.

If

$$\frac{\partial \rho}{\partial t} = 0, \tag{13.2}$$

then

© The Author(s), under exclusive license to Springer Nature Singapore Pte Ltd. 2022 221
F. Rahaman, *The Special Theory of Relativity*, La Matematica per il 3+2 136,
https://doi.org/10.1007/978-981-19-0497-4_13

$$\nabla \cdot \vec{J} = 0. \tag{13.3}$$

This expresses the fact that in case of stationary currents there is no net outward flux of current density.

13.3 Maxwell's Equations

Prior to Maxwell, there were four fundamental equations of electromagnetism prescribed by different scientists. Maxwell modified these equations and put them together in compact form known as Maxwell's equations in electromagnetism. Maxwell's equations represent mathematical expression of certain results. These equations cannot be verified directly, however their application to any situation can be verified. The electromagnetic theory of light is based on Maxwell's equations of electromagnetic field. These equations are differential forms of elementary laws of electricity and magnetism given by Gauss, Faraday and Ampere.

[1] Gauss's law in electrostatics

$$\nabla \cdot \vec{E} = \left(\frac{1}{\varepsilon_0}\right)\rho \tag{13.4}$$
$$= 0, \quad \text{for a region without charge.}$$

[2] Application of Gauss's law to static magnetic fields. It has no name

$$\nabla \cdot \vec{B} = 0. \tag{13.5}$$

[3] Faraday's law and Lenz's law of electromagnetic induction

$$\nabla \times \vec{E} = -\frac{\partial \vec{B}}{\partial t}. \tag{13.6}$$

[4] Ampere's law with Maxwell's corrections

$$\nabla \times \vec{B} - \mu_0 \varepsilon_0 \left(\frac{\partial \vec{E}}{\partial t}\right) = \mu_0 \vec{J}. \tag{13.7}$$

Ampere's law:

$$\nabla \times \vec{B} = \mu_0 \vec{J}, \tag{13.8}$$

where $\vec{E}$ = Electric field strength, $\vec{B}$ = Magnetic field strength,

$$\vec{J} = \text{current density}, \quad \rho = \text{charge density}.$$

$$\mu_0 = \text{magnetic permeability of free space} = 4\pi \times 10^{-7} \text{ Wb/A m},$$

$$\varepsilon_0 = \text{electric permittivity of free space} \approx 8.86 \times 10^{-12} C^2/\text{N m}^2.$$

Here,

$$\mu_0 \varepsilon_0 = \frac{1}{c^2}. \tag{13.9}$$

Maxwell's field equations can be solved for $\vec{B}$ and $\vec{E}$ in terms of a scalar function ϕ and a vector function $\vec{A}$ as

$$\vec{B} = \nabla \times \vec{A} \tag{13.10}$$

[it follows from (13.5) as $\nabla \cdot \nabla \times \vec{A} = 0$]

$$\vec{E} = -\left(\frac{\partial \vec{A}}{\partial t}\right) - \nabla \phi. \tag{13.11}$$

Faraday's law given in (13.6) implies

$$\nabla \times \vec{E} = -\left(\frac{\partial \vec{B}}{\partial t}\right) = -\left(\frac{\partial}{\partial t}\right)(\nabla \times \vec{A}) = -\nabla \times \left(\frac{\partial \vec{A}}{\partial t}\right).$$

This gives

$$\vec{E} = -\left(\frac{\partial \vec{A}}{\partial t}\right) - \nabla \phi$$

[$\nabla \phi$ is similar to integration constant as $\nabla \times \nabla \phi = 0$],
where $\vec{A}$ is known as *magnetic potential* and ϕ indicates *electric potential*.

13.4 Derivation of Equation of Continuity from Maxwell's Equations

If the charges move with velocity $\vec{u}$, then current density

$$\vec{j} = \rho \vec{u}. \tag{13.12}$$

Taking divergence of Eq. (13.7), we get

$$\nabla \cdot \nabla \times \vec{B} = \mu_0 \varepsilon_0 \nabla \cdot \left(\frac{\partial \vec{E}}{\partial t} \right) + \mu_0 \nabla \cdot \vec{j}$$

$$= \mu_0 \varepsilon_0 \left[\frac{\partial (\nabla \cdot \vec{E})}{\partial t} \right] + \mu_0 \nabla \cdot \vec{j}$$

As $\nabla \cdot \nabla \times \vec{B} = 0$, we have

$$\mu_0 \varepsilon_0 \left[\frac{\partial (\nabla \cdot \vec{E})}{\partial t} \right] + \mu_0 \nabla \cdot \vec{j} = 0.$$

Substituting, $\nabla \cdot \vec{E} = \left(\frac{1}{\varepsilon_0} \right) \rho$ in the above equation, we get

$$\nabla \cdot \vec{j} + \frac{\partial \rho}{\partial t} = 0,$$

which is the equation of continuity and expresses the law of conservation of total charge.

13.5 Displacement Current

In the set of Maxwell's equations, the following equation

$$\nabla \times \vec{B} - \mu_0 \varepsilon_0 \left(\frac{\partial \vec{E}}{\partial t} \right) = \mu_0 \vec{j}$$

is Ampere's law with Maxwell's corrections. The quantity $\left(\frac{\partial \vec{E}}{\partial t} \right)$ was introduced by Maxwell and is called *displacement current*.

Ampere's law was

$$\nabla \times \vec{B} = \mu_0 \vec{J}. \tag{13.13}$$

If one applies the divergence to (13.13), then one can obtain

$$\nabla \cdot \nabla \times \vec{B} = \mu_0 (\nabla \cdot \vec{J}).$$

This implies

$$\nabla \cdot \vec{J} = 0. \tag{13.14}$$

This equation is valid for steady-state problems, i.e. when charge density is not changing. The complete relation is given by the continuity equation

$$\nabla \cdot \vec{J} + \frac{\partial \rho}{\partial t} = 0. \tag{13.15}$$

Therefore, Maxwell realized that the definition of total current density is incomplete and he suggested that some quantity be added to the right-hand side of (13.13), i.e. to $\vec{J}$. Assuming it to be $\vec{J}^1$, Eq. (13.13) becomes

$$\nabla \times \vec{B} = \mu_0 (\vec{J} + \vec{J}^1). \tag{13.16}$$

Taking divergence of equation (13.16), one finds

$$\nabla \cdot \nabla \times \vec{B} = \mu_0 \nabla \cdot (\vec{J} + \vec{J}^1).$$

This implies

$$\nabla \cdot \vec{J}^1 = -\nabla \cdot \vec{J} = \frac{\partial \rho}{\partial t}. \tag{13.17}$$

From Gauss's law, one can see that electric field E is related to the charge density ρ by

$$\nabla \cdot \vec{E} = \left(\frac{1}{\varepsilon_0}\right) \rho. \tag{13.18}$$

Substituting the value of ρ from (13.18) into (13.17), we obtain

$$\nabla \cdot \vec{J}^1 = \frac{\partial [\varepsilon_0 (\nabla \cdot \vec{E})]}{\partial t} = \nabla \cdot \frac{\partial (\varepsilon_0 \vec{E})}{\partial t}.$$

This is true for any arbitrary volume, therefore we have

$$\vec{J}^1 = \varepsilon_0 \frac{\partial \vec{E}}{\partial t}. \tag{13.19}$$

Thus if we add $\varepsilon_0 \frac{\partial \vec{E}}{\partial t}$ to the right-hand side of (13.13), then its divergence will satisfy the continuity equation (13.15) so that the discrepancy is removed.

With $\varepsilon_0 \frac{\partial \vec{E}}{\partial t}$ included, we have generalized Ampere's law

$$\nabla \times \vec{B} - \mu_0 \varepsilon_0 \left(\frac{\partial \vec{E}}{\partial t} \right) = \mu_0 \vec{J}. \tag{13.20}$$

The new quantity first introduced by Maxwell and he called the added term $\varepsilon_0 \frac{\partial \vec{E}}{\partial t}$ in (13.20), the *displacement current*. After the inclusion of displacement current by Maxwell, the modified Ampere's law includes time-varying fields. According to Maxwell this added term is as effective as the conduction current $\vec{J}$ for producing magnetic field.

13.6 Transformation for Charge Density

Let us consider two frames S and S^1, S^1 is moving with velocity v relative to S along x-direction. Let a parallelepiped of volume V_0 containing charge density ρ_0 be at rest in S frame. Then,

$$V_0 = l_x l_y l_z$$

[l_x, l_y, l_z are the lengths of the edge of the body in S frame]

The volume of parallelepiped measured in S^1 frame is

$$V^1 = l_x \sqrt{1 - \frac{v^2}{c^2}} l_y l_z = V_0 \sqrt{1 - \frac{v^2}{c^2}} \tag{13.21}$$

[length contraction occurs only along x-direction]

Let charge density be measured in S^1 frame ρ^1 and charge contained in the volume V^1 is $\rho^1 V^1$. We know charge is invariant, which implies

$$\rho^1 V^1 = \rho_0 V_0. \tag{13.22}$$

Using (13.21), we obtain

$$\rho^1 = \frac{\rho_0}{\sqrt{1 - \frac{v^2}{c^2}}}. \tag{13.23}$$

This indicates that when a charged particle is moving then charge density will be increased.

13.7 Four Current Vector

Let us consider an elementary volume element

$$dV = dx_1 dx_2 dx_3,$$

containing rest charge density ρ_0 moving with velocity v. Then charge in this volume is

$$dq = \rho \, dx_1 dx_2 dx_3 \qquad (13.24)$$

[here, $\rho = \rho_0 / \sqrt{1 - \frac{v^2}{c^2}}$]

Now, we multiply both sides of (13.24) by a four vector dx_μ, which yields

$$dq \, dx_\mu = \rho \, dx_1 dx_2 dx_3 dx_\mu = \rho \frac{dx_\mu}{dt} dx_1 dx_2 dx_3 dt. \qquad (13.25)$$

We know charge is invariant and this implies the left-hand side of Eq. (13.25) is a four vector. Therefore, the right-hand side of this equation must be a four vector. Again, we know the four-dimensional volume element is an invariant quantity, i.e.

$$dx_1 dx_2 dx_3 dx_4 = \text{invariant}$$

or

$$dx_1 dx_2 dx_3 d(ict) = \text{invariant},$$

i.e.

$$dx_1 dx_2 dx_3 dt = \text{invariant}.$$

Hence, the right-hand side of Eq. (13.25) must be a four vector. In other words,

$$\rho \frac{dx_\mu}{dt}$$

should be a four vector.

This four vector is represented by J_μ and is known as current four vector, i.e.

$$J_\mu = \rho \frac{dx_\mu}{dt} = \frac{\rho_0}{\sqrt{1 - \frac{v^2}{c^2}}} \frac{dx_\mu}{dt} = \rho_0 \frac{dx_\mu}{d\tau} = \rho_0 u_\mu. \qquad (13.26)$$

Thus current four vector J_μ is defined by multiplying the invariant rest charge density ρ_0 by the four velocity vector u_μ so that

$$J_1 = \rho \frac{dx_1}{dt} = \frac{\rho_0}{\sqrt{1 - \frac{v^2}{c^2}}} v_1 = \rho_0 u_1, \quad J_2 = \rho \frac{dx_2}{dt} = \rho v_2 = \frac{\rho_0}{\sqrt{1 - \frac{v^2}{c^2}}} v_2 = \rho_0 u_2,$$

$$J_3 = \rho \frac{dx_3}{dt} = \frac{\rho_0}{\sqrt{1 - \frac{v^2}{c^2}}} v_3 = \rho_0 u_3, \quad J_4 = \rho \frac{dx_4}{dt} = \rho \frac{d(ict)}{dt} = ic\rho = \frac{ic\rho_0}{\sqrt{1 - \frac{v^2}{c^2}}}.$$

Thus,

$$J_\mu = [J_1, J_2, J_3, J_4 \equiv ic\rho] = [\rho_0 u_1, \rho_0 u_2, \rho_0 u_3, ic\rho]. \tag{13.27}$$

13.8   Equation of Continuity in Covariant Form

We write the continuity equation

$$\nabla \cdot \vec{J} + \frac{\partial \rho}{\partial t} = 0$$

explicitly as

$$\frac{\partial J_1}{\partial x_1} + \frac{\partial J_2}{\partial x_2} + \frac{\partial J_3}{\partial x_3} + \frac{\partial \rho}{\partial t} = 0$$

or

$$\frac{\partial J_1}{\partial x_1} + \frac{\partial J_2}{\partial x_2} + \frac{\partial J_3}{\partial x_3} + \frac{\partial (ic\rho)}{\partial (ict)} = 0$$

or

$$\frac{\partial J_1}{\partial x_1} + \frac{\partial J_2}{\partial x_2} + \frac{\partial J_3}{\partial x_3} + \frac{\partial J_4}{\partial x_4} = 0$$

or

$$\frac{\partial J_\mu}{\partial x_\mu} = 0. \tag{13.28}$$

This is the continuity equation in covariant form. Since the left-hand side of Eq. (13.28) is a four-dimensional scalar, it is invariant under Lorentz transformation. Also, this equation indicates that four divergence of the current four vector J_μ vanishes.

Proof of Invariance of Continuity Equation

Hint: We have to show $\frac{\partial J_\mu^1}{\partial x_\mu^1} = \frac{\partial J_\mu}{\partial x_\mu}$, i.e.

$$\frac{\partial J_1^1}{\partial x_1^1} + \frac{\partial J_2^1}{\partial x_2^1} + \frac{\partial J_3^1}{\partial x_3^1} + \frac{\partial J_4^1}{\partial x_4^1} = \frac{\partial J_1}{\partial x_1} + \frac{\partial J_2}{\partial x_2} + \frac{\partial J_3}{\partial x_3} + \frac{\partial J_4}{\partial x_4}.$$

We know that

$$x_1 = a(x_1^1 + vt^1); \quad x_2 = x_2^1; \quad x_3 = x_3^1; \quad t = a\left[t^1 + \left(\frac{vx_1^1}{c^2}\right)\right]$$

$[\beta = v/c \text{ and } a = 1/\sqrt{1 - \beta^2}]$

Applying chain rule, we have

$$\frac{\partial}{\partial x_1^1} = \frac{\partial}{\partial x_1}\frac{\partial x_1}{\partial x_1^1} + \frac{\partial}{\partial x_2}\frac{\partial x_2}{\partial x_1^1} + \frac{\partial}{\partial x_3}\frac{\partial x_3}{\partial x_1^1} + \frac{\partial}{\partial t}\frac{\partial t}{\partial x_1^1}.$$

Thus we get,

$$\frac{\partial}{\partial x_1^1} = a\left(\frac{\partial}{\partial x_1} + \frac{v}{c^2}\frac{\partial}{\partial t}\right). \tag{13.29}$$

Similarly,

$$\frac{\partial}{\partial x_2^1} = \frac{\partial}{\partial x_2}, \quad \frac{\partial}{\partial x_3^1} = \frac{\partial}{\partial x_3}, \quad \frac{\partial}{\partial t^1} = a\left(v\frac{\partial}{\partial x_1} + \frac{\partial}{\partial t}\right). \tag{13.30}$$

Now,

$$\frac{\partial J_1^1}{\partial x_1^1} + \frac{\partial J_2^1}{\partial x_2^1} + \frac{\partial J_3^1}{\partial x_3^1} + \frac{\partial J_4^1}{\partial x_4^1} = a\left(\frac{\partial[a(J_1 - v\rho)]}{\partial x_1} + \frac{v}{c^2}\frac{\partial[a(J_1 - v\rho)]}{\partial t}\right)$$

$$+ \frac{\partial J_2}{\partial x_2} + \frac{\partial J_3}{\partial x_3} + a\left(v\frac{\partial\left[a\left(\rho - \frac{v}{c^2}J_1\right)\right]}{\partial x_1}\right.$$

$$\left. + \frac{\partial\left[a\left(\rho - \frac{v}{c^2}J_1\right)\right]}{\partial t}\right)$$

$$= \frac{\partial J_1}{\partial x_1} + \frac{\partial J_2}{\partial x_2} + \frac{\partial J_3}{\partial x_3} + \frac{\partial J_4}{\partial x_4}.$$

13.9 Transformation of Four Current Vector

The four current vector can be written as

$$J_\mu = [J_1, J_2, J_3, J_4], \tag{13.31}$$

where $J_4 = ic\rho$.

We know Lorentz transformation of any four vector A_μ ($\mu = 1, 2, 3, 4$) may be expressed as

$$A_\mu^1 = a_{\mu\nu} A_\nu,$$

where $a_{\mu\nu}$ is given as

$$a_{\mu\nu} = \begin{bmatrix} a & 0 & 0 & ia\beta \\ 0 & 1 & 0 & 0 \\ 0 & 0 & 1 & 0 \\ -ia\beta & 0 & 0 & a \end{bmatrix}$$

$[\beta = v/c$ and $a = 1/\sqrt{1 - \beta^2}]$

Hence,

$$\begin{bmatrix} J_1^1 \\ J_2^1 \\ J_3^1 \\ J_4^1 \end{bmatrix} = \begin{bmatrix} a & 0 & 0 & ia\beta \\ 0 & 1 & 0 & 0 \\ 0 & 0 & 1 & 0 \\ -ia\beta & 0 & 0 & a \end{bmatrix} \begin{bmatrix} J_1 \\ J_2 \\ J_3 \\ J_4 \end{bmatrix}. \tag{13.32}$$

This implies

$$J_1^1 = aJ_1 + ia\beta J_4 = aJ_1 + ia\left(\frac{v}{c}\right)(ic\rho),$$

i.e.

$$J_1^1 = a\left[J_1 - v\rho\right] \tag{13.33}$$

$$J_2^1 = J_2, \quad J_3^1 = J_3 \tag{13.34}$$

$$J_4^1 = -ia\beta J_1 + aJ_4$$

or

$$\left(ic\rho^1\right) = -ia\left(\frac{v}{c}\right)J_1 + a\left(ic\rho\right),$$

i.e.

$$\rho^1 = a\left[\rho - \left(\frac{v}{c^2}\right)J_1\right]. \tag{13.35}$$

The inverse transformations are obtained by replacing v by $-v$ as

$$J_1 = a\left[J_1^1 + v\rho^1\right], \quad J_2 = J_2^1, \quad J_3 = J_3^1, \quad \rho = a\left[\rho^1 + \left(\frac{v}{c^2}\right)J_1^1\right]. \quad (13.36)$$

Let a body containing charge density be at rest in the S^1 frame so that its rest charge density is ρ_0. This means for observer in S frame the charged body is moving with velocity v. So, in S^1 frame,

$$J_1^1 = J_2^1 = J_3^1 = 0, \quad \rho^1 = \rho_0.$$

Hence,

$$J_1 = av\rho_0, \quad J_2 = 0, \quad J_3 = 0, \quad \rho = a\rho_0. \quad (13.37)$$

13.10 Maxwell's Equations in Covariant Form

In terms of magnetic potential $\vec{A}$ and electric potential ϕ, Maxwell's equations can be written as

$$\Box^2 \vec{A} = -\mu_0 \vec{J} \quad (13.38)$$

$$\Box^2 \phi = -\left(\frac{1}{\varepsilon_0}\right)\rho \quad (13.39)$$

with

$$\nabla \cdot \vec{A} + \frac{1}{c^2}\frac{\partial \phi}{\partial t} = 0. \quad (13.40)$$

Here, $\Box^2 = \nabla^2 - \frac{1}{c^2} \cdot \frac{\partial^2}{\partial t^2}$ is *d'Alembertian operator*.
Equation (13.40) is called *Lorentz-Gauge condition*.

Proof From Maxwell's equation (13.7), we have

$$\nabla \times \vec{B} - \mu_0\varepsilon_0\left(\frac{\partial \vec{E}}{\partial t}\right) = \mu_0 \vec{J}.$$

Using the $\vec{B}$ and $\vec{E}$ in terms of $\vec{A}$ and ϕ from (13.10) and (13.11), we obtain

$$\nabla \times \nabla \times \vec{A} = \mu_0 \varepsilon_0 \left(\frac{\partial \left[-\left(\frac{\partial \vec{A}}{\partial t} \right) - \nabla \phi \right]}{\partial t} \right) + \mu_0 \vec{J}$$

$$\implies \nabla(\nabla \cdot \vec{A}) - \nabla^2 \vec{A} = -\mu_0 \varepsilon_0 \left(\frac{\partial^2 \vec{A}}{\partial t^2} \right) - \mu_0 \varepsilon_0 \left(\frac{\partial(\nabla \phi)}{\partial t} \right) + \mu_0 \vec{J}$$

$$\implies \nabla^2 \vec{A} - \frac{1}{c^2} \frac{\partial^2 \vec{A}}{\partial t^2} = \nabla \left[\nabla \cdot \vec{A} + \frac{1}{c^2} \frac{\partial \phi}{\partial t} \right] - \mu_0 \vec{J}$$

$$\implies \nabla^2 \vec{A} - \frac{1}{c^2} \frac{\partial^2 \vec{A}}{\partial t^2} \equiv \square^2 \vec{A} = -\mu_0 \vec{J}$$

provided

$$\nabla \cdot \vec{A} + \frac{1}{c^2} \frac{\partial \phi}{\partial t} = 0.$$

Again from Maxwell equation (13.4), we have

$$\left(\frac{1}{\varepsilon_0} \right) \rho = \nabla \cdot \vec{E}$$

$$= \nabla \cdot \left[-\left(\frac{\partial \vec{A}}{\partial t} \right) - \nabla \phi \right]$$

$$= -\nabla^2 \phi - \frac{\partial(\nabla \cdot \vec{A})}{\partial t}$$

or

$$\square^2 \phi + \frac{1}{c^2} \frac{\partial^2 \phi}{\partial t^2} + \frac{\partial(\nabla \cdot \vec{A})}{\partial t} = -\left(\frac{1}{\varepsilon_0} \right) \rho$$

$$\implies \square^2 \phi + \frac{\partial}{\partial t} \left[\frac{1}{c^2} \frac{\partial \phi}{\partial t} + \nabla \cdot \vec{A} \right] = -\left(\frac{1}{\varepsilon_0} \right) \rho$$

$$\implies \square^2 \phi = -\left(\frac{1}{\varepsilon_0} \right) \rho$$

provided

$$\nabla \cdot \vec{A} + \frac{1}{c^2} \frac{\partial \phi}{\partial t} = 0.$$

13.10.1 The d'Alembertian Operator is Invariant Under Lorentz Transformation

Proof From (13.29) and (13.30), we have

$$\frac{\partial}{\partial x_1^1} = a\left(\frac{\partial}{\partial x_1} + \frac{v}{c^2}\frac{\partial}{\partial t}\right), \quad \frac{\partial}{\partial x_2^1} = \frac{\partial}{\partial x_2}, \quad \frac{\partial}{\partial x_3^1} = \frac{\partial}{\partial x_3}, \quad \frac{\partial}{\partial t^1} = a\left(v\frac{\partial}{\partial x_1} + \frac{\partial}{\partial t}\right).$$

From these, we can get

$$\frac{\partial^2}{\partial x_1^{1\,2}} = a^2\left(\frac{\partial^2}{\partial x_1^2} + \frac{v^2}{c^4}\frac{\partial^2}{\partial t^2} + \frac{2v}{c^2}\frac{\partial^2}{\partial x_1\partial t}\right) \tag{13.41}$$

$$\frac{\partial^2}{\partial x_2^{1\,2}} = \frac{\partial^2}{\partial x_2^2}, \quad \frac{\partial^2}{\partial x_3^{1\,2}} = \frac{\partial^2}{\partial x_3^2}, \quad \frac{\partial^2}{\partial t^{1\,2}} = a^2\left(v^2\frac{\partial^2}{\partial x_1^2} + \frac{\partial^2}{\partial t^2} + 2v\frac{\partial^2}{\partial x_1\partial t}\right). \tag{13.42}$$

Therefore,

$$\begin{aligned}
\Box^{1\,2} &= \frac{\partial^2}{\partial x_1^{1\,2}} + \frac{\partial^2}{\partial x_2^{1\,2}} + \frac{\partial^2}{\partial x_3^{1\,2}} - \frac{1}{c^2}\frac{\partial^2}{\partial t^{1\,2}} \\
&= a^2\left(\frac{\partial^2}{\partial x_1^2} + \frac{v^2}{c^4}\frac{\partial^2}{\partial t^2} + \frac{2v}{c^2}\frac{\partial^2}{\partial x_1\partial t}\right) + \frac{\partial^2}{\partial x_2^2} + \frac{\partial^2}{\partial x_3^2} \\
&\quad - \frac{a^2}{c^2}\left(v^2\frac{\partial^2}{\partial x_1^2} + \frac{\partial^2}{\partial t^2} + 2v\frac{\partial^2}{\partial x_1\partial t}\right) \\
&= \frac{\partial^2}{\partial x_1^2} + \frac{\partial^2}{\partial x_2^2} + \frac{\partial^2}{\partial x_3^2} - \frac{1}{c^2}\frac{\partial^2}{\partial t^2} = \Box^2.
\end{aligned}$$

It is shown that the d'Alembertian operator is the invariant four-dimensional Laplacian; therefore, the right-hand sides of (13.38) and (13.39) should be the components of a four vector. Hence, Lorentz covariance requires that the potential ϕ and A form a four vector potential.

13.10.2 Lorentz-Gauge Condition in Covariant Form

The Lorentz-Gauge condition is

$$\nabla \cdot \vec{A} + \frac{1}{c^2}\frac{\partial\phi}{\partial t} = 0$$

This can be written as

$$\nabla \cdot \vec{A} + \frac{\partial \left(\frac{i\phi}{c} \right)}{\partial (ict)} = 0$$

or

$$\frac{\partial A_1}{\partial x_1} + \frac{\partial A_2}{\partial x_2} + \frac{\partial A_3}{\partial x_3} + \frac{\partial A_4}{\partial x_4} = 0,$$

i.e.

$$\frac{\partial A_\mu}{\partial x_\mu} = 0.$$

This gives a clue about the fourth component of the four potential vector which is

$$A_4 = \frac{i\phi}{c}. \tag{13.43}$$

Therefore, the four potential vector can be written as

$$A_\mu = [A_1, A_2, A_3, A_4], \tag{13.44}$$

where $A_4 = \frac{i\phi}{c}$.

Now, the Maxwell equation (13.39) takes the following form:

$$\square^2 A_4 = -\mu_0 J_4. \tag{13.45}$$

Hence the total Maxwell's equations with Lorentz-Gauge condition can be written in covariant forms as

$$\square^2 A_\mu = -\mu_0 J_\mu \tag{13.46}$$

$$\frac{\partial A_\mu}{\partial x_\mu} = 0. \tag{13.47}$$

13.10.3   Gauge Transformations

From Eq. (13.10), we can construct a new vector potential keeping unchanged the magnetic field as

$$\vec{A}^1 = \vec{A} + \nabla\psi, \tag{13.48}$$

where ψ is any arbitrary scalar quantity.

$$[\vec{B}^1 = \nabla \times \vec{A}^1 = \nabla \times (\vec{A} + \nabla\psi) = \nabla \times \vec{A} = \vec{B} \quad \text{as} \quad \nabla \times \nabla \psi = 0]$$

If this addition keeps the electron field given in Eq. (13.11) unchanged, then we have to reconstruct the scalar potential ϕ in terms of a new choice of scalar potential ϕ^1 as

$$\vec{E}^1 = -\left(\frac{\partial \vec{A}^1}{\partial t}\right) - \nabla\phi^1 = -\left(\frac{\partial(\vec{A} + \nabla\psi)}{\partial t}\right) - \nabla\phi^1$$

$$= -\left(\frac{\partial \vec{A}}{\partial t}\right) - \nabla\left(\frac{\partial \psi}{\partial t}\right) - \nabla\phi^1$$

$$= -\left(\frac{\partial \vec{A}}{\partial t}\right) - \nabla\left(\frac{\partial \psi}{\partial t} + \phi^1\right) = -\left(\frac{\partial \vec{A}}{\partial t}\right) - \nabla\phi = \vec{E}$$

provided

$$\phi = \frac{\partial \psi}{\partial t} + \phi^1,$$

i.e.

$$\phi^1 = \phi - \frac{\partial \psi}{\partial t}. \tag{13.49}$$

Hence, electric field strength $\vec{E}$ and magnetic field strength $\vec{B}$ remain invariant under the transformations given in (13.48) and (13.49). Therefore, the set of transformations (13.48) and (13.49) leaves Maxwell's equations unchanged and the transformations are called *gauge transformations*.

The physical laws which remain invariant under gauge transformations are called *gauge invariants*.

When one applies gauge transformations to Maxwell's equations, the gauge function ψ should not be taken arbitrarily; rather Lorentz-Gauge condition (13.40) imposes the restriction on ψ.

Lorentz-Gauge condition for gauge transformations is

$$\nabla \cdot \vec{A}^1 + \frac{1}{c^2}\frac{\partial \phi^1}{\partial t} = 0.$$

Replacing the values from (13.48) and (13.49), we have

$$\nabla \cdot (\vec{A} + \nabla \psi) + \frac{1}{c^2} \frac{\partial \left(\phi - \frac{\partial \psi}{\partial t}\right)}{\partial t} = 0.$$

This implies

$$\nabla \cdot \vec{A} + \nabla^2 \psi + \frac{1}{c^2} \left(\frac{\partial \phi}{\partial t}\right) - \frac{1}{c^2} \left(\frac{\partial^2 \psi}{\partial t^2}\right) = 0.$$

Using Lorentz-Gauge condition

$$\nabla \cdot \vec{A} + \frac{1}{c^2} \frac{\partial \phi}{\partial t} = 0,$$

we obtain

$$\nabla^2 \psi - \frac{1}{c^2} \left(\frac{\partial^2 \psi}{\partial t^2}\right) = \Box^2 \psi = 0.$$

Thus the solution of this equation can be used to construct the gauge transformation functions A^1 and ϕ^1 from the old A and ϕ.

13.11   Transformation of Four Potential Vector

The four potential vector can be written as

$$A_\mu = [A_1, A_2, A_3, A_4], \tag{13.50}$$

where $A_4 = \frac{i\phi}{c}$. The first three components belong to magnetic potential and the fourth corresponds to electric potential.

We know Lorentz transformation of any four vector A_μ ($\mu = 1, 2, 3, 4$) may be expressed as

$$A^1_\mu = a_{\mu\nu} A_\nu,$$

where $a_{\mu\nu}$ is given as

$$a_{\mu\nu} = \begin{bmatrix} a & 0 & 0 & ia\beta \\ 0 & 1 & 0 & 0 \\ 0 & 0 & 1 & 0 \\ -ia\beta & 0 & 0 & a \end{bmatrix}$$

$[\beta = v/c$ and $a = 1/\sqrt{1 - \beta^2}]$

Hence,

$$\begin{bmatrix} A_1^1 \\ A_2^1 \\ A_3^1 \\ A_4^1 \end{bmatrix} = \begin{bmatrix} a & 0 & 0 & ia\beta \\ 0 & 1 & 0 & 0 \\ 0 & 0 & 1 & 0 \\ -ia\beta & 0 & 0 & a \end{bmatrix} \begin{bmatrix} A_1 \\ A_2 \\ A_3 \\ A_4 \end{bmatrix}. \tag{13.51}$$

This implies

$$A_1^1 = aA_1 + ia\beta A_4 = aA_1 + ia\left(\frac{v}{c}\right)\left(\frac{i\phi}{c}\right),$$

i.e.

$$A_1^1 = a\left[A_1 - \frac{v}{c^2}\phi\right] \tag{13.52}$$

$$A_2^1 = A_2, \quad A_3^1 = A_3$$
$$A_4^1 = -ia\beta A_1 + aA_4 \tag{13.53}$$

or

$$\left(\frac{i\phi^1}{c}\right) = -ia\left(\frac{v}{c}\right)A_1 + a\left(\frac{i\phi}{c}\right),$$

i.e.

$$\phi^1 = a\left[\phi - vA_1\right]. \tag{13.54}$$

The inverse transformations are obtained by replacing v by $-v$ as

$$A_1 = a\left[A_1^1 + \frac{v}{c^2}\phi^1\right], \quad A_2 = A_2^1, \quad A_3 = A_3^1, \quad \phi = a\left[\phi^1 + vA_1^1\right]. \tag{13.55}$$

13.12 The Electromagnetic Field Tensor

The electric field strength $\overrightarrow{E}$ and magnetic field strength $\overrightarrow{B}$ in Maxwell's equations are defined in terms of electromagnetic potentials $\overrightarrow{A}$ and ϕ by

$$\overrightarrow{B} = \nabla \times \overrightarrow{A} \text{ and } \overrightarrow{E} = -\left(\frac{\partial \overrightarrow{A}}{\partial t}\right) - \nabla \phi$$

In view of these equations, we define a second rank anti-symmetric tensor $F_{\mu\nu}$ which is constructed from the (four-dimensional) curl of the four potential A_μ as

$$F_{\mu\nu} = \text{curl}(A_\mu) = \frac{\partial A_\nu}{\partial x_\mu} - \frac{\partial A_\mu}{\partial x_\nu}. \tag{13.56}$$

Here, $F_{\mu\nu} = -F_{\nu\mu}$ and $F_{\mu\mu} = 0$. Interestingly, one can note that the components of $F_{\mu\nu}$ have remarkable correspondence with the components of electric field strength $\vec{E}$ and magnetic field strength $\vec{B}$, e.g.

$$F_{12} = \frac{\partial A_2}{\partial x_1} - \frac{\partial A_1}{\partial x_2} = (\nabla \times \vec{A})_3 = B_3$$

$$(B_1 i + B_2 j + B_3 k) = \begin{bmatrix} i & j & k \\ \frac{\partial}{\partial x_1} & \frac{\partial}{\partial x_2} & \frac{\partial}{\partial x_3} \\ A_1 & A_2 & A_3 \end{bmatrix}$$

$$F_{14} = \frac{\partial A_4}{\partial x_1} - \frac{\partial A_1}{\partial x_4} = \frac{\partial \left(\frac{i\phi}{c}\right)}{\partial x_1} - \frac{\partial A_1}{\partial (ict)}$$

$$= \left(\frac{i}{c}\right)\frac{\partial \phi}{\partial x_1} - \left(\frac{1}{ic}\right)\frac{\partial A_1}{\partial t}$$

$$= -\left(\frac{i}{c}\right)\left[-\frac{\partial \phi}{\partial x_1} - \frac{\partial A_1}{\partial t}\right]$$

$$= -\left(\frac{i}{c}\right) E_1 = -\left(\frac{i E_1}{c}\right), \text{ etc.}$$

The explicit values of the components of the above tensor may be depicted in the following form:

$$F_{\mu\nu} = \begin{bmatrix} 0 & B_3 & -B_2 & -\frac{iE_1}{c} \\ -B_3 & 0 & B_1 & -\frac{iE_2}{c} \\ B_2 & -B_1 & 0 & -\frac{iE_3}{c} \\ \frac{iE_1}{c} & \frac{iE_2}{c} & \frac{iE_3}{c} & 0 \end{bmatrix}. \tag{13.57}$$

This second rank anti-symmetric tensor $F_{\mu\nu}$ is called *electromagnetic field strength tensor*.

Again, we notice that the choice of the vector A_μ given in (13.10) may not be unique. One can choose the new one as

$$A_\mu^* = A_\mu + \nabla\psi, \quad (\psi \text{ is a scalar function}). \tag{13.58}$$

It is easy to verify that the replacement of A_μ by A_μ^* does not change the expression given in (13.10). However, Lorentz-Gauge condition $\frac{\partial A_\mu}{\partial x_\mu} = 0$ restricts the arbitrary choice of ψ. The function ψ should satisfy the

$$\nabla^2 \psi - \frac{1}{c^2}\left(\frac{\partial^2 \psi}{\partial t^2}\right) = \Box^2 \psi = 0. \tag{13.59}$$

Maxwell's equations given in (13.4)–(13.7) can be expressed in terms of electromagnetic field strength tensor $F_{\mu\nu}$ as

$$\frac{\partial F_{\mu\nu}}{\partial x_\nu} = \mu_0 J_\mu \tag{13.60}$$

$$\frac{\partial F_{\mu\nu}}{\partial x_\lambda} + \frac{\partial F_{\nu\lambda}}{\partial x_\mu} + \frac{\partial F_{\lambda\mu}}{\partial x_\nu} = 0. \tag{13.61}$$

The first and fourth equations in a set of Maxwell's equations can be expressed by Eq. (13.60). For example, when $\mu = 1$, Eq. (13.60) gives

$$\frac{\partial F_{12}}{\partial x_2} + \frac{\partial F_{13}}{\partial x_3} + \frac{\partial F_{14}}{\partial x_4} = \mu_0 J_1$$

or

$$\frac{\partial B_3}{\partial x_2} - \frac{\partial B_2}{\partial x_3} - \left(\frac{i}{c}\right)\left(\frac{1}{ic}\right)\frac{\partial E_1}{\partial t} = \mu_0 J_1$$

or

$$\frac{\partial B_3}{\partial x_2} - \frac{\partial B_2}{\partial x_3} - \left(\frac{1}{c^2}\right)\frac{\partial E_1}{\partial t} = \mu_0 J_1$$

or

$$\frac{\partial B_3}{\partial x_2} - \frac{\partial B_2}{\partial x_3} - \mu_0\varepsilon_0\frac{\partial E_1}{\partial t} = \mu_0 J_1$$

or

$$(\nabla \times \vec{B})_1 - \mu_0\varepsilon_0\frac{\partial E_1}{\partial t} = \mu_0 J_1, \text{ etc.}$$

When $\mu = 4$, Eq. (13.60) gives

$$\frac{\partial F_{42}}{\partial x_2} + \frac{\partial F_{43}}{\partial x_3} + \frac{\partial F_{41}}{\partial x_1} = \mu_0 J_4$$

or

$$\frac{\partial \left(\frac{iE_2}{c}\right)}{\partial x_2} + \frac{\partial \left(\frac{iE_3}{c}\right)}{\partial x_3} + \frac{\partial \left(\frac{iE_1}{c}\right)}{\partial x_1} = \mu_0 i c \rho$$

or

$$\frac{\partial E_2}{\partial x_2} + \frac{\partial E_3}{\partial x_3} + \frac{\partial E_1}{\partial x_1} = \frac{\rho}{\varepsilon_0} \left(\text{as } \varepsilon_0 \mu_0 = \frac{1}{c^2} \right)$$

or

$$\nabla \cdot \vec{E} = \frac{\rho}{\varepsilon_0}$$

The second and third equations in a set of Maxwell's equations can be expressed by Eq. (13.61). For example, if μ, ν and λ assume the values from any combination of (1, 2, 3), we always get from Eq. (13.61)

$$\frac{\partial F_{12}}{\partial x_3} + \frac{\partial F_{23}}{\partial x_1} + \frac{\partial F_{31}}{\partial x_2} = 0$$

or

$$\frac{\partial B_3}{\partial x_3} + \frac{\partial B_1}{\partial x_1} + \frac{\partial B_2}{\partial x_2} = 0 \implies \nabla \cdot \vec{B} = 0$$

Again, if we take $\mu = 2$, $\nu = 4$ and $\lambda = 1$, then Eq. (13.61) gives

$$\frac{\partial F_{24}}{\partial x_1} + \frac{\partial F_{41}}{\partial x_2} + \frac{\partial F_{12}}{\partial x_4} = 0$$

or

$$-\frac{\partial \left(\frac{iE_2}{c}\right)}{\partial x_1} + \frac{\partial \left(\frac{iE_1}{c}\right)}{\partial x_2} + \frac{\partial B_3}{\partial (ict)} = 0$$

or

$$(\nabla \times \vec{E})_3 = -\frac{\partial B_3}{\partial t}, \text{ etc.}$$

Equation (13.61) can also be written in terms of dual electromagnetic field tensor $^*F_{\mu\nu}$ as

$$\frac{\partial ^*F_{\mu\nu}}{\partial x_\nu} = 0.$$

Here, dual electromagnetic field tensor $^*F_{\mu\nu}$ is defined as

$$^*F_{\mu\nu} = \frac{1}{2}\epsilon_{\mu\nu\alpha\beta}F_{\alpha\beta},$$

where $\epsilon_{\mu\nu\alpha\beta}$ is Levi-Civita tensor defined as

$$\epsilon_{\mu\nu\alpha\beta} = 0, \quad \text{if any two indices are equal}$$
$$= +1 \quad \text{if } \mu\nu\,\alpha\beta \text{ is an even permutation of } 0, 1, 2, 3$$
$$= -1 \quad \text{if } \mu\nu\,\alpha\beta \text{ is an odd permutation of } 0, 1, 2, 3.$$

The components of $^*F_{\mu\nu}$ are given by

$$^*F_{\mu\nu} = \begin{bmatrix} 0 & -\frac{iE_3}{c} & -\frac{iE_2}{c} & B_1 \\ \frac{iE_3}{c} & 0 & -\frac{iE_1}{c} & B_2 \\ -\frac{iE_2}{c} & \frac{iE_1}{c} & 0 & B_3 \\ -B_1 & -B_2 & -B_3 & 0 \end{bmatrix}.$$

Maxwell's equations given in (13.60) and (13.61) are in tensor form. We know a tensor equation must be the same in all coordinate systems, i.e. invariant. Therefore, the Maxwell equations are invariant under Lorentz transformation. Thus, it is self-evident the covariance of Maxwell's equations under Lorentz transformation.

13.13 Lorentz Transformation of Electromagnetic Fields

The components in the electromagnetic field strength tensor $F_{\mu\nu}$ are electric and magnetic fields $\vec{E}$ and $\vec{B}$. Here $F_{\mu\nu}$ is a second rank anti-symmetric tensor. Therefore, one can use the Lorentz transformation matrix $a_{\mu\nu}$ to get the required Lorentz transformation of the electromagnetic field strength tensor $F_{\mu\nu}$ as

$$F_{\mu\nu}^1 = a_{\mu\alpha}a_{\nu\beta}F_{\alpha\beta}, \tag{13.62}$$

where

$$a_{\mu\nu} = \begin{bmatrix} a & 0 & 0 & ia\beta \\ 0 & 1 & 0 & 0 \\ 0 & 0 & 1 & 0 \\ -ia\beta & 0 & 0 & a \end{bmatrix}$$

$[\beta = v/c \text{ and } a = 1/\sqrt{1 - \beta^2}]$

Using the above transformation equations, one can get the transformation of electric and magnetic field components as

$$E_1^1 = E_1, \quad E_2^1 = a[E_2 - c\beta B_3], \quad E_3^1 = a[E_3 + c\beta B_2] \qquad (13.63)$$

$$B_1^1 = B_1, \quad B_2^1 = a\left[B_2 + \left(\frac{\beta}{c}\right)E_3\right], \quad B_3^1 = a\left[B_3 - \left(\frac{\beta}{c}\right)E_2\right]. \qquad (13.64)$$

The above relations can be obtained from (13.62) as follows:

[1] *For magnetic field components:*

Let us choose $\mu = 3$ and $\nu = 1$ because $F_{31} = B_2$. Writing explicitly the components of (13.62), we get

$$F_{31}^1 = a_{3\alpha}a_{1\beta}F_{\alpha\beta} = a_{33}[a_{11}F_{31} + a_{14}F_{34}]$$

or

$$B_2^1 = aB_2 + ia\beta\left(\frac{-iE_3}{c}\right) = a\left[B_2 + \left(\frac{\beta}{c}\right)E_3\right].$$

Now we choose $\mu = 2$ and $\nu = 3$ because $F_{23} = B_1$. Equation (13.62) implies

$$F_{23}^1 = a_{2\alpha}a_{3\beta}F_{\alpha\beta} = a_{22}a_{33}F_{23} \implies B_1^1 = B_1$$

The choice $\mu = 1$ and $\nu = 2$ gives $F_{12} = B_3$. Hence, (13.62) yields

$$F_{12}^1 = a_{1\alpha}a_{2\beta}F_{\alpha\beta} = a_{11}a_{22}F_{12} + a_{14}a_{22}F_{42}$$

or

$$B_3^1 = aB_3 + ia\beta\left(\frac{iE_2}{c}\right) = a\left[B_3 - \left(\frac{\beta}{c}\right)E_2\right].$$

[2] *For electric field components:*

$$F_{24}^1 = a_{2\alpha}a_{4\beta}F_{\alpha\beta} = a_{22}a_{44}F_{24} + a_{22}a_{41}F_{21}$$

or

$$-\left(\frac{E_2^1}{c}\right) = a - \left(\frac{E_2}{c}\right) + ia\beta B_3$$

$$\implies E_2^1 = a(E_2 - c\beta B_3)$$

$$F_{14}^1 = a_{1\alpha}a_{4\beta}F_{\alpha\beta} = a_{11}a_{44}F_{14} + a_{14}a_{41}F_{41}$$

$$\implies E_1^1 = E_1$$

$$F_{34}^1 = a_{3\alpha}a_{4\beta}F_{\alpha\beta} = a_{33}a_{44}F_{34} + a_{33}a_{41}F_{31}$$

$$\implies E_3^1 = a(E_3 + c\beta B_2).$$

Alternative

The electric field strength $\vec{E}$ and magnetic field strength $\vec{B}$ in Maxwell's equations are defined in terms of electromagnetic potentials $\vec{A}$ and ϕ by

$$\vec{B} = \nabla \times \vec{A} \text{ and } \vec{E} = -\left(\frac{\partial \vec{A}}{\partial t}\right) - \nabla \phi.$$

We know the transformation of electromagnetic potentials $\vec{A}$ and ϕ

$$A_1^1 = a\left[A_1 - \frac{v}{c^2}\phi\right], \quad A_2^2 = A_2, \quad A_3^1 = A_3, \quad \phi^1 = a\left[\phi - vA_1\right]$$

$[\beta = v/c \text{ and } a = 1/\sqrt{1-\beta^2}]$
We also know

$$\frac{\partial}{\partial x_1^1} = a\left(\frac{\partial}{\partial x_1} + \frac{v}{c^2}\frac{\partial}{\partial t}\right), \quad \frac{\partial}{\partial x_2^1} = \frac{\partial}{\partial x_2}, \quad \frac{\partial}{\partial x_3^1} = \frac{\partial}{\partial x_3}, \quad \frac{\partial}{\partial t^1} = a\left(v\frac{\partial}{\partial x_1} + \frac{\partial}{\partial t}\right).$$

Now the Lorentz transformations of electric field strength $\vec{E}$ and magnetic field strength $\vec{B}$ will be

$$B^1 = \nabla A^1 \tag{13.65}$$

$$\vec{E}^1 = -\left(\frac{\partial \vec{A}^1}{\partial t}\right) - \nabla \phi^1. \tag{13.66}$$

In terms of components, the transformations of magnetic field strength are written as

$$iB_1^1 + jB_2^1 + kB_3^1 = i\left(\frac{\partial A_3^1}{\partial x_2^1} - \frac{\partial A_2^1}{\partial x_3^1}\right) + j\left(\frac{\partial A_1^1}{\partial x_3^1} - \frac{\partial A_3^1}{\partial x_1^1}\right) + k\left(\frac{\partial A_2^1}{\partial x_1^1} - \frac{\partial A_1^1}{\partial x_2^1}\right).$$

Now,

$$B_1^1 = \left(\frac{\partial A_3^1}{\partial x_2^1} - \frac{\partial A_2^1}{\partial x_3^1}\right) = \left(\frac{\partial A_3}{\partial x_2} - \frac{\partial A_2}{\partial x_3}\right) = B_1$$

$$B_2^1 = \left(\frac{\partial A_1^1}{\partial x_3^1} - \frac{\partial A_3^1}{\partial x_1^1}\right) = \frac{\partial\left[a\left(A_1 - \frac{v}{c^2}\phi\right)\right]}{\partial x_3} - a\left(\frac{\partial}{\partial x_1} + \frac{v}{c^2}\frac{\partial}{\partial t}\right)A_3$$

$$\implies B_2^1 = a\left[B_2 + \left(\frac{\beta}{c}\right)E_3\right]$$

$$B_3^1 = \left(\frac{\partial A_2^1}{\partial x_1^1} - \frac{\partial A_1^1}{\partial x_2^1}\right) = a\left(\frac{\partial}{\partial x_1} + \frac{v}{c^2}\frac{\partial}{\partial t}\right)A_2 - \frac{\partial\left[a\left(A_1 - \frac{v}{c^2}\phi\right)\right]}{\partial x_2}$$

$$\implies B_3^1 = a\left[B_3 - \left(\frac{\beta}{c}\right)E_2\right].$$

Similarly, the transformation components of electric field strength are given by

$$i E_1^1 + j E_2^1 + k E_3^1 = i \left(\frac{\partial A_1^1}{\partial t^1} - \frac{\partial \phi^1}{\partial x_1^1} \right) + j \left(\frac{\partial A_2^1}{\partial t^1} - \frac{\partial \phi^1}{\partial x_2^1} \right) + k \left(\frac{\partial A_3^1}{\partial t^1} - \frac{\partial \phi^1}{\partial x_3^1} \right).$$

Now,

$$
\begin{aligned}
E_1^1 &= \left(\frac{\partial A_1^1}{\partial t^1} - \frac{\partial \phi^1}{\partial x_1^1} \right) = a \left(v \frac{\partial}{\partial x_1} + \frac{\partial}{\partial t} \right) \left[a \left(A_1 - \frac{v}{c^2} \phi \right) \right] \\
&\quad - a \left(\frac{\partial}{\partial x_1} + \frac{v}{c^2} \frac{\partial}{\partial t} \right) \left[a \left(\phi - v A_1 \right) \right] \\
\Longrightarrow E_1^1 &= E_1 \\
E_2^1 &= \left(\frac{\partial A_2^1}{\partial t^1} - \frac{\partial \phi^1}{\partial x_2^1} \right) = a \left(v \frac{\partial}{\partial x_1} + \frac{\partial}{\partial t} \right) A_2 - \frac{\partial \left[a \left(\phi - v A_1 \right) \right]}{\partial x_2} \\
\Longrightarrow E_2^1 &= a(E_2 - v B_3) \\
E_3^1 &= \left(\frac{\partial A_3^1}{\partial t^1} - \frac{\partial \phi^1}{\partial x_3^1} \right) = a \left(v \frac{\partial}{\partial x_1} + \frac{\partial}{\partial t} \right) A_3 - \frac{\partial \left[a \left(\phi - v A_1 \right) \right]}{\partial x_3} \\
\Longrightarrow E_3^1 &= a(E_3 + v B_2).
\end{aligned}
$$

The inverse transformation of electric and magnetic field components can be obtained by replacing v by $-v$ and interchanging primes and unprimes.

$$E_1 = E_1^1, \quad E_2 = a[E_2^1 + c\beta B_3^1], \quad E_3 = a[E_3^1 - c\beta B_2^1] \tag{13.67}$$

$$B_1 = B_1^1, \quad B_2 = a \left[B_2^1 - \left(\frac{\beta}{c} \right) E_3^1 \right], \quad B_3 = a \left[B_3^1 + \left(\frac{\beta}{c} \right) E_2^1 \right]. \tag{13.68}$$

It is seen that electromagnetic field components are unaffected in the direction of motion; however, transverse components are modified.

Here, the transformations of electric and magnetic field components can be written in terms of the components of the fields parallel ($_\parallel$) and perpendicular ($_\perp$) to the direction of the relative velocity between S and S^1.

$$E_\parallel^1 = E_\parallel, \quad E_\perp^1 = a \left[E_\perp + (\vec{v} \times \vec{B}_\perp) \right] \tag{13.69}$$

$$B_\parallel^1 = B_\parallel, \quad B_\perp^1 = a \left[B_\perp - \frac{1}{c^2} (\vec{v} \times \vec{E}_\perp) \right]. \tag{13.70}$$

However, one can write the above transformation equations for electric and magnetic field components in more compact forms as

$$\vec{B}^{\,1} = a\left[\vec{B} - \frac{1}{c^2}(\vec{v} \times \vec{E}) + \frac{(\vec{v} \cdot \vec{B})\vec{v}}{v^2}\left(\frac{1}{a} - 1\right)\right] \tag{13.71}$$

$$\vec{E}^{\,1} = a\left[\vec{E} + \frac{1}{c^2}(\vec{v} \times \vec{B}) + \frac{(\vec{v} \cdot \vec{E})\vec{v}}{v^2}\left(\frac{1}{a} - 1\right)\right]. \tag{13.72}$$

If one substitutes $\vec{v} = (v, 0, 0)$ in Eqs. (13.71) and (13.72), then Eqs. (13.69) and (13.70) will recover.

It is interesting to note that a purely electric field or a purely magnetic field in one frame of reference will appear to be a combination of electric and magnetic fields in another frame of reference except the components in the direction of relative motion between the frames. Thus, if one of these fields is absent in one frame of reference, then that field may appear in another reference frame. For example, suppose in S frame there is no magnetic field. The transformation equation (13.71) implies that in S^1 frame, magnetic field exists which is given as $\vec{B}^{\,1} = -\frac{1}{c^2}(\vec{v} \times \vec{E})$.

13.14 Maxwell's Equations Are Invariant Under Lorentz Transformations

Maxwell's equations can be expressed in the covariant form with the help of electromagnetic four potentials A_μ and current four potentials J_μ along with the Lorentz-Gauge condition as

$$\Box^2 A_\mu = -\mu_0 J_\mu, \quad \frac{\partial A_\mu}{\partial x_\mu} = 0.$$

Now, to prove the invariance of Maxwell's equations under Lorentz transformations, we will have to show that the above equations must retain the same form in S^1 frame. Thus Maxwell's equations will be invariant if

$$\Box^{1\,2} A_\mu^1 = -\mu_0 J_\mu^1, \quad \frac{\partial A_\mu^1}{\partial x_\mu^1} = 0.$$

We know that d'Alembertian operator is invariant under Lorentz transformations, i.e. $\Box^{1\,2} = \Box^2$.

(i) $\mu = 1$:

$$\Box'^2 A_1^1 = \Box^2 A_1^1$$

$$= \Box^2 \left[a \left(A_1 - \frac{v}{c^2} \phi \right) \right]$$

$$= a \Box^2 A_1 - a \frac{v}{c^2} \Box^2 \phi$$

$$= -a \mu_0 J_1 - a \frac{v}{c^2} \left(-\frac{\rho}{\varepsilon_0} \right)$$

$$= -a \mu_0 J_1 + a \rho \mu_0 v$$

$$= -a \mu_0 (J_1 - v \rho) = -\mu_0 J_1^1.$$

Thus

$$\Box'^2 A_1^1 = -\mu_0 J_1^1.$$

(ii) $\mu = 2$:

$$\Box'^2 A_2^1 = \Box^2 A_2^1$$

$$= \Box^2 A_2 = -\mu_0 J_2^1.$$

Hence

$$\Box'^2 A_2^1 = -\mu_0 J_2^1, \quad \text{as } J_2^1 = J_2.$$

(iii) $\mu = 3$:

$$\Box'^2 A_3^1 = \Box^2 A_3^1$$

$$= \Box^2 A_3 = -\mu_0 J_3^1.$$

That is

$$\Box'^2 A_3^1 = -\mu_0 J_3^1, \quad \text{as } J_3^1 = J_3.$$

(iv) $\mu = 4$:

$$\Box'^2 A_4^1 = \Box^2 A_4^1$$

$$= \Box^2 \left[a \left(A_4 - i \beta A_1 \right) \right]$$

$$= a \Box^2 A_4 - i \beta \Box^2 A_1$$

$$= -a \mu_0 J_4 - a i \beta (-\mu_0 J_1)$$

$$= -a \mu_0 (J_4 - i \beta J_1) = -\mu_0 J_4^1.$$

Therefore

$$\Box'^2 A_4^1 = -\mu_0 J_4^1.$$

The cases (i)–(iv) indicate that

$$\Box'^2 A_\mu^1 = -\mu_0 J_\mu^1.$$

To show the invariance of Lorentz-Gauge condition $\frac{\partial A_\mu^1}{\partial x_\mu^1} = 0$, we use the transformation of A_μ^1 which is given by

$$A_\mu^1 = a_{\mu\nu} A_\nu.$$

This implies

$$A_\mu^1 = \frac{\partial x_\mu^1}{\partial x_\nu} A_\nu$$

or

$$\frac{\partial A_\mu^1}{\partial x_\mu^1} = \frac{\partial}{\partial x_\mu^1} \frac{\partial x_\mu^1}{\partial x_\nu} A_\nu$$

or

$$\frac{\partial A_\mu^1}{\partial x_\mu^1} = \frac{\partial A_\nu}{\partial x_\nu}.$$

We know Lorentz-Gauge condition $\frac{\partial A_\nu}{\partial x_\nu} = 0$, hence

$$\frac{\partial A_\mu^1}{\partial x_\mu^1} = 0.$$

Thus, Maxwell's equations are invariant under Lorentz transformations.

Alternative

We know Maxwell's equations are given by

$$\frac{\partial F^{\mu\nu}}{\partial x_\nu} = \mu_0 J^\mu, \quad \frac{\partial F_{\mu\nu}}{\partial x_\lambda} + \frac{\partial F_{\nu\lambda}}{\partial x_\mu} + \frac{\partial F_{\lambda\mu}}{\partial x_\nu} = 0.$$

Therefore to prove its invariance, we will have to show

$$\frac{\partial F^{1\mu\nu}}{\partial x_\nu^1} = \mu_0 J^{1\mu}, \quad \frac{\partial F_{\mu\nu}^1}{\partial x_\lambda^1} + \frac{\partial F_{\nu\lambda}^1}{\partial x_\mu^1} + \frac{\partial F_{\lambda\mu}^1}{\partial x_\nu^1} = 0.$$

Note that we have used the first equation in contravariant form of electromagnetic field tensor. This tensor transforms according to the rule given as

$$F'^{\mu\nu} = \frac{\partial x'_\mu}{\partial x_\beta} \frac{\partial x'_\nu}{\partial x_\sigma} F^{\beta\sigma}.$$

Now, we have

$$
\begin{aligned}
\frac{\partial F'^{\mu\nu}}{\partial x'_\nu} &= \frac{\partial}{\partial x'_\nu} \left[\frac{\partial x'_\mu}{\partial x_\beta} \frac{\partial x'_\nu}{\partial x_\sigma} F^{\beta\sigma} \right] \\
&= \frac{\partial x_\alpha}{\partial x'_\nu} \frac{\partial}{\partial x_\alpha} \left[\frac{\partial x'_\mu}{\partial x_\beta} \frac{\partial x'_\nu}{\partial x_\sigma} F^{\beta\sigma} \right] \\
&= \frac{\partial x_\alpha}{\partial x'_\nu} \frac{\partial x'_\mu}{\partial x_\beta} \frac{\partial x'_\nu}{\partial x_\sigma} \frac{\partial}{\partial x_\alpha} \left[F^{\beta\sigma} \right] \\
&= \frac{\partial x'_\mu}{\partial x_\beta} \delta^\alpha_\sigma \frac{\partial}{\partial x_\alpha} \left[F^{\beta\sigma} \right] \\
&= \frac{\partial x'_\mu}{\partial x_\beta} \frac{\partial}{\partial x_\alpha} \left[F^{\beta\alpha} \right] \\
&= \frac{\partial x'_\mu}{\partial x_\beta} \mu_0 J^\beta.
\end{aligned}
$$

Therefore, we get

$$\frac{\partial F'^{\mu\nu}}{\partial x'_\nu} = \mu_0 J'^\mu.$$

For the second Maxwell equation, we have used the covariant form of electromagnetic field tensor. This tensor transforms according to the following rule:

$$F'_{\mu\nu} = \frac{\partial x_\beta}{\partial x'_\mu} \frac{\partial x_\lambda}{\partial x'_\nu} F_{\beta\lambda}.$$

Now, we have

$$
\begin{aligned}
\frac{\partial F'_{\mu\nu}}{\partial x'_\sigma} &= \frac{\partial}{\partial x'_\sigma} \left[\frac{\partial x_\beta}{\partial x'_\mu} \frac{\partial x_\lambda}{\partial x'_\nu} F_{\beta\lambda} \right] \\
&= \frac{\partial x_\alpha}{\partial x'_\sigma} \frac{\partial}{\partial x_\alpha} \left[\frac{\partial x_\beta}{\partial x'_\mu} \frac{\partial x_\lambda}{\partial x'_\nu} F_{\beta\lambda} \right].
\end{aligned}
$$

Thus

$$\frac{\partial F^1{}_{\mu\nu}}{\partial x^1_\sigma} = \frac{\partial x_\alpha}{\partial x^1_\sigma}\frac{\partial x_\beta}{\partial x^1_\mu}\frac{\partial x_\lambda}{\partial x^1_\nu}\frac{\partial}{\partial x_\alpha}\left[F_{\beta\lambda}\right].$$

Similarly,

$$\frac{\partial F^1{}_{\nu\sigma}}{\partial x^1_\mu} = \frac{\partial x_\beta}{\partial x^1_\mu}\frac{\partial x_\lambda}{\partial x^1_\nu}\frac{\partial x_\alpha}{\partial x^1_\sigma}\frac{\partial}{\partial x_\beta}\left[F_{\lambda\alpha}\right]$$

$$\frac{\partial F^1{}_{\sigma\mu}}{\partial x^1_\nu} = \frac{\partial x_\lambda}{\partial x^1_\nu}\frac{\partial x_\alpha}{\partial x^1_\sigma}\frac{\partial x_\beta}{\partial x^1_\mu}\frac{\partial}{\partial x_\lambda}\left[F_{\alpha\beta}\right].$$

After adding these three results, we get

$$\frac{\partial F^1_{\mu\nu}}{\partial x^1_\sigma} + \frac{\partial F^1_{\nu\sigma}}{\partial x^1_\mu} + \frac{\partial F^1_{\sigma\mu}}{\partial x^1_\nu} = \frac{\partial x_\alpha}{\partial x^1_\sigma}\frac{\partial x_\beta}{\partial x^1_\mu}\frac{\partial x_\lambda}{\partial x^1_\nu}\left[\frac{\partial F_{\beta\lambda}}{\partial x_\alpha} + \frac{\partial F_{\lambda\alpha}}{\partial x_\beta} + \frac{\partial F_{\alpha\beta}}{\partial x_\lambda}\right] = 0.$$

13.15 Lorentz Force on a Charged Particle

Now we determine the electromagnetic forces experienced by charged particle moving in an electromagnetic field. Let us consider two frames S and S^1, S^1 is moving with velocity v relative to S along x-direction. Let a charged particle of charge q be rest in system S^1. This means the charged particle is moving with velocity $\vec{v} = v\vec{i} = (v, 0, 0)$ relative to system S. Since the particle is at rest in system S^1, no magnetic field exists in S^1, i.e. field is purely electrostatic. In other words,

$$B^1_x = B^1_y = B^1_z = 0. \tag{13.73}$$

But the particle is moving relative to S, therefore to an observer in S frame the purely electric field appears both as an electric and a magnetic field. We know the Lorentz transformation formulae of force components as

$$F^1_x = F_x, \quad F^1_y = \frac{F_y}{\sqrt{1-\beta^2}}, \quad F^1_z = \frac{F_z}{\sqrt{1-\beta^2}} \tag{13.74}$$

[see the results given in Eq. (12.27)]

Note that if $\vec{F}$ is the electric field measured in system S due to charge stationary in it and $\vec{F}^1$ is the electric field measured in system S^1, moving with velocity v relative to S in any direction, then resolving the electric field along and transverse to velocity v is

$$F^1_\parallel = F_\parallel \quad \text{and} \quad F^1_\perp = \frac{F_\perp}{\sqrt{1-\beta^2}},$$

where $\parallel$ and $\perp$ are along and perpendicular to the direction of velocity.

The force F^1 in S^1 system due to the electric field strength $\vec{E}^1$ is $\vec{F}^1 = q\vec{E}^1$ since q is invariant under Lorentz transformation. Writing explicitly we get

$$F_x^1 = qE_x^1, \quad F_y^1 = qE_y^1, \quad F_z^1 = qE_z^1 \tag{13.75}$$

Using (13.74), we get

$$F_x = F_x^1 = qE_x^1, \quad F_y = F_y^1\sqrt{1 - \beta^2} = qE_y^1\sqrt{1 - \beta^2},$$

$$F_z = F_z^1\sqrt{1 - \beta^2} = qE_z^1\sqrt{1 - \beta^2}. \tag{13.76}$$

However, we know Lorentz transformations of electric field components as

$$E_x^1 = E_x, \quad E_y^1 = \frac{E_y - vB_z}{\sqrt{1 - \beta^2}}, \quad E_z^1 = \frac{E_z + vB_y}{\sqrt{1 - \beta^2}}. \tag{13.77}$$

Using (13.77), components of force from (13.76) are

$$F_x = qE_x, \quad F_y = q(E_y - vB_z), \quad F_z = q(E_z + vB_y). \tag{13.78}$$

Therefore, we obtain

$$\vec{F} = F_x\vec{i} + F_y\vec{j} + F_z\vec{k} = qE_x\vec{i} + q(E_y - vB_z)\vec{j} + q(E_z + vB_y)\vec{k}$$

or

$$\vec{F} = q\vec{E} + qv(B_y\vec{k} - B_z\vec{j})$$
$$= q\vec{E} + q[v\vec{i} \times (B_x\vec{i} + B_y\vec{j} + B_z\vec{k})]$$
$$= q\vec{E} + q(\vec{v} \times \vec{B}),$$

i.e.

$$\vec{F} = q[\vec{E} + \vec{v} \times \vec{B}]. \tag{13.79}$$

The force experienced by charged particle moving in an electromagnetic field is known as *Lorentz force* on a particle of charge q. Note that this Lorentz force is a direct consequence of the relativity principle.

Lorentz Force in Covariant Form

We have seen that the equation of motion of a particle of rest mass m_0 and charge q in an electromagnetic field ($\vec{E}$ and $\vec{B}$) is given by Lorentz force equation as

$$\vec{F} = q[\vec{E} + \vec{v} \times \vec{B}], \quad \text{where } \vec{v} \text{ is the velocity of the charge.}$$

Let the force experienced per unit volume of charge density ρ is $\vec{f}$. Then the above expression gives

$$\vec{f} = \rho[\vec{E} + \vec{v} \times \vec{B}].$$

This implies

$$\vec{f} = \rho\vec{E} + \rho\vec{v} \times \vec{B} = \rho\vec{E} + \vec{J} \times \vec{B}. \tag{13.80}$$

Writing explicitly this equation, we get

$$f_1 = \rho E_1 + J_2 B_3 - J_3 B_2, \quad f_2 = \rho E_2 + J_3 B_1 - J_1 B_3, \quad f_3 = \rho E_3 + J_1 B_2 - J_2 B_1. \tag{13.81}$$

We know electromagnetic field strength tensor $F_{\mu\nu}$ can be expressed in terms of the components of electric field strength $\vec{E}$ and magnetic field strength $\vec{E}$ as

$$F_{\mu\nu} = \begin{bmatrix} 0 & B_3 & -B_2 & -\frac{iE_1}{c} \\ -B_3 & 0 & B_1 & -\frac{iE_2}{c} \\ B_2 & -B_1 & 0 & -\frac{iE_3}{c} \\ \frac{iE_1}{c} & \frac{iE_2}{c} & \frac{iE_3}{c} & 0 \end{bmatrix}.$$

Thus, Eq. (13.81) can be written in terms of the components of electromagnetic field strength tensor $F_{\mu}\nu$ as

$$f_1 = F_{11} J_1 + F_{12} J_2 + F_{13} J_3 + F_{14} J_4$$
$$f_2 = F_{21} J_1 + F_{22} J_2 + F_{23} J_3 + F_{14} J_4$$
$$f_3 = F_{31} J_1 + F_{32} J_2 + F_{33} J_3 + F_{34} J_4.$$

Here J_μ is the current four vector with the fourth component $J_4 = ic\rho$. Note that the right-hand sides of the above equations are the components of four vectors of electromagnetic field and current density. Therefore, it is expected that the above equations could be written in tensor form as

$$f_\mu = m_0 \frac{du_\mu}{d\tau} = F_{\mu\nu} J_\nu. \tag{13.82}$$

Thus, we have the expression of *four-dimensional Lorentz force* or the Lorentz force in covariant form. Here the first three components in (13.82) represent the ordinary three-dimensional Lorentz force and the fourth component represents the rate of work done on the particle.

Here the fourth component is

$$f_4 = F_{41} J_1 + F_{42} J_2 + F_{43} J_3 + F_{44} J_4$$
$$= \frac{i E_1}{c} J_1 + \frac{i E_2}{c} J_2 + \frac{i E_3}{c} J_3 + 0 = \frac{i}{c} \vec{E} \cdot \vec{J} = \frac{i \rho}{c} \vec{E} \cdot \vec{v}.$$

Thus, the fourth component represents the amount of work done by the electric field.

The expression given in (13.82) can also be written using Maxwell's equation in the following form:

$$f_\mu = \frac{1}{\mu_0} F_{\mu\nu} \frac{\partial F_{\nu\lambda}}{\partial x_\lambda} \quad \text{as} \quad \frac{\partial F_{\nu\lambda}}{\partial x_\lambda} = \mu_0 J_\nu. \tag{13.83}$$

13.16   Electromagnetic Field Produced by a Moving Charge

Consider a particle of charge q moving with uniform velocity v in S system. We choose another system S^1 moving with velocity v relative to S system in which this charge is at rest at the origin (see Fig. 13.1). Since the charge is at rest in S^1, it produces only electric field $\vec{E}^1$ in S^1. Let P be any point whose position vector is $\vec{r}^1$ relative to S^1. Therefore, from Coulomb's law, the electric field at P in S^1 is given as

$$\vec{E}^1 = \frac{q \vec{r}^1}{4 \pi \epsilon_0 r^{13}} \quad \text{here,} \quad \vec{B}^1 = 0 \quad r^1 = \left| \vec{r}^1 \right|. \tag{13.84}$$

Writing explicitly, we have

$$E_x^1 = \frac{q x^1}{4 \pi \epsilon_0 r^{13}}, \quad E_y^1 = \frac{q y^1}{4 \pi \epsilon_0 r^{13}}, \quad E_z^1 = \frac{q z^1}{4 \pi \epsilon_0 r^{13}} \tag{13.85}$$

$$B_x^1 = 0, \quad B_y^1 = 0, \quad B_z^1 = 0. \tag{13.86}$$

Fig. 13.1 Charge q and reference frame S^1 are moving with velocity v along x-direction

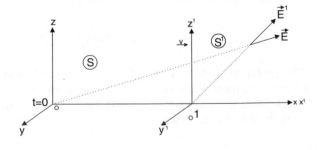

The reverse transformation relations give the components of electric and magnetic fields ($\vec{E}$ and $\vec{B}$) at point P in S which are given as

$$E_x = E_x^1, \quad E_y = \frac{E_y^1 + vB_z^1}{\sqrt{1 - \beta^2}} = \frac{E_y^1}{\sqrt{1 - \beta^2}}, \quad E_z = \frac{E_z^1 - vB_y^1}{\sqrt{1 - \beta^2}} = \frac{E_z^1}{\sqrt{1 - \beta^2}}$$

(13.87)

$$B_x = B_x^1 = 0, \quad B_y = \frac{B_y^1 - \frac{v}{c^2}E_z^1}{\sqrt{1 - \beta^2}} = -\frac{\frac{v}{c^2}E_z^1}{\sqrt{1 - \beta^2}}, \quad B_z = \frac{B_z^1 - \frac{v}{c^2}E_y^1}{\sqrt{1 - \beta^2}} = \frac{\frac{v}{c^2}E_y^1}{\sqrt{1 - \beta^2}}.$$

(13.88)

Let position vector of P be $\vec{r}$ relative to S. Then the components of the radius vector $\vec{r}$ are related to the components of the radius vector $\vec{r}^1$ as

$$x^1 = \frac{x - vt}{\sqrt{1 - \beta^2}}; \quad y^1 = y; \quad z^1 = z; \quad t^1 = \frac{t - \frac{vx}{c^2}}{\sqrt{1 - \beta^2}}.$$

At time $t = 0$, when q is at origin O of S, then the above equations give

$$x^1 = \frac{x}{\sqrt{1 - \beta^2}}; \quad y^1 = y; \quad z^1 = z; \quad \text{and} \quad t^1 = -\frac{\frac{vx}{c^2}}{\sqrt{1 - \beta^2}}.$$

Let θ be the angle made by $\vec{r}$ with x-axis, then

$$x = r\cos\theta \quad \text{where } r^2 = x^2 + y^2 + z^2.$$

This implies

$$y^2 + z^2 = r^2 \sin^2\theta.$$

We know that

$$r^{12} = x^{12} + y^{12} + z^{12}.$$

Therefore, at $t = 0$

$$r^{12} = \frac{x^2}{1 - \beta^2} + y^2 + z^2 = \frac{r^2}{1 - \beta^2}(1 - \beta^2 \sin^2\theta).$$

Hence

$$r^1 = \frac{r}{\sqrt{1-\beta^2}} \left(1 - \beta^2 \sin^2\theta\right)^{\frac{1}{2}}.$$

Thus the components of electric field of charge q at $t = 0$ as seen in S are given as

$$E_x = E_x^1 = \frac{1}{4\pi\epsilon_0}\frac{qx^1}{r^{13}} = \frac{1}{4\pi\epsilon_0}\frac{\frac{qx}{\sqrt{1-\beta^2}}(1-\beta^2)^{\frac{3}{2}}}{\left[r^3(1-\beta^2\sin^2\theta)^{\frac{3}{2}}\right]} \tag{13.89}$$

$$E_y = \frac{E_y^1}{\sqrt{1-\beta^2}} = \frac{1}{\sqrt{1-\beta^2}}\frac{qy^1}{4\pi\epsilon_0 r^{13}} = \frac{1}{4\pi\epsilon_0}\frac{qy(1-\beta^2)}{r^3(1-\beta^2\sin^2\theta)^{\frac{3}{2}}} \tag{13.90}$$

$$E_z = \frac{E_z^1}{\sqrt{1-\beta^2}} = \frac{1}{\sqrt{1-\beta^2}}\frac{1}{4\pi\epsilon_0}\frac{qz^1}{r^{13}} = \frac{1}{4\pi\epsilon_0}\frac{qz(1-\beta^2)}{r^3(1-\beta^2\sin^2\theta)^{\frac{3}{2}}}. \tag{13.91}$$

Writing in compact form, we get the electric field strength of charge q at $t = 0$ as observed in S

$$\vec{E} = i E_x + j E_y + k E_z = \frac{1}{4\pi\epsilon_0}\frac{q(1-\beta^2)\vec{r}}{r^3(1-\beta^2\sin^2\theta)^{\frac{3}{2}}}. \tag{13.92}$$

This indicates that $\vec{E}$ is a radial field directed out along $\vec{r}$. We can write Eq. (13.88) in the following form:

$$\vec{E} = i E_x + j E_y + k E_z = \frac{1}{4\pi\epsilon_0}\frac{q(1-\beta^2)}{r^2(1-\beta^2\sin^2\theta)^{\frac{3}{2}}}\frac{\vec{r}}{r}. \tag{13.93}$$

This implies the field is an inverse square one.

Similarly, the components of magnetic field $\vec{B}$ in S follow explicitly from substituting $\vec{B}^1$ and $\vec{E}^1$ as

$$B_x = B_x^1 = 0 \tag{13.94}$$

$$B_y = -\frac{\frac{v}{c^2}E_z^1}{\sqrt{1-\beta^2}} = -\frac{1}{4\pi\epsilon_0}\frac{\frac{v}{c^2}}{\sqrt{1-\beta^2}}\frac{qz^1}{r^{13}}$$

$$= -\frac{1}{4\pi\epsilon_0 c^2}\frac{v}{\sqrt{1-\beta^2}}\frac{qz(1-\beta^2)^{\frac{3}{2}}}{r^3(1-\beta^2\sin^2\theta)^{\frac{3}{2}}}$$

or

$$B_y = -\frac{\mu_0}{4\pi r^3}\frac{qvz(1-\beta^2)}{r^3(1-\beta^2\sin^2\theta)^{\frac{3}{2}}} \quad \left(\text{since } \mu_0\epsilon_0 = \frac{1}{c^2}\right) \tag{13.95}$$

$$B_z = \frac{\frac{v}{c^2}E_y^1}{\sqrt{1-\beta^2}} = \frac{1}{4\pi\epsilon_0}\frac{\frac{v}{c^2}}{\sqrt{1-\beta^2}}\frac{qy^1}{r^{13}}$$

$$= \frac{1}{4\pi\epsilon_0 c^2}\frac{v}{\sqrt{1-\beta^2}}\frac{qy(1-\beta^2)^{\frac{3}{2}}}{r^3(1-\beta^2\sin^2\theta)^{\frac{3}{2}}}$$

or

$$B_z = \frac{\mu_0}{4\pi r^3}\frac{qvy(1-\beta^2)}{r^3(1-\beta^2\sin^2\theta)^{\frac{3}{2}}}. \tag{13.96}$$

Writing in compact form, we get the magnetic field strength of charge q at $t = 0$ as observed in S

$$\vec{B} = \vec{i}B_x + \vec{j}B_y + \vec{k}B_z = \frac{\mu_0}{4\pi}\frac{q(1-\beta^2)(\vec{v}\times\vec{r})}{(1-\beta^2\sin^2\theta)r^3}. \tag{13.97}$$

This is *relativistic Biot–Savart Law* for magnetic field of a moving charge.

Note that $\vec{v} = (v, 0, 0)$, i.e. the moving direction is x-axis, so that x-component of the magnetic field is zero. For classical limit, $\beta \ll 1$, Eq. (13.97) implies

$$\vec{B} = \frac{\mu_0}{4\pi}\frac{q(\vec{v}\times\vec{r})}{r^3}. \tag{13.98}$$

This is the *classical Biot–Savart Law* for magnetic field of a moving charge.

13.17 Relativistic Lagrangian and Hamiltonian Functions of a Charged Particle in an Electromagnetic Field

Assume a particle of mass m_0 and charge q moving with velocity $\vec{v}$ in an electromagnetic field ($\vec{E}$ and $\vec{B}$). The force on the charged particle is given by Lorentz force

$$\vec{F} = q[\vec{E} + \vec{v}\times\vec{B}].$$

Magnetic field $\vec{B}$ and electric field $\vec{E}$ can be written in terms of electric potential ϕ and magnetic potential $\vec{A}$ as

$$\vec{B} = \nabla \times \vec{A}, \quad \vec{E} = -\left(\frac{\partial \vec{A}}{\partial t}\right) - \nabla\phi.$$

Using these values, the Lorentz force takes the form

$$\vec{F} = q\left[-\left(\frac{\partial \vec{A}}{\partial t}\right) - \nabla\phi + \vec{v} \times (\nabla \times \vec{A})\right]$$

or

$$\vec{F} = q\left[-\nabla\phi - \left(\frac{\partial \vec{A}}{\partial t} + (\vec{v} \cdot \nabla)\vec{A}\right) + \nabla(\vec{v} \cdot \vec{A})\right]. \tag{13.99}$$

The total derivative of $\vec{A}$ can be expressed as

$$\frac{d\vec{A}}{dt} = \frac{\partial \vec{A}}{\partial t} + \sum_{i=1}^{3} \frac{\partial \vec{A}}{\partial x_i}\frac{dx_i}{dt} = (\vec{v} \cdot \nabla)\vec{A} + \frac{\partial \vec{A}}{\partial t}. \tag{13.100}$$

Using (13.100), Eq. (13.99) yields

$$\vec{F} = q\left[-\nabla\phi - \frac{d\vec{A}}{dt} + \nabla(\vec{v} \cdot \vec{A})\right] = q\left[-\nabla(\phi - \vec{v} \cdot \vec{A}) - \frac{d\vec{A}}{dt}\right]. \tag{13.101}$$

Now, we write the x-component of both sides of Eq. (13.101) as

$$F_1 = q\left[-\frac{\partial}{\partial x}(\phi - \vec{v} \cdot \vec{A}) - \frac{d\vec{A}_1}{dt}\right]. \tag{13.102}$$

One can write $\frac{d\vec{A}_1}{dt}$ in the following form:

$$\frac{d\vec{A}_1}{dt} = \frac{d}{dt}\left[\frac{\partial}{\partial \dot{x}}(\vec{v} \cdot \vec{A})\right] \text{ as } \frac{\partial}{\partial \dot{x}}(\vec{v} \cdot \vec{A}) = \frac{\partial}{\partial v_1}(\vec{v} \cdot \vec{A}) = A_1.$$

Again electric potential ϕ is independent of the velocity, so $\frac{\partial \phi}{\partial \dot{x}} = 0$. Hence Eq. (13.102) can be written in the following form:

$$F_1 = -\frac{\partial U}{\partial x} + \frac{d}{dt}\left[\frac{\partial U}{\partial \dot{x}}\right], \tag{13.103}$$

where

$$U = q(\phi - \vec{v} \cdot \vec{A}) \tag{13.104}$$

U is the generalized potential.

Thus, the relativistic Lagrangian L_{em} for a charged particle in an electromagnetic field is $L_{em} = L_m - U$, where $L_m = m_0 c^2 \left(1 - \sqrt{1 - \frac{v^2}{c^2}}\right)$ is the relativistic Lagrangian for the free particle. The terms in U describe the interaction of the charge with the field. Hence, we have

$$L_{em} = m_0 c^2 \left(1 - \sqrt{1 - \frac{v^2}{c^2}}\right) - q\phi + q\vec{v} \cdot \vec{A}. \tag{13.105}$$

For classical limit ($v \ll c$), we have

$$L_{em} = \frac{1}{2} m_0 v^2 - q\phi + q\vec{v} \cdot \vec{A}. \tag{13.106}$$

This Lagrangian indicates that the potential is velocity dependent.

The generalized momentum with components (p_1, p_2, p_3) of the charged particle in an electromagnetic field is defined as $p_i = \frac{\partial L_{em}}{\partial \dot{x}_i}$. Thus, we get

$$p_1 = \frac{\partial L_{em}}{\partial \dot{x}} \equiv \frac{\partial L_{em}}{\partial v_1} = \left(\frac{m_0 \dot{x}}{\sqrt{1 - \frac{v^2}{c^2}}} + q A_1\right).$$

Hence, the generalized momentum is given by the vector equation

$$\vec{p} = \left(\frac{m_0 \vec{v}}{\sqrt{1 - \frac{v^2}{c^2}}} + q\vec{A}\right). \tag{13.107}$$

For classical limit ($v \ll c$), we have

$$\vec{p} = \left(m_0 \vec{v} + q\vec{A}\right). \tag{13.108}$$

The Hamiltonian for a charged particle in an electromagnetic field is defined as

$$H_{em} = p_i \dot{x}_i - L_{em} = \vec{p} \cdot \vec{v} - L_{em}.$$

Putting the values of $\vec{p}$ and L_{em}, we get

$$H_{em} = \left(\frac{m_0 \vec{v} \cdot \vec{v}}{\sqrt{1 - \frac{v^2}{c^2}}} + q\vec{A} \cdot \vec{v}\right) - m_0 c^2 \left(1 - \sqrt{1 - \frac{v^2}{c^2}}\right) + q\phi - q\vec{v} \cdot \vec{A}$$

or

$$H_{em} = m_0 c^2 \left(\frac{1}{\sqrt{1 - \frac{v^2}{c^2}}} - 1 \right) + q\phi. \tag{13.109}$$

For classical limit ($v \ll c$), we have

$$H_{em} = \frac{1}{2} m_0 v^2 + q\phi. \tag{13.110}$$

One can also write the Hamiltonian in terms of momentum as

$$\left[(H_{em} - q\phi) + m_0 c^2 \right]^2 - [(\vec{p} - q\vec{A}) \cdot (\vec{p} - q\vec{A})]c^2 = m_0^2 c^4$$

or

$$H_{em} = c\sqrt{(\vec{p} - q\vec{A})^2 + m_0 c^2} + q\phi - m_0 c^2. \tag{13.111}$$

Example 13.1 Show that the charge of an electron is relativistically invariant.

Hint: Let us consider two frames S and S^1, S^1 is moving with velocity v relative to S along x-direction. Then the transformation equations for current and charge density are

$$J_1^1 = a[J_1 - v\rho], \quad J_2^1 = J_2, \quad J_3^1 = J_3, \quad \rho^1 = a\left[\rho - \left(\frac{v}{c^2}\right)J_1\right],$$

where $a = 1/\sqrt{1 - \frac{v^2}{c^2}}$.

Let us consider the charged electron to be at rest in S frame, then obviously current density J is zero. Hence, the above transformations become

$$J_1^1 = -av\rho, \quad J_2^1 = J_2, \quad J_3^1 = J_3, \quad \rho^1 = \frac{\rho}{\sqrt{1 - \frac{v^2}{c^2}}}.$$

It is apparent that the charge contained in the elementary volume $dV^1 = dx_1^1 dx_2^1 dx_3^1$ in S^1 frame is

$$dq^1 = \rho^1 dV^1 = \rho^1 dx_1^1 dx_2^1 dx_3^1$$

$$= \rho^1 \left(dx_1 \sqrt{1 - \frac{v^2}{c^2}} \right) dx_2 dx_3$$

$$= \rho dx_1 dx_2 dx_3 = dq.$$

Thus, both frames of reference S and S^1 measure the same charge. In other words, charge of an electron is relativistically invariant.

Example 13.2 Show that $A^2 - \frac{\phi^2}{c^2}$ is invariant under Lorentz transformation, where $A^2 = A_1^2 + A_2^2 + A_3^2$.

Hint: One can note that

$$A_\mu \cdot A_\mu = A_1 A_1 + A_2 A_2 + A_3 A_3 + A_4 A_4 = A^2 - \frac{\phi^2}{c^2}.$$

Now,

$$A_\mu^1 \cdot A_\mu^1 = A_1^1 A_1^1 + A_2^1 A_2^1 + A_3^1 A_3^1 + \frac{i\phi^1}{c} \frac{i\phi^1}{c}$$

$$= a^2 \left[A_1 - \frac{v}{c^2}\phi \right]^2 + A_2^2 + A_3^2 - \frac{1}{c^2} a^2 [\phi - vA_1]^2$$

$$= A^2 - \frac{\phi^2}{c^2} = A_\mu \cdot A_\mu \quad \left(\text{putting the value, } a = \frac{1}{\sqrt{1 - \frac{v^2}{c^2}}} \right).$$

Example 13.3 Show that $J^2 - \rho^2 c^2$ is invariant under Lorentz transformation, where $J^2 = J_1^2 + J_2^2 + J_3^2$.

Hint: As in Example (13.2), we note that

$$J_\mu \cdot J_\mu = J_1 J_1 + J_2 J_2 + J_3 J_3 + J_4 J_4 = J^2 - \rho^2 c^2, \text{ etc.}$$

Example 13.4 Show that $AJ - \phi\rho$ is invariant under Lorentz transformation, where $J^2 = J_1^2 + J_2^2 + J_3^2$.

Hint: Here,

$$A_\mu \cdot J_\mu = A_1 J_1 + A_2 J_2 + A_3 J_3 + A_4 J_4 = AJ - \phi\rho, \text{ etc.}$$

Example 13.5 Find the non-relativistic limit of the inverse transformation of four current vector. Interpret the results.

Hint: The inverse transformations of four current vector are given as

$$J_1 = a \left[J_1^1 + v\rho^1 \right], \quad J_2 = J_2^1, \quad J_3 = J_3^1, \quad \rho = a \left[\rho^1 + \left(\frac{v}{c^2} \right) J_1^1 \right].$$

For non-relativistic limit $v \ll c$, the above results become

$$J_1 \approx J_1 + v\rho^1, \quad J_2 = J_2^1, \quad J_3 = J_3^1, \quad \rho \approx \rho^1, \text{ etc.}$$

Example 13.6 Show that d'Alembertian operator $\Box^2$ is invariant under Lorentz transformations but not Laplacian ∇^2.

Hint: See the text.

Example 13.7 If total charge is an invariant quantity, how can it be that a neutral wire in one frame appears to be charged in another frame?

Hint: Let us consider a long straight current-carrying wire at rest in S frame. In this wire, the free electrons move with velocity u to the right and the positive ions are at rest.

$$+ \qquad + \qquad + \qquad + \qquad +$$

$$\ominus \longrightarrow u \quad \ominus \longrightarrow u \quad \ominus \longrightarrow u \quad \ominus \longrightarrow u \quad \ominus \longrightarrow u$$

Here, the number of free electrons and the number of ions per unit volume are the same, say n, so that the net charge in any volume of the wire is zero. In this consideration, the positive charges (ions) are at rest in S frame, whereas electrons are in motion. Now, the negative charge density (due to electrons) is $\rho^- = -ne$, where e is the magnitude of an electronic charge and the positive charge density (due to ions) is $\rho^+ = ne$. Therefore, the net charge density measured in S frame is

$$\rho = \rho^- + \rho^+ = 0$$

For current density, $J_x^- = -neu$, $J_x^+ = 0$, so that

$$J_x = J_x^- + J_x^+ = u\rho^-.$$

Suppose S^1 frame is moving with velocity v relative to S frame along x-direction. Now, observer in S^1 frame measures the current density as

$$\rho^{-1} = \frac{\rho^- - \frac{vJ_x^-}{c^2}}{\sqrt{1 - \frac{v^2}{c^2}}}, \quad \rho^{+1} = \frac{\rho^+ - \frac{vJ_x^+}{c^2}}{\sqrt{1 - \frac{v^2}{c^2}}}.$$

Therefore, he obtains the net charge as (putting the values of J_x^-, J_x^+, ρ^-, ρ^+)

$$\rho^1 = \rho^{-1} + \rho^{+1} = \frac{\frac{neuv}{c^2}}{\sqrt{1 - \frac{v^2}{c^2}}},$$

which is positive and non-zero. Thus, observer in S^1 frame finds the wire to be positively charged.

Example 13.8 Show that $c^2 B^2 - E^2$ is invariant under Lorentz transformation.

Hint: We will have to show

$$c^2 B'^2 - E'^2 = c^2(B_1'^2 + B_2'^2 + B_3'^2) - (E_1'^2 + E_2'^2 + E_3'^2) = c^2 B^2 - E^2.$$

Use the transformations of electric and magnetic field components as

$$E_1' = E_1, \quad E_2' = a[E_2 - c\beta B_3], \quad E_3' = a[E_3 + c\beta B_2]$$

$$B_1' = B_1, \quad B_2' = a\left[B_2 + \left(\frac{\beta}{c}\right)E_3\right], \quad B_3' = a\left[B_3 - \left(\frac{\beta}{c}\right)E_2\right], \text{ etc.}$$

Example 13.9 Show that $\vec{E} \cdot \vec{B}$ is invariant under Lorentz transformation.

Hint: Here one has to show

$$\vec{E}^1 \cdot \vec{B}^1 = E_1' B_1' + E_2' B_2' + E_3' B_3' = \vec{E} \cdot \vec{B}.$$

Example 13.10 Show that the continuity equation is self-contained in the homogeneous pair of Maxwell's equation.

Hint: Maxwell's equation in terms of electromagnetic tensor $F_{\mu\gamma}$ is

$$\frac{\partial F_{\mu\gamma}}{\partial x_\gamma} = \mu_0 J_\mu. \tag{1}$$

Differentiating (1) with respect to x_μ, we get

$$\frac{\partial^2 F_{\mu\gamma}}{\partial x_\mu \partial x_\gamma} = \mu_0 \frac{\partial J_\mu}{\partial x_\mu}. \tag{2}$$

Since $F_{\gamma\mu}$ is anti-symmetric, i.e. $F_{\gamma\mu} = -F_{\gamma\mu}$, then from Eq. (2), we get

$$-\frac{\partial^2 F_{\gamma\mu}}{\partial x_\mu \partial x_\gamma} = \mu_0 \frac{\partial J_\mu}{\partial x_\mu}. \tag{3}$$

Interchanging the dummy indices μ and γ in the above equation, we get

$$-\frac{\partial^2 F_{\gamma\mu}}{\partial x_\gamma \partial x_\mu} = \mu_0 \frac{\partial J_\mu}{\partial x_\mu}. \tag{4}$$

Using the property of perfect differential,

$$\text{i.e.} \left(\frac{\partial^2}{\partial x_\mu \partial x_\gamma} = \frac{\partial^2}{\partial x_\gamma \partial x_\mu}\right)$$

$$-\frac{\partial^2 F_{\gamma\mu}}{\partial x_\mu \partial x_\gamma} = \mu_0 \frac{\partial J_\mu}{\partial x_\mu}. \tag{5}$$

Equations $(2) + (5)$ give

$$2\mu_0 \frac{\partial J_\mu}{\partial x_\mu} = 0,$$

i.e.

$$\frac{\partial J_1}{\partial x_1} + \frac{\partial J_2}{\partial x_2} + \frac{\partial J_3}{\partial x_3} + \frac{\partial J_4}{\partial x_4} = 0$$

or

$$\frac{\partial J_1}{\partial x_1} + \frac{\partial J_2}{\partial x_2} + \frac{\partial J_3}{\partial x_3} + \frac{\partial (ic\rho)}{\partial (ict)} = 0$$

This implies

$$\nabla \cdot \vec{J} + \frac{\partial \rho}{\partial t} = 0,$$

which is the equation of continuity.

Example 13.11 Show that a purely electric field in one frame appears both as an electric and a magnetic field to an observer moving relative to the first.

Hint: use Eqs. (13.69) and (13.70).

Example 13.12 Show that a purely magnetic field in one frame appears both as an electric and a magnetic field to an observer moving relative to the first.

Hint: use Eqs. (13.69) and (13.70).

Example 13.13 Starting from the four-dimensional form of homogeneous Maxwell's equation $\frac{\partial F^{\mu\nu}}{\partial x_\nu} = 0$ obtain the wave equation for the field in a vacuum in the four-dimensional form.

Hint:

$$\frac{\partial F^{\mu\nu}}{\partial x_\nu} = 0 \implies \frac{\partial \left[\frac{\partial A_\nu}{\partial x_\mu} - \frac{\partial A_\mu}{\partial x_\nu} \right]}{\partial x_\nu} = 0$$

or

$$\frac{\partial^2 A_\nu}{\partial x_\nu \partial x_\mu} - \frac{\partial^2 A_\mu}{\partial x_\nu \partial x_\nu} = 0$$

We know,

$$\frac{\partial^2 A_\nu}{\partial x_\nu \partial x_\mu} = \frac{\partial}{\partial x_\mu} \frac{\partial A_\nu}{\partial x_\nu} = 0$$

[The Lorentz-Gauge condition gives $\frac{\partial A_\nu}{\partial x_\nu} = 0$.]
 Thus, we get

$$\frac{\partial^2 A_\mu}{\partial x_\nu \partial x_\nu} = 0$$

In other words,

$$\Box^2 A_\mu = 0 \quad \text{here,} \quad \frac{\partial^2}{\partial x_\nu \partial x_\nu} \equiv \Box^2 = \nabla^2 - \frac{1}{c^2} \frac{\partial^2}{\partial t^2}.$$

Example 13.14 Assuming the Lorentz-Gauge condition $\frac{\partial A_\nu}{\partial x_\nu} = 0$ show that Maxwell's equation gives $\frac{\partial^2 A_\mu}{\partial x_\nu \partial x_\nu} = -\mu_0 J_\mu$.

Hint: Maxwell's equation gives

$$\frac{\partial F^{\mu\nu}}{\partial x_\nu} = \mu_0 J_\mu \implies \frac{\partial \left[\frac{\partial A_\nu}{\partial x_\mu} - \frac{\partial A_\mu}{\partial x_\nu} \right]}{\partial x_\nu} = \mu_0 J_\mu$$

or

$$\frac{\partial^2 A_\nu}{\partial x_\nu \partial x_\mu} - \frac{\partial^2 A_\mu}{\partial x_\nu \partial x_\nu} = \mu_0 J_\mu$$

We know,

$$\frac{\partial^2 A_\nu}{\partial x_\nu \partial x_\mu} = \frac{\partial}{\partial x_\mu} \frac{\partial A_\nu}{\partial x_\nu} = 0.$$

[Using the Lorentz-Gauge condition gives $\frac{\partial A_\nu}{\partial x_\nu} = 0$.]
 Thus, we get

$$\frac{\partial^2 A_\mu}{\partial x_\nu \partial x_\nu} = -\mu_0 J_\mu.$$

Example 13.15 Starting with Maxwell's equation show that $\frac{\partial J_\mu}{\partial x_\mu} = 0$.

Hint: See Example 13.10.

Example 13.16 Using the transformation laws of electric E and magnetic B field components find the transformation laws for force.

Hint: The force experienced by charged particle moving with velocity $\vec{v}$ in an electromagnetic field is known as *Lorentz force* on a particle of charge q which is given by $\vec{F} = q[\vec{E} + \vec{v} \times \vec{B}]$. Writing explicitly, we have

$$F_1 = qE_1, \quad F_2 = q(E_2 - vB_3), \quad F_3 = q(E_3 + vB_2).$$

The components of the force in S^1 system are due to electrostatic force

$$F_1^1 = qE_1^1, \quad F_2^1 = qE_2^1, \quad F_3^1 = qE_3^1.$$

We also know the transformations of electric field components as

$$E_1^1 = E_1, \quad E_2^1 = a[E_2 - c\beta B_3], \quad E_3^1 = a[E_3 + c\beta B_2].$$

Therefore, we have

$$F_1^1 = qE_1^1 = qE_1, \quad F_2^1 = qE_2^1 = qa[E_2 - c\beta B_3] = aF_2,$$
$$F_3^1 = qE_3^1 = aq[E_3 + c\beta B_2] = aF_3.$$

Example 13.17 Show that $F_{\mu\nu} F_{\mu\nu} = 2[B^2 - \frac{E^2}{c^2}]$.

Hint:

$$F_{\mu\nu} F_{\mu\nu} = F_{1\nu}^2 + F_{2\nu}^2 + F_{3\nu}^2 + F_{4\nu}^2$$
$$= [F_{11}^2 + F_{21}^2 + F_{31}^2 + F_{41}^2] + [F_{12}^2 + F_{22}^2 + F_{32}^2 + F_{42}^2]$$
$$+ [F_{13}^2 + F_{23}^2 + F_{33}^2 + F_{43}^2] + [F_{14}^2 + F_{24}^2 + F_{34}^2 + F_{44}^2].$$

We know the explicit values of the components of the electromagnetic field tensor $F_{\mu\nu}$ as

$$F_{\mu\nu} = \begin{bmatrix} 0 & B_3 & -B_2 & -\frac{iE_1}{c} \\ -B_3 & 0 & B_1 & -\frac{iE_2}{c} \\ B_2 & -B_1 & 0 & -\frac{iE_3}{c} \\ \frac{iE_1}{c} & \frac{iE_2}{c} & \frac{iE_3}{c} & 0 \end{bmatrix}.$$

Putting the values of the components of the electromagnetic field tensor, one can get

$$F_{\mu\nu} F_{\mu\nu} = 2\left[B_1^2 + B_2^2 + B_3^2 - \frac{E_1^2}{c^2} - \frac{E_2^2}{c^2} - \frac{E_3^2}{c^2} \right] = 2\left[B^2 - \frac{E^2}{c^2} \right].$$

Example 13.18 Show that $\frac{1}{64\pi^2}(E^2 + B^2)^2 - \frac{1}{16\pi^2}(\vec{E} \times \vec{B}) \cdot (\vec{E} \times \vec{B})$ is invariant under Lorentz transformation.

Hint: Using this result

$$(\vec{E} \times \vec{B}) \cdot (\vec{E} \times \vec{B}) = E^2 B^2 - (\vec{E} \cdot \vec{B})^2,$$

the given expression implies

$$\frac{1}{64\pi^2}(E^2 + B^2)^2 - \frac{1}{16\pi^2}(\vec{E} \times \vec{B}) \cdot (\vec{E} \times \vec{B}) = \frac{1}{64\pi^2}[(E^2 - B^2)^2 + 4(\vec{E} \cdot \vec{B})^2].$$

Here, $(E^2 - B^2)$ and $(\vec{E} \cdot \vec{B})$ are invariant, hence, etc.

Example 13.19 Show that if $\vec{E} \geq c\vec{B}$ in one inertial frame, then $\vec{E}^1 \geq c\vec{B}^1$ in all other inertial frames.

Hint: We know $c^2 B^2 - E^2$ is invariant quantity, therefore

$$c^2 B^{12} - E^{12} = c^2 B^2 - E^2 \leq 0 \implies \vec{E}^1 \leq c\vec{B}^1.$$

Example 13.20 Show that if electric field $\vec{E}$ and magnetic field $\vec{B}$ are perpendicular in one inertial frame, then they are perpendicular in all other inertial frames.

Hint: We know, $\vec{E} \cdot \vec{B}$ is invariant quantity, therefore

$$\vec{E}^1 \cdot \vec{B}^1 = \vec{E} \cdot \vec{B} = 0 \text{ since } \vec{E} \text{ and } \vec{B} \text{ are perpendicular.}$$

Example 13.21 Show that for a given electromagnetic field, we can find an inertial frame in which either $\vec{B} = 0$ if $\vec{E} \geq c\vec{B}$ or $\vec{E} = 0$ if $\vec{E} \leq c\vec{B}$ at a given point if $\vec{E} \cdot \vec{B} = 0$ at that point.

Hint:
Case I: Let S^1 be any frame in which no magnetic field is present. Therefore,

$$c^2 B^2 - E^2 = c^2 B^{12} - E^{12} = -E^{12} < 0, \quad \text{i.e. } \vec{E} \geq c\vec{B}.$$

We know from Eq. (13.13)

$$B_{\parallel}^1 = B_{\parallel} = 0, \quad B_{\perp}^1 = a\left[B_{\perp} - \frac{1}{c^2}(\vec{v} \times \vec{E}_{\perp})\right] = 0.$$

These give

$$\vec{B} = \frac{1}{c^2}(\vec{v} \times \vec{E}).$$

From this expression, we see that

$$\vec{v} \cdot \vec{B} = \vec{v} \cdot \frac{1}{c^2}(\vec{v} \times \vec{E}) = 0.$$

Therefore, velocity vector of S^1 frame is perpendicular to the magnetic field. Since electric field exists in S^1, so the velocity vector of S^1 frame is also perpendicular to the electric field. This implies the relative velocity vector between $\vec{E}$ and $\vec{B}$ frames is perpendicular to the plane containing S and S^1 frames. Also,

$$\vec{E} \cdot \vec{B} = \vec{E} \cdot \frac{1}{c^2}(\vec{v} \times \vec{E}) = 0.$$

Now,

$$\vec{E} \times \vec{B} = \vec{E} \times \frac{1}{c^2}(\vec{v} \times \vec{E}) = \frac{E^2}{c^2}\vec{v}.$$

This implies

$$\vec{v} = \frac{c^2(\vec{E} \times \vec{B})}{E^2}.$$

Note that this type of frame is not unique.

Case II: Let S^1 be any frame in which no electric field is present. Therefore,

$$c^2 B^2 - E^2 = c^2 B^{1^2} - E^{1^2} = c^2 B^{1^2} > 0 \text{ i.e. } \vec{E} \le c\vec{B}.$$

We know from Eq. (13.12)

$$E_\parallel^1 = E_\parallel = 0, \quad E_\perp^1 = a[E_\perp + (\vec{v} \times \vec{B}_\perp)] = 0.$$

These give

$$\vec{E} = -(\vec{v} \times \vec{B}).$$

From this expression, we see that

$$\vec{v} \cdot \vec{E} = \vec{v} \cdot (\vec{v} \times \vec{B}) = 0.$$

Therefore, velocity vector of S^1 frame is perpendicular to the electric field. Since magnetic field exists in S^1, the velocity vector of S^1 frame is also perpendicular to the magnetic field. This implies the relative velocity vector between $\vec{E}$ and $\vec{B}$ frames is perpendicular to the plane containing S and S^1 frames. Also,

$$\vec{E} \cdot \vec{B} = -(\vec{v} \times \vec{B}) \cdot \vec{B} = 0.$$

Now,

$$\vec{B} \times \vec{E} = \vec{B} \times [-(\vec{v} \times \vec{B})] = -B^2 \vec{v}.$$

This implies

$$\vec{v} = \frac{c^2(\vec{E} \times \vec{B})}{B^2}.$$

Note that this type of frame is not unique.

Example 13.22 Find the trajectory of a particle of charge q, rest mass m_0 moving in a uniform electrostatic field E along x-axis. Given that the initial momentum in the y-direction is p_0 and initial momentum in the x-direction is zero.

Hint: Since the electrostatic field E is in x-direction, we assume the initial velocity of the charged particle to be in the y-direction. Then the charged particle moves in the xy-plane. Here the equation of motion is

$$\vec{F} = \frac{d\vec{p}}{dt} = q\vec{E}.$$

From the given condition,

$$\frac{dp_x}{dt} = qE, \quad \frac{dp_y}{dt} = 0.$$

These yield

$$p_x = qEt, \quad p_y = p_0$$

[at $t = 0$, $p_x = 0$, $p_y = p_0$]. The total energy of the particle

$$T_E = \sqrt{p^2 c^2 + m_0^2 c^4} = \sqrt{c^2(p_x^2 + p_y^2) + m_0^2 c^4} = \sqrt{(c^2 p_0^2 + c^2 q^2 E^2 t^2) + m_0^2 c^4}.$$

Again, we know

$$\vec{p} = m\vec{v} \text{ and } T_E = mc^2, \quad \text{therefore, } \vec{v} = \frac{\vec{p}}{T_E}c^2.$$

This implies

$$v_x = \frac{dx}{dt} = \frac{p_x}{T_E}c^2 = \frac{qEtc^2}{\sqrt{(c^2 p_0^2 + c^2 q^2 E^2 t^2) + m_0^2 c^4}}$$

and

$$v_y = \frac{dy}{dt} = \frac{p_y}{T_E}c^2 = \frac{p_0 c^2}{\sqrt{(c^2 p_0^2 + c^2 q^2 E^2 t^2) + m_0^2 c^4}}.$$

Solving these equations, we get

$$x = \frac{1}{qE}\sqrt{(c^2 p_0^2 + c^2 q^2 E^2 t^2) + m_0^2 c^4}$$

$$y = \frac{p_0 c}{qE} \sinh^{-1}\left(\frac{cqEt}{\sqrt{c^2 p_0^2 + m_0^2 c^4}}\right).$$

Eliminating t from these equations, we get the required trajectory of the charged particle as

$$x = \frac{\sqrt{c^2 p_0^2 + m_0^2 c^4}}{qE} \cosh\left(\frac{qEy}{p_0 c}\right).$$

Chapter 14
Electromagnetic Waves

14.1 Introduction

Maxwell equations indicate that the electric and magnetic fields can travel in space in form of waves. These waves are known as **electromagnetic waves**.

Magnetic field strength $\vec{B}$ can be written in terms of magnetic intensity $\vec{H}$ as

$$\vec{B} = \mu \vec{H}, \tag{14.1}$$

where μ is magnetic permeability.

Electric field strength $\vec{E}$ can be written in terms of electric displacement $\vec{D}$ as

$$\vec{D} = \epsilon \vec{E}, \tag{14.2}$$

where ϵ is electric permittivity.

If electrical conductivity be σ, then the conduction current density is

$$\vec{J} = \sigma \vec{E}. \tag{14.3}$$

With this new quantities, Maxwell's equations read as

$$\vec{\nabla} \times \vec{H} = \vec{J} + \frac{\partial \vec{D}}{\partial t}, \tag{14.4}$$

$$\vec{\nabla} \times \vec{E} = -\frac{\partial \vec{B}}{\partial t}, \tag{14.5}$$

$$\vec{\nabla} \cdot \vec{D} = \rho, \tag{14.6}$$

F. Rahaman, *The Special Theory of Relativity*, La Matematica per il 3+2 136,
https://doi.org/10.1007/978-981-19-0497-4_14

$$\vec{\nabla} \cdot \vec{H} = 0. \tag{14.7}$$

Note 1: An insulator or alternatively known as a **dielectric** is a material in which the electrons are firmly bound to the nucleus of the atom so that there are no free electrons available for the conduction of electric current. The dielectric is termed as **isotropic** if the variation in activities of the dielectric is independent of the direction of the applied field. If behaviour of the dielectric is changed depending on the direction of applied field, then it is **anisotropic**. In the presence of dielectric, the electric field exerts a force on each charged particle. The relative displacement of the charges is called **polarization**. Here the dielectric is known to be **polarized** and acts like a dipole.

Note 2: Suppose a medium filled with simple material, then

    a. **linear:** μ, ϵ are constants;

    b. **isotropic:** properties of the medium are same in all directions;

    c. **homogeneous:** no special position in the space, i.e. there is translator symmetry in all directions;

    d. **source free:** no charge density, i.e. the charge density ρ is zero;

    e. **non-conducting:** the conductivity σ is zero. In other words, current density $\vec{J} = \sigma \vec{E}$ is also zero.

14.2   Wave Equation for Magnetic Intensity $\vec{H}$

Let us consider the medium is homogeneous. In this case, the magnetic permeability μ, electric permittivity ϵ and electric conductivity σ are constants. Now we take curl for both sides of Eq. (14.4) to yield

$$\vec{\nabla} \times (\vec{\nabla} \times \vec{H}) = \vec{\nabla} \times \vec{J} + \vec{\nabla} \times \frac{\partial \vec{D}}{\partial t}. \tag{14.8}$$

Using (14.2) and (14.3), we get

$$\vec{\nabla} \times \vec{\nabla} \times \vec{H} = \sigma \vec{\nabla} \times \vec{E} + \epsilon \frac{\partial}{\partial t} (\vec{\nabla} \times \vec{E}). \tag{14.9}$$

Using (14.1) and (14.5), we obtain

$$\vec{\nabla} \times \vec{\nabla} \times \vec{H} = -\mu \sigma \frac{\partial \vec{H}}{\partial t} - \mu \epsilon \frac{\partial^2 \vec{H}}{\partial t^2}$$

$$\Rightarrow \vec{\nabla}(\vec{\nabla} \cdot \vec{H}) - \nabla^2 \vec{H} = -\sigma\mu\frac{\partial \vec{H}}{\partial t} - \epsilon\mu\frac{\partial^2 \vec{H}}{\partial t^2}. \qquad (14.10)$$

We know $\vec{\nabla} \cdot \vec{H} = 0$, [from (14.7)], and therefore, finally, we get

$$\vec{\nabla}^2 \vec{H} - \epsilon\mu\frac{\partial^2 \vec{H}}{\partial t^2} - \sigma\mu\frac{\partial \vec{H}}{\partial t} = 0. \qquad (14.11)$$

This is the wave equation for magnetic intensity $\vec{H}$.

14.3 Wave Equation for Electric Field Strength $\vec{E}$

As before we take curl for both sides of Eq. (14.5) to yield

$$\vec{\nabla} \times \vec{\nabla} \times \vec{E} = -\vec{\nabla} \times \frac{\partial \vec{B}}{\partial t}. \qquad (14.12)$$

Using Eqs. (14.1), (14.2), (14.3), we obtain from the above equation as

$$\vec{\nabla} \times \vec{\nabla} \times \vec{E} = -\mu\sigma\frac{\partial \vec{E}}{\partial t} - \epsilon\mu\frac{\partial^2 \vec{E}}{\partial t^2}$$

$$\Rightarrow \vec{\nabla}(\vec{\nabla} \cdot \vec{E}) - \vec{\nabla}^2 \vec{E} = -\mu\sigma\frac{\partial \vec{E}}{\partial t} - \epsilon\mu\frac{\partial^2 \vec{E}}{\partial t^2}. \qquad (14.13)$$

If the medium is charge free, i.e. $\rho = 0$, then $\vec{\nabla} \cdot \vec{E} = (\frac{1}{\epsilon})\vec{\nabla} \cdot \vec{D} = 0$. Hence, we get

$$\vec{\nabla}^2 \vec{E} - \epsilon\mu\frac{\partial^2 \vec{E}}{\partial t^2} - \sigma\mu\frac{\partial \vec{E}}{\partial t} = 0. \qquad (14.14)$$

This is the wave equation for electric field strength $\vec{E}$.
One can note that the wave equations for $\vec{E}$ & $\vec{H}$ are same form.

14.4   Electromagnetic Waves in a Non-conducting Dielectric Medium

For non-conducting medium, the conductivity vanishes, i.e. $\sigma = 0$. Now from Eqs. (14.11) and (14.14), we have

$$\vec{\nabla}^2 \vec{H} = \epsilon\mu \frac{\partial^2 \vec{H}}{\partial t^2}, \tag{14.15}$$

$$\vec{\nabla}^2 \vec{E} = \epsilon\mu \frac{\partial^2 \vec{E}}{\partial t^2}. \tag{14.16}$$

These equations have the form of standard wave equation as

$$\vec{\nabla}^2 \psi = \frac{1}{v^2} \frac{\partial^2 \psi}{\partial t^2}, \tag{14.17}$$

where

$$v = \frac{1}{\sqrt{\epsilon\mu}} \tag{14.18}$$

is the velocity of electromagnetic wave propagation.

Equation (14.17) indicates that both electric and magnetic fields propagate with the same velocity as given in (14.18). In free space $\epsilon = \epsilon_0$ and $\mu = \mu_0$. Therefore, electromagnetic wave propagates in free space with velocity $\frac{1}{\sqrt{\epsilon_0\mu_0}}$.

We know, $\epsilon_0 = 8.854 \times 10^{-12}$ F/m and $\mu_0 = 4\pi \times 10^{-7}$ H/m, and therefore

$$\frac{1}{\sqrt{\epsilon_0\mu_0}} = 2.9978 \times 10^8 \text{ m/sec},$$

which is the velocity of light in free space as measured experimentally. Thus,

$$\frac{1}{\sqrt{\epsilon_0\mu_0}} = c, \quad i.e. \ \epsilon_0\mu_0 = \frac{1}{c^2}. \tag{14.19}$$

This proves that light is a form of electromagnetic radiation. Thus, the proof of light as an electromagnetic radiation is a great achievement of Maxwell's theory. Similar to visible light, γ-rays, X-rays, ultraviolet rays, infrared radiation and radio waves are all electromagnetic in nature. However, the wave lengths are different. All these waves have the same velocity $c = \frac{1}{\sqrt{\mu_0\epsilon_0}} = 2.9978 \times 10^8$ m/sec in free space.

Suppose the electromagnetic wave is propagating in z-direction, then the solution of the wave equation (14.17) will be considerably simplified. Here the field depends only on z-coordinate. In this case, the wave is known as **plane**.

Here

$$\nabla = \vec{k}\frac{\partial}{\partial z}, \quad \nabla^2 = \vec{\nabla} \cdot \vec{\nabla} = \frac{\partial^2}{\partial z^2} \quad \left(as \; \frac{\partial}{\partial x} = \frac{\partial}{\partial y} = 0\right). \quad (14.20)$$

Hence, Eqs. (14.15) and (14.16) assume the forms

$$\frac{\partial^2 \vec{H}}{\partial z^2} - \frac{1}{v^2}\frac{\partial^2 \vec{H}}{\partial t^2} = 0, \quad (14.21)$$

$$\frac{\partial^2 \vec{E}}{\partial z^2} - \frac{1}{v^2}\frac{\partial^2 \vec{E}}{\partial t^2} = 0. \quad (14.22)$$

The solutions of Eqs. (14.21) and (14.22) are

$$E = E_0 e^{i(kz-\omega t)} \quad (14.23)$$

$$H = H_0 e^{i(kz-\omega t)}. \quad (14.24)$$

Here ω is angular frequency of the wave and $k = \frac{2\pi}{\lambda}$ is the wave number.

Now differentiating (14.23) w.r.t z and t, we obtain

$$\frac{\partial \vec{E}}{\partial z} = ik E_0 e^{i(kz-\omega t)} = ik\vec{E},$$

$$\frac{\partial \vec{E}}{\partial t} = -i\omega E_0 e^{i(kz-\omega t)} = -i\omega\vec{E}.$$

Therefore, it is clear that $\frac{\partial}{\partial z}$ is equivalent to (ik) while $\frac{\partial}{\partial t}$ is equivalent to $(-i\omega)$.

In isotropic medium, Maxwell's equation (14.6) implies

$$\vec{\nabla} \cdot \vec{D} = 0 \; or \; \vec{\nabla} \cdot \vec{E} = 0. \quad (14.25)$$

Using, $\vec{\nabla} = \vec{k}\frac{\partial}{\partial z}$, we get

$$\vec{k} \frac{\partial}{\partial z} \cdot \vec{E} = 0, \tag{14.26}$$

$$\Rightarrow \vec{k}\,(ik) \cdot \vec{E} = 0, \qquad \left(as\ \frac{\partial}{\partial z} \rightarrow ik\right) \tag{14.27}$$

$$\Rightarrow \vec{k} \cdot \vec{E} = 0. \tag{14.28}$$

This indicates that $\vec{k}$, the unit vector in the direction of wave propagation, is perpendicular to $\vec{E}$.

Similarly, Eq. (14.7) yields $\vec{k} \cdot \vec{H} = 0$.

Thus, the vectors $\vec{E}$, $\vec{H}$ are transverse to the direction of propagation. In other words, electromagnetic wave is transverse in nature.

Equation (14.4) implies

$$\vec{\nabla} \times \vec{H} = -i\omega\epsilon \vec{E}, \qquad \left(as\ \frac{\partial}{\partial t} \rightarrow -i\omega\right) \tag{14.29}$$

$$\Rightarrow \vec{k} \frac{\partial}{\partial z} \times \vec{H} = -i\omega\epsilon \vec{E}, \qquad \left(as\ \vec{\nabla} \rightarrow \vec{k} \frac{\partial}{\partial z}\right) \tag{14.30}$$

$$\Rightarrow ik\vec{k} \times \vec{H} = -i\omega\epsilon \vec{E}. \qquad \left(as\ \frac{\partial}{\partial z} \rightarrow ik\right). \tag{14.31}$$

The last equation yields

$$\vec{H} \cdot \vec{E} = 0$$

Therefore, $\vec{E}$ and $\vec{H}$ are mutually perpendicular to each other.
The general case will be discussed in Section (14.7).

14.5   Poynting's Theorem (Energy Conservation)

Electromagnetic energy propagates through space by means of electromagnetic waves. In 1884, John Henry Poynting derived an important result about the conservation of energy for a configuration consisting of electric and magnetic fields acting on charges. He showed that Maxwell's equations lead to conservation of energy. This is Poynting theorem. This conservation theorem comprises with the effects of both displacement current and of magnetic induction.

Recall Maxwell's equations which are given by

$$\vec{\nabla} \times \vec{H} = \vec{J} + \frac{\partial \vec{D}}{\partial t}; \quad \vec{\nabla} \times \vec{E} = -\frac{\partial \vec{B}}{\partial t}; \quad \vec{\nabla} \cdot \vec{D} = \rho; \quad \vec{\nabla} \cdot \vec{B} = 0 \qquad (14.32)$$

$$\vec{B} = \mu \vec{H}; \quad D = \epsilon \vec{E}; \quad \vec{J} = \sigma \vec{E}. \qquad (14.33)$$

Taking the dot product of the first equation in (14.32) with $\vec{E}$, we get

$$\vec{E}.(\vec{\nabla} \times \vec{H}) = \vec{E} \cdot \vec{J} + \vec{E} \cdot \frac{\partial \vec{D}}{\partial t}. \qquad (14.34)$$

Also, taking dot product of second equation in (14.32) with $\vec{H}$, we get

$$\vec{H} \cdot (\vec{\nabla} \times \vec{E}) = -\vec{H} \cdot \frac{\partial \vec{B}}{\partial t}. \qquad (14.35)$$

Now (14.34)–(14.35), we get

$$\vec{E} \cdot \vec{J} + \left(\vec{E} \cdot \frac{\partial \vec{D}}{\partial t} + \vec{H} \cdot \frac{\partial \vec{B}}{\partial t} \right) = -\vec{\nabla} \cdot (\vec{E} \times \vec{H}). \qquad (14.36)$$

Now integrating over a fixed finite volume V yields

$$\int_V \vec{E} \cdot \vec{J} dV + \int_V \left(\vec{E} \cdot \frac{\partial \vec{D}}{\partial t} + \vec{H} \cdot \frac{\partial \vec{B}}{\partial t} \right) dV$$

$$= -\int_V \vec{\nabla} \cdot (\vec{E} \times \vec{H}) dV$$

$$= -\int_S (\vec{E} \times \vec{H}) \cdot \vec{n} dS. \qquad \text{(by divergence theorem)} \qquad (14.37)$$

Here volume V is bounded by the surface S and $\vec{n}$ is unit normal vector on S.

This equation is known as **Poynting theorem**. We will show that this equation expresses the law of conservation of energy.

The volume integral of $\vec{E} \cdot \vec{J}$ signifies the power given to electric charges by the electric field. The right-hand side of Eq. (14.37) represents a net energy flux carried through the boundary surface by the electromagnetic fields.

The vector

$$\vec{Y} = \vec{E} \times \vec{H} \tag{14.38}$$

represents the direction of the energy flow and its magnitude is equal to the rate of the flux of energy through a unit area normal to itself which is known as **Poynting's vector**.

The unit of $\vec{Y}$ is watts per metre square. The second term in Eq. (14.37) represents the time rate of change of the energy stored in the electromagnetic fields within V.

The corresponding time rate of change of the energy density u is

$$\frac{\partial u}{\partial t} = \left(\vec{E} \cdot \frac{\partial \vec{D}}{\partial t} + \vec{H} \cdot \frac{\partial \vec{B}}{\partial t} \right). \tag{14.39}$$

For an electromagnetically linear, isotropic material (ϵ, μ are constants), we have

$$\frac{\partial u}{\partial t} = \frac{\partial}{\partial t} \left(\frac{1}{2} \epsilon E^2 + \frac{1}{2} \frac{B^2}{\mu} \right). \tag{14.40}$$

This gives

$$u = \left(\frac{1}{2} \epsilon E^2 + \frac{1}{2} \frac{B^2}{\mu} \right) = u_e + u_m, \tag{14.41}$$

where $u_e = \frac{1}{2} \epsilon E^2$ and $u_m = \frac{1}{2} \frac{B^2}{\mu}$. This is the sum of electric and magnetic energies. The following equation

$$\int_V \sigma E^2 dV + \frac{\partial}{\partial t} \int_V \frac{1}{2} \epsilon E^2 dV + \frac{\partial}{\partial t} \int_V \frac{1}{2} \mu H^2 dV = - \int_S (\vec{E} \times \vec{H}) \cdot \vec{n} dS \tag{14.42}$$

is the **integral form of Poynting's theorem**. Equation (14.36) yields the **differential (point) form of Poynting's theorem** as

$$-\sigma E^2 - \frac{\partial}{\partial t} \left(\frac{1}{2} \epsilon E^2 + \frac{1}{2} \mu H^2 \right) = \vec{\nabla} \cdot (\vec{E} \times \vec{H}). \tag{14.43}$$

The first term in the left-hand side of Eq. (14.42) indicates the power dissipation as heat (Joule's law), the second term implies the rate of change of stored electric

Fig. 14.1 Pillbox-shaped
surface at the interface of
two media

energy and the last term represents the rate of change of stored magnetic energy. The right-hand side term indicates the power flowing into the region V.

Thus, the **statement of Poynting's theorem** can be made as 'The total work done on the charges by the electromagnetic force within V is equal to the net energy flowing out through the surface S per unit time'.

Hence, Poynting's theorem in Eq. (14.37) is an equation expressing energy conservation.

14.6 Boundary Conditions

Electromagnetic waves within material media are affected due to polarization and magnetization effects of that medium. In other words, electric and magnetic field vectors experience a modification while passing from one medium to the other. In general, the fields $\vec{E}, \vec{B}, \vec{D}, \vec{H}$ will be discontinuous at a boundary surface between two distinct media. The explicit form of these discontinuities could be estimated from Maxwell's equations in their integral forms. The differential forms of Maxwell's equations can be cast in integral forms as

1. $\int_S \vec{D} \cdot \vec{n} dS = 4\pi \int_V \rho dV = Q$, Q = total charge within a volume V and S is a closed surface bounding it. $\vec{n}$ is a unit outward normal vector to the surface S.
2. $\int_S \vec{B} \cdot \vec{n} dS = 0$.
3. $\oint_P \vec{E} \cdot d\vec{l} = -\frac{\partial}{\partial t} \int_S \vec{B} \cdot d\vec{S}$, for any surface S bounded by the closed loop P.
4. $\oint_P \vec{H} \cdot d\vec{l} = \int_S \vec{J} \cdot \vec{n} dS + \frac{\partial}{\partial t} \int_S \vec{D} \cdot \vec{n} dS$

Now we will discuss the boundary conditions of the electromagnetic fields $\vec{E}, \vec{B}, \vec{D}, \vec{H}$ at the boundary surface between different media.

Fig. 14.2 Rectangular loop
at the interface between two
media. This rectangular loop
is bounding a surface S

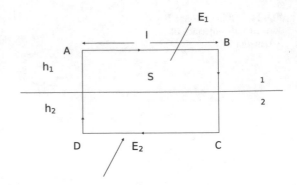

a. Boundary condition for magnetic field $\vec{B}$

Consider an infinitesimal pillbox (popularly known as Gaussian pillbox) connects the
boundary surface between two distinct media comprising different electromagnetic
fields (see Fig. 14.1).

The volume enclosed by the pillbox-shaped surface S contains three surfaces S_1,
S_2 and S_3 as shown in Fig. 14.1. Now, we obtain from the integral form of Maxwell's
equation (14.2) as

$$\int_S \vec{B}.\vec{n}dS = \int_{S_1} \vec{B}.\vec{n_1}dS + \int_{S_2} \vec{B}.\vec{n_2}dS + \int_{S_3} \vec{B}.\vec{n_3}dS = 0. \qquad (14.44)$$

When the height of the pillbox is very small, i.e. when S_1 and S_2 approach to each
other towards the interface, then the side surface area S_3 is negligible. As a result,
for bounded $\vec{B}$, the last term of the equation vanishes. Since the directions of $\vec{n_1}$ and
$\vec{n_2}$ are opposite, we get

$$\vec{B}.\vec{n_1} = \vec{B}.\vec{n_2}. \qquad (14.45)$$

Thus, the normal component of the magnetic induction is continuous across the
boundary.

b. Boundary condition for electric field $\vec{E}$
Now we will find the boundary conditions on the tangential component of the electric
field from Maxwell equation

$$\vec{\nabla} \times \vec{E} = -\frac{\partial \vec{B}}{\partial t}. \qquad (14.46)$$

Consider a rectangular loop across the interface between the two media (see
Fig. 14.2).

Fig. 14.3 Pillbox of cross-sectional area S with small thickness h

Now integrating over the loop of the above equation, we get

$$\int_S (\vec{\nabla} \times \vec{E}) \cdot \vec{n} dS = -\int_S \frac{\partial \vec{B}}{\partial t} \cdot \vec{n} dS.$$

For very small h_1 and h_2, i.e. when the loop is shrunk, the sides of the rectangle will vanish. Due to the vanishing of the area of integration, the right-hand side of the above equation also vanishes. Applying Stoke's theorem to the left-hand side, we obtain

$$\oint_{ABCD} \vec{E} \cdot d\vec{l} = \int_{AB} \vec{E}_1 \cdot d\vec{l} + \int_{CD} \vec{E}_1 \cdot d\vec{l} + contribution \ from \ sides \ BC \ and \ DA. \quad (14.47)$$

As before, for vanishing BC and DA, the contribution from sides BC and DA will vanish. Hence, we get

$$\int_{AB} \vec{E}_1 \cdot d\vec{l} + \int_{CD} \vec{E}_2 \cdot d\vec{l} = lE_{1t} - lE_{2t} = 0 \Rightarrow E_{1t} = E_{2t}. \quad (14.48)$$

Thus, the tangential components of the electric fields (E_{1t}, E_{2t}) are continuous across the interface.

c. Boundary condition for electric displacement $\vec{D}$

The boundary condition for the normal component of the electric displacement can be obtained from $\vec{\nabla}.\vec{D} = 4\pi\rho$. Let us assume that the electric displacement vector $\vec{D}$ is directed from medium 2 into medium 1. Consider a pillbox-shaped surface (see Fig. 14.3) with cross-sectional area S and very small thickness.

Now, integrating over the volume V enclosed by the pillbox, we have

$$\int_V \vec{\nabla} \cdot \vec{D} dV = 4\pi \int_V \rho dV = 4\pi\sigma S, \quad (14.49)$$

where σ = surface charge density (charge per unit area).

$(4\pi \int_V \rho dV = 4\pi \rho h S, h = \text{thickness}; \lim_{h\to 0} 4\pi \rho h S = 4\pi \sigma S)$

Now, applying the divergence theorem, we have

$$\int_S \vec{D}_1.\vec{n}_1 dS + \int_S \vec{D}_2 \cdot \vec{n}_2 dS + contributions\ from\ walls = 4\pi\sigma S, \quad (14.50)$$

where $\vec{n}_1$ and $\vec{n}_2$ are outwardly drawn opposite directed unit normal vectors. Now neglecting the side surface area of the pillbox, we have

$$\vec{D}_1 \cdot \vec{n}_1 S + \vec{D}_2.\vec{n}_2 S = 4\pi\sigma S \qquad (14.51)$$

or

$$D_{1n} - D_{2n} = 4\pi\sigma. \qquad (14.52)$$

Thus component of D, which is perpendicular to the inter-surface of two different media, is discontinuous. Actually the normal component of electric displacement changes by an amount equal to the free surface density of the charge at the interface.

d. Boundary condition for current density $\vec{J}$

The boundary condition for the current density $\vec{J}$ can be obtained from the Maxwell equation

$$\vec{\nabla} \times \vec{H} = \vec{J} + \frac{\partial \vec{D}}{\partial t}. \qquad (14.53)$$

This equation leads to the continuity equation by taking divergence both sides as

$$\nabla \cdot J + \frac{\partial \rho}{\partial t} = 0. \qquad (14.54)$$

So, we try to find the boundary condition for the normal component of current density $\vec{J}$.

Let us assume that the current density vector $\vec{J}$ is directed from medium 2 into medium 1. Consider a pillbox-shaped surface (see Fig. 14.4) with cross-sectional area S and very small thickness.

Now, integrating over the volume V enclosed by the pillbox, we have

$$-\int_V \vec{\nabla}.\vec{J} dV = \int_V \frac{\partial \rho}{\partial t} dV = \frac{\partial \sigma}{\partial t} S, \qquad (14.55)$$

where σ = surface charge density.

Now, applying the divergence theorem, we have

Fig. 14.4 Pillbox of
cross-sectional area S with
small thickness *h*

Fig. 14.5 Pillbox of
cross-sectional area *S* with
small thickness *h*

$$-\int_S \vec{J}_1 . \vec{n}_1 dS - \int_S \vec{J}_2 . \vec{n}_2 dS + contributions\ from\ walls = \frac{\partial \sigma}{\partial t} S, \quad (14.56)$$

where $\vec{n}_1$ *and* $\vec{n}_2$ are outwardly drawn opposite directed unit normal vectors. Now neglecting the side surface area of the pillbox, we have

$$-\vec{J}_1 \cdot \vec{n}_1 S - \vec{J}_2 \cdot \vec{n}_2 S = \frac{\partial \sigma}{\partial t} S \quad (14.57)$$

or

$$J_{2n} - J_{1n} = \frac{\partial \sigma}{\partial t}. \quad (14.58)$$

Thus, component of J that is perpendicular to the inter-surface of two different media is discontinuous. Actually the normal component of current density changes by an amount equal to the rate of change of free surface density of the charge at the interface.

e. Boundary condition for magnetic intensity vector $\vec{H}$

To find the boundary conditions for the tangential component of the magnetic intensity at a boundary between two different media, we use the following Maxwell
equation:

$$\vec{\nabla} \times \vec{H} = \vec{J} + \frac{\partial \vec{D}}{\partial t}. \tag{14.59}$$

Consider a rectangular loop across the interface between the two media (see
Fig. 14.5).

Now integrating over the loop of the above equation, we get

$$\int_S (\vec{\nabla} \times \vec{H}) \cdot \vec{n} dS = \int_S \vec{J}.\vec{n} dS + \int_S \frac{\partial \vec{D}}{\partial t} \cdot \vec{n} dS. \tag{14.60}$$

For very small h_1 and h_2, i.e. when the loop is shrunk, the sides of the rectangle will
vanish. Applying Stoke's theorem to the left-hand side, we obtain

$$\oint_{ABCD} \vec{H}.\vec{dl} = \int_{AB} \vec{H}_1.\vec{dl} + \int_{CD} \vec{H}_2 \cdot \vec{dl} + contribution\ from\ sides\ BC\ and DA. \tag{14.61}$$

As before, for vanishing BC and DA, the contribution from sides BC and DA will
vanish. Hence, we get

$$\oint_{ABCD} \vec{H} \cdot \vec{dl} = \int_{AB} \vec{H}_1 \cdot \vec{dl} + \int_{CD} \vec{H}_2 \cdot \vec{dl} = l H_{1t} - l H_{2t}. \tag{14.62}$$

Here the tangential components of the magnetic fields are H_{1t} and H_{2t}.

The second term on the right-hand side of (14.60) will vanish, since h_1 and h_2
will be very very small, i.e. the area is infinitely small. Here the first term on the
right-hand side of (14.60) may not vanish since, although the area is infinitely small,
$\vec{J}$ could be infinitely large. As a result, a surface current could exist on the boundary,
normal to the area ABCD. Hence Eq. (14.60) yields

$$l H_{1t} - l H_{2t} = J_S l \quad \Rightarrow \quad H_{1t} - H_{2t} = J_S, \tag{14.63}$$

where J_S denotes a surface current density perpendicular to the direction of the
tangential component of H which will be matched. It is evident that tangential
component of H is continuous only if there is no surface current at the interface of
the media. In this case

$$H_{1t} = H_{2t}, \tag{14.64}$$

i.e. tangential component of H is continuous at the interface of the media.

14.7 Plane Electromagnetic Waves in a Non-conducting Isotropic Medium

A plane wave is defined as a wave whose amplitude is the same at any point in plane perpendicular to a specified direction.

The wave equations for electric field strength $\vec{E}$ and magnetic intensity $\vec{H}$ are

$$\nabla^2 \vec{E} - \frac{1}{v^2} \frac{\partial^2 \vec{E}}{\partial t^2} = 0, \tag{14.65}$$

$$\nabla^2 \vec{H} - \frac{1}{v^2} \frac{\partial^2 \vec{H}}{\partial t^2} = 0. \tag{14.66}$$

Note 3: In free space phase, velocity v can be replaced by c and $\mu \to \mu_0, \epsilon \to \epsilon_0$, i.e.

$$v = \frac{1}{\sqrt{\epsilon \mu}} \to c = \frac{1}{\sqrt{\epsilon_0 \mu_0}}. \tag{14.67}$$

The plane wave solution of these equations can be written as

$$\vec{E}(\vec{r}, t) = \vec{E}_0 e^{i\vec{k} \cdot \vec{r} - i\omega t}, \tag{14.68}$$

$$\vec{H}(\vec{r}, t) = \vec{H}_0 e^{i\vec{k} \cdot \vec{r} - i\omega t}, \tag{14.69}$$

where $\vec{E}_0$, $\vec{H}_0$ are complex amplitudes which are constants in space and time.

$\vec{k}$ is wave propagation vector denoted as

$$\vec{k} = k\hat{n} = \frac{2\pi}{\lambda}\hat{n} = \frac{2\pi\nu}{v}\hat{n} = \frac{\omega}{v}\hat{n} \tag{14.70}$$

$[k = \frac{\omega}{v}, T = \frac{1}{\nu} \to$ time period, ν=frequency, $\frac{v}{\nu} = \lambda$=wave length, $\omega = \frac{2\pi}{T} \to$ angular frequency.]

Here $\hat{n}$ is a unit vector in the direction of wave propagation.

Now we try to find $\vec{\nabla} \cdot \vec{H}$ and $\vec{\nabla} \cdot \vec{E}$:

$$\vec{\nabla} \cdot \vec{E} = \left(i\frac{\partial}{\partial x} + j\frac{\partial}{\partial y} + k\frac{\partial}{\partial z} \right) \cdot \vec{E}_0 e^{k_x x + k_y y + k_z z - i\omega t} \tag{14.71}$$

$[\vec{k} = k_x \hat{i} + k_y \hat{j} + k_z \hat{k}, \vec{r} = x\hat{i} + y\hat{j} + z\hat{k}]$, i.e.

Fig. 14.6 A general plane
wave propagates in
$+z$-direction

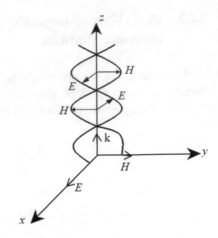

$$\vec{\nabla} \cdot \vec{E} = i\vec{k} \cdot \vec{E}_0 e^{i\vec{k}\cdot\vec{r}-i\omega t}.$$ (14.72)

Therefore,

$$\vec{\nabla} \cdot \vec{E} = i\vec{k} \cdot \vec{E}.$$ (14.73)

Similarly,

$$\vec{\nabla} \cdot \vec{H} = i\vec{k} \cdot \vec{H}.$$ (14.74)

For isotropic non-conducting medium

$$\vec{D} = \epsilon\vec{E}, \vec{B} = \mu\vec{H}, \vec{J} = \sigma\vec{E} = 0, \rho = 0, v = \frac{1}{\sqrt{\mu\epsilon}}.$$ (14.75)

Therefore, Maxwell's equations imply

$$\vec{\nabla} \cdot \vec{H} = 0 \; as \; well \; as \; \vec{\nabla} \cdot \vec{E} = 0$$

. These imply

$$\vec{k} \cdot \vec{E} = 0,$$ (14.76)

$$\vec{k} \cdot \vec{H} = 0.$$ (14.77)

This indicates that the field vectors $\vec{E}$ and $\vec{H}$ are both perpendicular to the direction
of propagation vector $\vec{k}$ (see Fig. 14.6), i.e. no components of electric field $\vec{E}$ and
magnetic field intensity $\vec{H}$ in the direction of wave propagation.

Thus, electromagnetic waves in isotropic non-conducting medium are transverse in nature. It is known as **Transverse Electromagnetic Wave [TEM]**.

From other Maxwell equations,

$$\nabla \times \vec{E} = -\mu \frac{\partial \vec{H}}{\partial t} \text{ and } \nabla \times \vec{H} = \epsilon \frac{\partial \vec{E}}{\partial t}, \tag{14.78}$$

we obtain

$$\vec{k} \times \vec{E} = \mu \omega \vec{H}, \quad \vec{k} \times \vec{H} = -\epsilon \omega \vec{E}. \tag{14.79}$$

This equation implies that

$$\vec{E} \cdot \vec{H} = 0. \tag{14.80}$$

Thus, $\vec{E}$ and $\vec{H}$ are mutually perpendicular and they are also perpendicular to the direction of the wave propagation $\vec{k}$. Thus, $\vec{E}, \vec{H}, \vec{k}$ form a set of orthogonal vector which constitute a right-handed coordinate system.

From Eq. (14.79), we have

$$\vec{H} = \frac{1}{\mu \omega}(\vec{k} \times \vec{E})$$

$$= \frac{k}{\mu \omega}(\vec{n} \times \vec{E}) \qquad (as \ \vec{k} = k\vec{n}),$$

$$= \frac{1}{\mu v}(\vec{n} \times \vec{E}) \quad \left(as \ k = \frac{\omega}{v}, \ v = \frac{1}{\sqrt{\epsilon \mu}} \right).$$

Hence, we get that

$$\vec{H} = \sqrt{\frac{\epsilon}{\mu}}(\vec{n} \times \vec{E}). \tag{14.81}$$

The ratio of magnitude of $\vec{E}$ and $\vec{H}$ is given by

$$\frac{|\vec{E}|}{|\vec{H}|} = z = \sqrt{\frac{\mu}{\epsilon}}. \tag{14.82}$$

This implies that the field vectors $\vec{E}$ and $\vec{H}$ are in the same phase. They take some relative magnitudes at all points at all time. Thus, unit of z is ohm and is known as **impedance** of isotropic non-conducting medium. Impedance is actually a measure of the total opposition of a circuit to current. In free space $z_0 = \sqrt{\frac{\mu_0}{\epsilon_0}} = 377$ Ohms as v can be replaced by c, i.e. $c^2 = \frac{1}{\mu_0 \epsilon_0}$.

The plane in which the electric field $\vec{E}$ lies is called the **plane of polarization** of the electromagnetic wave. For a TEM wave, the plane of polarization is perpendicular to the direction of propagation. If the direction of $\vec{E}$ does not change with time, the wave is said to be **linearly polarized.** Poynting vector for a plane electromagnetic wave in a isotropic non-conducting medium can be represented as

$$\vec{S} = \vec{E} \times \vec{H} = E \times \sqrt{\frac{\epsilon}{\mu}} \vec{n} \times \vec{E} \qquad (14.83)$$

$$= \frac{\vec{E} \times (\vec{n} \times \vec{E})}{z} = \frac{(\vec{E}\vec{E})\vec{n} - (\vec{E} \cdot \vec{n})\vec{E}}{z}. \qquad (14.84)$$

Hence, **Poynting vector for a plane electromagnetic wave** is given by

$$\vec{S} = \frac{E^2}{z}\vec{n}, \qquad (14.85)$$

where $\vec{E} \cdot \vec{n} = 0$ as $\vec{E}$ is perpendicular to $\vec{n}$.

The ratio of electrostatic and magnetostatic energy densities wave field is

$$\frac{u_e}{u_m} = \frac{\frac{1}{2}\epsilon E^2}{\frac{1}{2}\mu H^2} = \frac{\epsilon}{\mu}z^2 = \frac{\epsilon}{\mu} \cdot \frac{\mu}{\epsilon} = 1,$$

where $z = \frac{|\vec{E}|}{|\vec{H}|} = \sqrt{\frac{\mu}{\epsilon}}$.

Thus, electrostatic and magnetostatic energies are equal.

Therefore, total electromagnetic energy density is

$$u = u_m + u_e = 2u_e = 2\frac{1}{2}\epsilon E^2 = \epsilon E^2. \qquad (14.86)$$

In free space

$$u = \epsilon_0 E^2. \qquad (14.87)$$

The **Average Energy Density** for a plane electromagnetic wave is obtained as

$$< u > = \epsilon < E^2 > = \epsilon < (E_0 e^{i\vec{k}\cdot\vec{r} - i\omega t})^2 >_{real}$$

$$= \frac{\int_0^T \epsilon E_0^2 cos^2(\omega t - \vec{k} \cdot \vec{r})dt}{\int_0^T dt}$$

$$= \frac{1}{2}\epsilon E_0^2. \qquad [using\ \omega T = 2\pi]$$

Also the **Average Poynting Vector** for a plane electromagnetic wave is obtained as

$$< S >=< \vec{E} \times \vec{H} >=< \frac{E^2}{z}\vec{n} >$$

$$= \frac{1}{z} < (E_0 e^{i\vec{k}\cdot\vec{r}-i\omega t})^2 >_{real} \vec{n}$$

$$= \frac{1}{z}\vec{n} \frac{\int_0^T E_0^2 cos^2(\omega t - \vec{k}.\vec{r})dt}{\int_0^T dt} = \frac{E_0^2 \vec{n}}{2z}. \qquad (14.88)$$

Note 4: For finding actual Physical fields we often take real parts of complex exponentials.

14.8 Plane Electromagnetic Waves in a Conducting Medium

In a conducting medium, there is an induced current density $\vec{J}$ which is linearly proportional to the electric field E as $\vec{J} = \sigma\vec{E}$. σ is a constant of proportionality, known as conductivity. From Eqs. (14.11) and (14.14), one can write the wave equations for electric field strength $\vec{E}$ and magnetic intensity $\vec{H}$ in a conducting medium as

$$\nabla^2 \vec{E} - \sigma\mu\frac{\partial\vec{E}}{\partial t} - \epsilon\mu\frac{\partial^2\vec{E}}{\partial t^2} = 0 \qquad (14.89)$$

$$\nabla^2 \vec{H} - \sigma\mu\frac{\partial\vec{H}}{\partial t} - \epsilon\mu\frac{\partial^2\vec{H}}{\partial t^2} = 0. \qquad (14.90)$$

Let us assume the wave to be monochromatic and the fields vary as $e^{i\vec{k}.\vec{r}-i\omega.t}$. Therefore, the solutions of the above two equations are

$$\vec{E} = \vec{E}_0 e^{i\vec{k}.\vec{r}-i\omega t}, \tag{14.91}$$

$$\vec{H} = \vec{H}_0 e^{i\vec{k}.\vec{r}-i\omega t}, \tag{14.92}$$

where $\vec{k}$ is a wave vector may be complex while $\vec{E}_0$, $\vec{H}_0$ are complex amplitudes which are constants in space and time.

Putting Eq. (14.91) in Eq. (14.89), we get

$$(-k^2 + i\sigma\mu\omega + \mu\epsilon\omega^2)\vec{E} = 0. \tag{14.93}$$

This implies that

$$-k^2 + i\sigma\mu\omega + \mu\epsilon\omega^2 = 0 \tag{14.94}$$

$$k^2 = \mu\epsilon\omega^2 \left(1 + \frac{i\sigma}{\epsilon\omega}\right). \tag{14.95}$$

Here, the first term corresponds to the displacement current and the second term to the conduction current contribution. This shows that propagation constant $\vec{k}$ is complex.

Let us assume, $k = \alpha + i\beta$, i.e.

$$k^2 = \alpha^2 - \beta^2 + 2i\alpha\beta. \tag{14.96}$$

Comparing Eq. (14.96) with Eq. (14.95) we get

$$\alpha = \sqrt{\mu\epsilon}\omega \left[\frac{\sqrt{1 + (\frac{\sigma}{\omega\epsilon})^2} + 1}{2}\right]^{\frac{1}{2}} \tag{14.97}$$

$$\beta = \sqrt{\mu\epsilon}\omega \left[\frac{\sqrt{1 + (\frac{\sigma}{\omega\epsilon})^2} - 1}{2}\right]^{\frac{1}{2}}. \tag{14.98}$$

The quantity β characterizes the rate at which the amplitude of the wave decays with distances.

The quantity β is known as **attenuation constant** of the wave. Actually, attenuation defines the rate of loss of amplitude during the propagation of an electromagnetic wave. The quantity α gives the rate of change of phase with distance. α is known as **phase constant** of the wave.

Now in terms of α and β, the field vector $\vec{E}$ takes the form

$$\vec{E} = \vec{E}_0 e^{i(\alpha+i\beta)\hat{n}\cdot\vec{r}-i\omega t} \qquad (using \ \vec{k} = k\vec{n}) \qquad (14.99)$$

or

$$\vec{E} = \vec{E}_0 e^{-\beta\hat{n}\cdot\vec{r}} e^{i(\alpha\hat{n}\cdot\vec{r}-\omega t)}. \qquad (14.100)$$

Similarly, the magnetic intensity $\vec{H}$ can be obtained as

$$\vec{H} = \vec{H}_0 e^{-\beta\hat{n}\cdot\vec{r}} e^{i(\alpha\hat{n}\cdot\vec{r}-\omega t)}. \qquad (14.101)$$

The above equations suggest that due to presence of the term $e^{-\beta\hat{n}\cdot\vec{r}}$, the field amplitudes are damping. The quantity β is a measure of attenuation and is called as **absorption coefficient**. Also in the last exponential terms in (14.100) and (14.101), one can note that α plays the role of k. Hence, one can conclude that the field vectors are propagated in conducting medium with speed, i.e. phase velocity

$$v = \frac{\omega}{\alpha} = \frac{1}{\sqrt{\mu\epsilon}} \left[\frac{\sqrt{(1 + (\frac{\sigma}{\omega\epsilon})^2)} + 1}{2} \right]^{-\frac{1}{2}}. \qquad (14.102)$$

Maxwell's equations

$$\vec{\nabla} \cdot \vec{H} = 0 \ \& \ \vec{\nabla} \cdot \vec{E} = 0 \qquad (14.103)$$

imply

$$\vec{k} \cdot \vec{E} = 0 \ \& \ \vec{k} \cdot \vec{H} = 0. \qquad (14.104)$$

These mean that both field vectors are perpendicular to the direction of propagation of vector $\vec{k}$.

This indicates **electromagnetic wave in a conducting medium is transverse.**

Maxwell's equations

$$\vec{\nabla} \times \vec{E} = -\mu\frac{\partial \vec{H}}{\partial t} \ \& \ \vec{\nabla} \times \vec{H} = \vec{J} + \epsilon\frac{\partial \vec{E}}{\partial t} \qquad (14.105)$$

imply

$$i \vec{k} \times \vec{E} = i\mu\omega \vec{H} \quad i.e. \quad \vec{k} \times \vec{E} = \mu\omega \vec{H} \tag{14.106}$$

and

$$i \vec{k} \times \vec{H} = (\sigma - i\epsilon\omega) \vec{E} \quad i.e. \quad \vec{k} \times \vec{H} = -(\epsilon\omega + i\sigma).\vec{E} \tag{14.107}$$

These equations imply that electromagnetic field vectors $\vec{E}$, $\vec{H}$ are mutually perpendicular and also perpendicular to propagation vector $\vec{k}$.

Equation (14.106) yields

$$\vec{H} = \frac{\vec{k} \times \vec{E}}{\mu\omega} = \frac{k(\vec{n} \times \vec{E})}{\mu\omega} = \frac{(\alpha + i\beta)(\vec{n} \times \vec{E})}{\mu\omega}. \tag{14.108}$$

Therefore, the ratio of the magnitude of $\vec{E}$ and $\vec{H}$ is given by

$$\frac{|H|}{|E|} = \frac{(\alpha + i\beta)}{\mu\omega}. \tag{14.109}$$

Poynting vector for a plane electromagnetic wave is given by

$$\vec{S} = \vec{E} \times \vec{H} = \frac{(\alpha + i\beta)[\vec{E} \times (\vec{n} \times \vec{E})]}{\mu\omega},$$

$$\vec{S} = \frac{(\alpha + i\beta)E^2 \vec{n}}{\mu\omega}. \qquad (using \ \vec{E} \cdot \vec{n} = 0) \tag{14.110}$$

Note 5: Good conductors are characterized by $\frac{\sigma}{\omega\epsilon} \gg 1$. In this case, Eqs. (14.97) and (14.98) yield

$$\beta \cong \sqrt{\frac{\mu\sigma\omega}{2}} \cong \alpha. \tag{14.111}$$

Thus for good conductor, real or imaginary part of the wave vector $\vec{k}$ will be equal. This indicates that the decay of the wave will be very fast. For a good conductor, the phase velocity is

$$v = \frac{\omega}{\alpha} \cong \sqrt{\frac{2\omega}{\mu\sigma}}. \tag{14.112}$$

Here, the phase velocity depends on the frequency, i.e. there is dispersion.

Fig. 14.7 Skin depth is determined by the depth at which the amplitude of the wave has been decayed by a factor by $\frac{1}{e}$ of its initial amplitude at the surface

Poor conductors are characterized by $\frac{\sigma}{\omega\epsilon} << 1$. In this case, Eqs. (14.97) and (14.98) yield

$$\beta \cong \sqrt{\frac{\mu\sigma^2}{4\epsilon}} \quad , \quad \alpha \cong \omega\sqrt{\mu\epsilon}. \tag{14.113}$$

Thus

$$\beta = \frac{\alpha\sigma}{2\omega\epsilon}. \tag{14.114}$$

It indicates that $\beta << \alpha$. Thus for poor conductor, the decay of the wave will be very slow as one can see in the case of non-conducting material. For a poor conductor, the phase velocity is

$$v = \frac{\omega}{\alpha} \cong \frac{1}{\sqrt{\mu\epsilon}}. \tag{14.115}$$

Here, the phase velocity does not depend on the frequency, i.e. there is no dispersion.

14.9 Skin Depth

The waves given by Eqs. (14.100) and (14.101) show an exponential damping or attenuation with distance. For large value of β, one can get large attenuation.

The reciprocal of the attenuation constant, $\frac{1}{\beta}$, measures the depth at which electromagnetic wave must travel within a conductor before its amplitude has decayed by a factor $\frac{1}{e} = 0.369$ of its initial amplitude at the surface. This depth is called **skin depth** (see Fig. 14.7) and usually characterized by

$$\delta = \frac{1}{\beta} = \frac{1}{\omega\sqrt{\mu\epsilon}} \left[\frac{\sqrt{\{1 + (\frac{\sigma}{\omega\epsilon})^2\}} - 1}{2} \right]^{-\frac{1}{2}} . \qquad (14.116)$$

For good conductors ($\frac{\sigma}{\omega\epsilon} \gg 1$), Eq. (14.116) yields

$$\delta \cong \sqrt{\frac{2}{\mu\sigma\omega}}. \qquad (14.117)$$

Note that skin depth depends on the frequency and it decreases with increasing frequency. Copper is a good conductor. Its skin depth is 0.86 cm at 60 Hz, however, at 1 MHz, it has decayed to 0.00667 cm. As a result, circuit current flows only on the surface of the conductors at high frequency. Skin depth plays a significant role to measure the depth to which an electromagnetic wave can penetrate a conducting medium. Hence, one should be aware to use the conducting sheets for electromagnetic shields that must be thicker than the skin depth. Equation (14.117) indicates that the skin depth decreases as the conductivity σ, magnetic permeability μ and frequency ω increase. For poor conductor ($\frac{\sigma}{\omega\epsilon} \ll 1$), Eq. (14.116) yields

$$\delta \cong \sqrt{\frac{4\epsilon}{\mu\sigma^2}}. \qquad (14.118)$$

Here, the skin depth does not depend on the frequency; however, the skin depth increases with the increase of the electric permittivity ϵ.

14.10   Wave Guides

A hollow metallic pipe of infinite extent with the uniform cross section for propagating electromagnetic waves by consecutive reflections from the inner walls of the pipe is called as a **wave guide**. Actually, wave guides are a specific type of transmission passage that is used to monitor the waves along the length of the tube. Since the good conductors have very small skin depth, a tinny glazing of a good conductor stimulates the performance of an ideal wave guide. To study the propagation of electromagnetic waves within a wave guide, we follow Maxwell's equations subject to the boundary conditions at the bouncing walls on the electromagnetic field vectors $\vec{E}$ and $\vec{H}$.

We now consider the transmission of electromagnetic waves along a hollow conducting pipe (wave guide) (see Fig. 14.8) of arbitrary cross section. We assume also that wave propagates along z-axis and the wave guide is a perfect conductor, $\sigma = \infty$, so that $\vec{E} = 0$ and $\vec{B} = 0$ inside the material itself. The boundary conditions at the inner wall are

Fig. 14.8 A hollow pipe of
infinite extent with uniform
cross section as wave guide

$$E^{\parallel} = 0, \tag{14.119}$$

$$B^{\perp} = 0, \tag{14.120}$$

where $E^{\parallel}$ and $B^{\perp}$ are tangential component of $\vec{E}$ and normal component of $\vec{B}$, respectively.

For simplicity, the interior of the wave guide is assumed to be vacuum. Therefore, each component of $\vec{E}$ & $\vec{B}$ should satisfy the wave equations in vacuum

$$\vec{\nabla}^2 \vec{E} - \frac{1}{c^2} \frac{\partial^2 \vec{E}}{\partial t^2} = 0, \tag{14.121}$$

$$\vec{\nabla}^2 \vec{B} - \frac{1}{c^2} \frac{\partial^2 \vec{B}}{\partial t^2} = 0. \tag{14.122}$$

Apart from these wave equations, the Maxwell equations and the boundary conditions must be satisfied.

As the wave propagates within wave guide along z-axis, the solutions of (14.119) and (14.120) assume the form

$$\vec{E}(x, y, z, t) = \vec{E}_0(x, y)e^{i(kz - \omega t)}, \quad \vec{B}(x, y, z, t) = \vec{B}_0(x, y)e^{i(kz - \omega t)}. \tag{14.123}$$

The electric field and magnetic field also satisfy Maxwell equations in the interior of the wave guide:

$$\vec{\nabla} \cdot \vec{E} = 0 \tag{14.124}$$

$$\vec{\nabla} \times \vec{E} = -\frac{\partial \vec{B}}{\partial t} \tag{14.125}$$

$$\vec{\nabla} \cdot \vec{B} = 0 \tag{14.126}$$

$$\vec{\nabla} \times \vec{B} = \frac{1}{c^2} \frac{\partial \vec{E}}{\partial t}. \tag{14.127}$$

Now one can determine the values of $\vec{E}_0(x, y)$, $\vec{B}_0(x, y)$ from the differential Eqs. (14.124)–(14.127) using the boundary conditions (14.119) and (14.120).

Let us assume the components of $\vec{E}_0(x, y)$, $\vec{B}_0(x, y)$ are given by

$$\vec{E}_0 = E_x i + E_y j + E_z k \quad ; \quad \vec{B}_0 = B_x i + B_y j + B_z k. \tag{14.128}$$

Now, putting these in Maxwell's equations (14.126) and (14.127), we get

$$\frac{\partial E_y}{\partial x} - \frac{\partial E_x}{\partial y} = i\omega B_z \quad ; \quad \frac{\partial B_y}{\partial x} - \frac{\partial B_x}{\partial y} = -\frac{i\omega}{c^2} E_z,$$

$$\frac{\partial E_z}{\partial y} - ik E_y = i\omega B_x \quad ; \quad \frac{\partial B_z}{\partial y} - ik B_y = -\frac{i\omega}{c^2} E_x,$$

$$ik E_x - \frac{\partial E_z}{\partial x} = i\omega B_y \quad ; \quad ik B_x - \frac{\partial B_z}{\partial x} = -\frac{i\omega}{c^2} B_y. \tag{14.129}$$

Solving these equations, we get the transverse components

$$E_x = \frac{1}{\frac{\omega^2}{c^2} - k^2} \left(k \frac{\partial E_z}{\partial x} + \omega \frac{\partial B_z}{\partial y} \right), \tag{14.130}$$

$$E_y = \frac{1}{\frac{\omega^2}{c^2} - k^2} \left(k \frac{\partial E_z}{\partial y} - \omega \frac{\partial B_z}{\partial x} \right), \tag{14.131}$$

$$B_x = \frac{1}{\frac{\omega^2}{c^2} - k^2} \left(k \frac{\partial B_z}{\partial x} - \frac{\omega}{c^2} \frac{\partial E_z}{\partial y} \right), \tag{14.132}$$

$$B_y = \frac{1}{\frac{\omega^2}{c^2} - k^2} \left(k \frac{\partial B_z}{\partial y} + \frac{\omega}{c^2} \frac{\partial E_z}{\partial x} \right). \tag{14.133}$$

Now we have to determine the longitudinal components E_z and B_z. Inserting Eqs. (14.130)–(14.133) in (14.124) and (14.126), we get

$$\left[\frac{\partial^2}{\partial x^2} + \frac{\partial^2}{\partial y^2} + \left(\frac{\omega^2}{c^2} - k^2\right)\right] E_z = 0 \tag{14.134}$$

$$\left[\frac{\partial^2}{\partial x^2} + \frac{\partial^2}{\partial y^2} + \left(\frac{\omega^2}{c^2} - k^2\right)\right] B_z = 0. \tag{14.135}$$

Note that boundary conditions for $\vec{E}$ *and* $\vec{B}$ are different and cannot be satisfied simultaneously. Subsequently, the electromagnetic fields may be assigned into two different categories.

a. If $E_z = 0$, we call these **TE waves** (transverse electric waves).

b. If $B_z = 0$, we call these **TM waves** (transverse magnetic waves).

c. If both $E_z = 0$, $B_z = 0$, we call them **TEM waves** (transverse electromagnetic waves).

For hollow wave guide, either TE or TM waves can propagate but TEM cannot exist.

Note that in case of wave guide, we will have to solve only Eqs. (14.134) and (14.135) by using the boundary conditions (14.119) and (14.120). So for different wave guides, e.g. rectangular wave guide, circular wave guide, etc., one can get different solutions.

Note 6: The metal tube with rectangular cross section (with lengths a and b) is known as rectangular wave guide. The conductivity of this type of wave guide is zero.

For TE wave in rectangular wave guide:

i. $E_z(x, y) = 0$ everywhere.

ii. $B_z(x, y)$ satisfies the wave equation (14.135).

iii. The boundary conditions for $B_z(x, y)$ are

$$\frac{\partial B_z(x, y)}{\partial x} = 0 \quad at \quad x = 0 \, and \, x = a \tag{14.136}$$

and

$$\frac{\partial B_z(x, y)}{\partial y} = 0 \quad at \quad y = 0 \, and \, x = b. \tag{14.137}$$

To solve Eq. (14.135), one can use the method of separation of variables. The solution for $B_z(x, y)$ is

$$B_z(x, y) = B_0 \cos \frac{m\pi x}{a} \cos \frac{n\pi y}{b}, \tag{14.138}$$

where m, n are two parameters that define a wave guide mode and $\frac{\omega^2}{c^2} - k^2 = \frac{m^2\pi^2}{a^2} + \frac{n^2\pi^2}{b^2}$.

For TM wave in rectangular wave guide:

i. $B_z(x, y) = 0$ everywhere.
ii. $E_z(x, y)$ satisfies the wave equation (14.134).
iii. The boundary conditions for $E_z(x, y)$ are

$$\frac{\partial E_z(x, y)}{\partial x} = 0 \quad at \quad x = 0 \ and \ x = a \qquad (14.139)$$

and

$$\frac{\partial E_z(x, y)}{\partial y} = 0 \quad at \quad y = 0 \ and \ x = b. \qquad (14.140)$$

To solve Eq. (14.134), one can also use the method of separation of variables. The solution for $E_z(x, y)$ is

$$E_z(x, y) = E_0 \sin \frac{m'\pi x}{a} \sin \frac{n'\pi y}{b}, \qquad (14.141)$$

where m', n' are two parameters that define a wave guide mode and $\frac{\omega^2}{c^2} - k^2 = \frac{m'^2\pi^2}{a^2} + \frac{n'^2\pi^2}{b^2}$.

TE and TM waves in circular wave guide:

Consider a cylindrical wave guide with circular cross section of radius a. In cylindrical coordinate (r, θ, ϕ), the wave equations (14.134) and (14.135) take the following form:

$$\frac{1}{r}\frac{\partial}{\partial r}\left(r\frac{\partial \psi}{\partial r}\right) + \frac{1}{r^2}\frac{\partial^2 \psi}{\partial \theta^2} + \left(\frac{\omega^2}{c^2} - k^2\right)\psi = 0, \qquad (14.142)$$

where ψ stands for $E_z(r, \theta) \ and \ B_z(r, \theta)$.

For the TE mode in circular wave guide:

i. $E_z(r, \theta) = 0$ everywhere.
ii. $B_z(r, \theta)$ satisfies the wave equation (14.142).
iii. The boundary condition for $B_z(r, \theta)$ is

$$\frac{\partial B_z(r, \theta)}{\partial r} = 0 \quad at \quad r = a, \ \forall \ \theta. \qquad (14.143)$$

To solve Eq. (14.142), one can use the method of separation of variables. The solution for $B_z(r, \theta)$ is

$$B_z(r, \theta) = B_n J_n(\alpha r) \cos n\theta, \tag{14.144}$$

where J_n is the Bessel function of order n and $\alpha^2 = \frac{\omega^2}{c^2} - k^2$.

For the TM mode in circular wave guide:

i. $B_z(r, \theta) = 0$ everywhere.
ii. $E_z(r, \theta)$ satisfies the wave equation (14.142).
iii. The boundary condition for $E_z(r, \theta)$ is

$$E_z(r, \theta) = 0, \quad at \ r = a, \ \forall \ \theta. \tag{14.145}$$

To solve Eq. (14.142), one can use the method of separation of variables. The solution for $E_z(r, \theta)$ is

$$E_z(r, \theta) = E_n J_n(\alpha r) \cos n\theta, \tag{14.146}$$

where J_n is the Bessel function of order n and $\alpha^2 = \frac{\omega^2}{c^2} - k^2$.

14.11 Coulomb Gauge

The wave equations in terms of electromagnetic potentials are written as (see Eq. (13.10))

$$\nabla^2 \vec{A} - \mu\epsilon \frac{\partial^2 \vec{A}}{\partial t^2} - \vec{\nabla}\left(\vec{\nabla} \cdot \vec{A} + \mu\epsilon \frac{\partial \phi}{\partial t} \right) = -\mu \vec{J} \tag{14.147}$$

and

$$\nabla^2 \phi - \mu\epsilon \frac{\partial^2 \phi}{\partial t^2} + \frac{\partial}{\partial t}\left(\vec{\nabla} \cdot \vec{A} + \mu\epsilon \frac{\partial \phi}{\partial t} \right) = -\frac{\rho}{\epsilon}. \tag{14.148}$$

Equation (14.148) yields

$$\nabla^2 \phi + \frac{\partial}{\partial t}(\vec{\nabla} \cdot \vec{A}) = -\frac{\rho}{\epsilon}. \tag{14.149}$$

The **Coulomb gauge** is defined by the gauge condition that restricts the divergence of A as

$$\vec{\nabla} \cdot \vec{A} = 0. \tag{14.150}$$

Using the Coulomb gauge, Eq. (14.149) reduces to Poisson's equation

$$\nabla^2 \phi = -\frac{\rho}{\epsilon}. \tag{14.151}$$

The solution of this Poisson's equation is given by

$$\phi(r, t) = \frac{1}{4\pi\epsilon} \int \frac{\rho(r', t)}{|r - r'|} dv'. \tag{14.152}$$

Thus, the scalar potential can be expressed in terms of instantaneous Coulomb potential due to charge density $\rho(r, t)$.

14.12 Hertz Vector

The entire electromagnetic field may be characterized by a sole vector $\vec{z}$, known as the **Hertz vector**. H R Hertz showed that it is feasible under usual circumstances to describe an electromagnetic field in terms of a sole vector function. Actually, it can be shown that Maxwell's equations is equivalent to a single wave equation in terms of $\vec{z}$.

Recall Maxwell's equations:

$$\vec{\nabla} \cdot \vec{E} = 4\pi\rho \;\;, \;\;\; \vec{\nabla} \cdot \vec{B} = 0, \tag{14.153}$$

$$\vec{\nabla} \times \vec{E} = -\frac{\partial \vec{B}}{\partial t} \;\;, \;\;\; \vec{\nabla} \times \vec{H} = \vec{J} + \frac{\partial \vec{D}}{\partial t}. \tag{14.154}$$

Let us consider a homogeneous linear non-conducting medium, then

$$\rho = 0, \;\; \vec{J} = \sigma \vec{E} = 0, \;\; \vec{B} = \mu \vec{H}, \;\; \vec{D} = \epsilon \vec{E}. \tag{14.155}$$

Using these conditions, the Maxwell equations will take the form as

$$\vec{\nabla} \cdot \vec{E} = 0 \;\;, \;\;\; \vec{\nabla} \cdot \vec{B} = 0 \tag{14.156}$$

$$\vec{\nabla} \times \vec{E} = -\frac{\partial \vec{B}}{\partial t} \quad , \quad \vec{\nabla} \times \vec{B} = \mu\epsilon \frac{\partial \vec{E}}{\partial t}. \tag{14.157}$$

Now,

$$\vec{\nabla} \cdot \vec{B} = 0 \Rightarrow \vec{B} = \vec{\nabla} \times \vec{A}, \tag{14.158}$$

where $\vec{A}$ is electromagnetic vector potential.

We assume that the vector potential $\vec{A}$ is proportional to the time derivative of a vector $\vec{z}$ (motive will become understandable),

$$\vec{A} = \mu\epsilon \frac{\partial \vec{z}}{\partial t}, \tag{14.159}$$

where $\vec{z}$ is known as Hertz vector.

Now, one can get $\vec{B}$ in terms of $\vec{z}$ as

$$\vec{B} = \mu\epsilon \frac{\partial}{\partial t} (\vec{\nabla} \times \vec{z}). \tag{14.160}$$

The Maxwell equation $\vec{\nabla} \times \vec{B} = \mu\epsilon \frac{\partial \vec{E}}{\partial t}$ implies

$$\frac{\partial \vec{E}}{\partial t} = \frac{\partial}{\partial t} (\vec{\nabla} \times \vec{\nabla} \times \vec{z}) \tag{14.161}$$

or

$$\vec{E} = \vec{\nabla}(\vec{\nabla} \cdot \vec{z}) - \nabla^2 \vec{z}. \tag{14.162}$$

Here, the integration constant does not contribute anything for the evaluation of the field, and therefore we set it equal to zero.

From equation $\vec{\nabla} \times \vec{E} = -\frac{\partial \vec{B}}{\partial t}$, we obtain

$$\vec{\nabla} \times \vec{E} = -\frac{\partial}{\partial t} \left(\mu\epsilon \frac{\partial}{\partial t} (\vec{\nabla} \times \vec{z}) \right) = -\vec{\nabla} \times \left(\mu\epsilon \frac{\partial^2 \vec{z}}{\partial t^2} \right). \tag{14.163}$$

This equation implies

$$\vec{E} = -\mu\epsilon \frac{\partial^2 \vec{z}}{\partial t^2} - \vec{\nabla}\phi, \tag{14.164}$$

where the scalar function ϕ is purely arbitrary.

Equations (14.162) and (14.164) are identical if

$$\phi = -\vec{\nabla} \cdot \vec{z} \tag{14.165}$$

and

$$\nabla^2 \vec{z} - \mu\epsilon \frac{\partial^2 \vec{z}}{\partial t^2} = 0 \tag{14.166}$$

or

$$\nabla^2 \vec{z} - \frac{1}{v^2} \frac{\partial^2 \vec{z}}{\partial t^2} = 0, \tag{14.167}$$

where $\mu\epsilon = \frac{1}{v^2}$, v being phase velocity of the wave.

If we write the velocity of the electromagnetic wave as c, then Eq. (14.167) can be written as

$$\Box^2 \vec{z} = 0, \tag{14.168}$$

where $\Box^2 = \nabla^2 - \frac{1}{c^2} \frac{\partial^2}{\partial t^2}$ is the d'Alembertian operator.

Thus, we get a single equation in terms of Hertz vector $\vec{z}$ in place of a set of four Maxwell's equations.

Note 7: In homogeneous linear non-conducting medium, these fields lie outside to the region. So to get the most general physical implication of the Hertz vectors, we have to relate them to their sources. Hence, we have to consider the inhomogeneous equations from which Eq. (14.158) has been established.

Let us assume the electric and magnetic polarization vectors $\vec{P}$ and $\vec{M}$ associated with matter and vanish in free space. Then we rewrite the Maxwell equations as

$$\vec{\nabla} \cdot \vec{E} = \frac{1}{\epsilon} \vec{\nabla} \cdot \vec{P} \quad , \quad \vec{\nabla} \cdot \vec{H} = -\vec{\nabla} \cdot \vec{M}, \tag{14.169}$$

$$\vec{\nabla} \times \vec{E} + \mu \frac{\partial \vec{H}}{\partial t} = -\mu \frac{\partial \vec{M}}{\partial t} \quad , \quad \vec{\nabla} \times \vec{H} = \frac{\partial \vec{P}}{\partial t} + \epsilon \frac{\partial \vec{E}}{\partial t}. \tag{14.170}$$

The electric Hertz vector from these sources is given by

$$\nabla^2 \vec{z} - \mu\epsilon \frac{\partial^2 \vec{z}}{\partial t^2} = -\frac{1}{\epsilon} \vec{P}. \tag{14.171}$$

The magnetic Hertz vector from these sources is given by

$$\nabla^2 \vec{z} - \mu\epsilon \frac{\partial^2 \vec{z}}{\partial t^2} = -\mu \vec{M}. \tag{14.172}$$

14.13   A Brief Introduction of Relativistic Wave Equation

In 1926, Schrödinger proposed non-relativistic wave equation for a free particle. Here, the non-relativistic equation is given by

$$H = \frac{p^2}{2m} + V = E, \tag{14.173}$$

where E is the energy of the particle having mass m and momentum $\vec{p}$ moving in a potential V. Here H stands for Hamiltonian of a free particle.

Considering $\vec{p}$ and E as operators,

$$\vec{p} = -i\hbar\vec{\nabla}, \tag{14.174}$$

$$E = i\hbar\frac{\partial}{\partial t}. \tag{14.175}$$

One can find the **Schrödinger equation** as

$$\left(-\frac{\hbar^2}{2m}\nabla^2 + V\right)\psi(\vec{r}, t) = i\hbar\frac{\partial}{\partial t}\psi(\vec{r}, t), \tag{14.176}$$

which can be expressed as

$$H\psi(\vec{r}, t) = i\hbar\frac{\partial}{\partial t}\psi(\vec{r}, t). \tag{14.177}$$

Here $\psi(\vec{r}, t)$ represents a wave function of the particle which is normalized to unity. It is also understood as the probability of finding a particle at the time t with position $\vec{r}$ is $|\psi(\vec{r}, t)|^2$.

For a free particle of spin 0, the relativistic relationship between the energy and momentum $\vec{p}$ is

$$E = \pm(m^2c^4 + c^2p^2)^{\frac{1}{2}}, \tag{14.178}$$

where m is the mass.

As before one can substitute the energy and momentum as differential operators (given in Eqs. (14.174) and (14.175)), one can find

$$-\hbar^2\frac{\partial^2\psi}{\partial t^2} = (-c^2\hbar^2\nabla^2 + m^2c^4)\psi. \tag{14.179}$$

This equation is known as **Klein–Gordon equation**, taking positive sign.

This equation can be written as

$$\left[\nabla^2 - \frac{1}{c^2}\frac{\partial^2}{\partial t^2} - \frac{m^2 c^2}{\hbar^2} \right] \psi = 0, \tag{14.180}$$

which is represented as

$$\left(\Box^2 - \frac{m^2 c^2}{\hbar^2} \right) \psi = 0, \tag{14.181}$$

where $\Box^2 = \nabla^2 - \frac{1}{c^2}\frac{\partial^2}{\partial t^2} = \frac{\partial}{\partial x^\mu}\frac{\partial}{\partial x^\nu}$ is called the D'Alembertian operator, where $x^4 = ict$.

The complex conjugate of (14.180)) is given by

$$\left[\nabla^2 - \frac{1}{c^2}\frac{\partial}{\partial t^2} - \frac{m^2 c^2}{\hbar^2} \right] \psi^* = 0. \tag{14.182}$$

Multiplying (14.180) by ψ^* and (14.182) by ψ from the left and subtracting, we get

$$\psi^* \nabla^2 \psi - \psi \nabla^2 \psi^* + \frac{1}{c^2}\left(\psi\frac{\partial^2 \psi^*}{\partial t^2} - \psi^*\frac{\partial^2 \psi}{\partial t^2} \right) = 0 \tag{14.183}$$

or, $\vec{\nabla}\cdot(\psi^*\vec{\nabla}\psi - \psi\vec{\nabla}\psi^*) + \frac{1}{c^2}\frac{\partial}{\partial t}\left(\psi\frac{\partial \psi^*}{\partial t} - \psi^*\frac{\partial \psi}{\partial t} \right) = 0. \tag{14.184}$

Here, we have used the vector identity $\vec{\nabla}\cdot(\phi\vec{\nabla}\psi - \psi\vec{\nabla}\phi) = \phi\nabla^2\psi - \psi\nabla^2\phi$. Now the (14.184) can be seen as continuity equation

$$\vec{\nabla}\cdot\vec{J} + \frac{\partial\rho}{\partial t} = 0, \tag{14.185}$$

where the current density $\vec{J}$ and probability density ρ are defined as

$$\vec{J} = \frac{\hbar}{2mi}\left(\psi^*\vec{\nabla}\psi - \psi\vec{\nabla}\psi^* \right), \tag{14.186}$$

$$\rho = \frac{\hbar}{2mic^2}\left(\psi\frac{\partial\psi^*}{\partial t} - \psi^*\frac{\partial\psi}{\partial t} \right). \tag{14.187}$$

Unlike in Schrödinger equation, the probability density in Klein–Gordon equation given in (14.187) contains arbitrary functions ψ and $\frac{\partial\psi}{\partial t}$ which can be both positive and negative values. However, the probability density ρ must be a positive finite quantity and therefore initially ignored by the physics community until Pauli and Weisskopf realized that it can described charged particles, by multiplying the probability density with $\pm e$. Another problem arises due to the second-order differential equation in time t that leads to negative values in ρ as well. Also, since the Klein–Gordon equation comes from Einstein's $E - p$ relationship, there exists a negative energy solution even for a free particle as $E = \pm\sqrt{p^2c^2 + m^2c^4}$. This negative energy solution arises a serious problem as there exist infinite negative energy states, which is due to non-bounded values of $|\vec{p}|$. Hence, wave function interpretation of Klein–Gordon wave equation is not possible.

It is difficult to interpret the positive and negative signs. In classical mechanics, negative could be ignored. In quantum mechanics, the energy of the particle could be changed discontinuously.

To avoid the difficulties arising in Klein–Gordon equation, in 1928, Dirac proposed a new formulation which demands that an equation is linear in H, and therefore E must be linear in $\vec{p}$.

Keeping in mind the relativistic relation, $E^2 = p^2c^2 + m^2c^4$, Dirac chose a linear relation as

$$E = c\vec{\alpha} \cdot \vec{p} + \beta mc^2. \tag{14.188}$$

Here β and components of $\vec{\alpha}$ will be determined. Notice that $\vec{\alpha}$ and β do not contain $\vec{r}, t, \vec{p}, E$, and therefore commute with all of them to preserve linearity.

Squaring Eq. (14.188) yields

$$E^2 = c^2(\vec{\alpha} \cdot \vec{p})^2 + \beta^2 m^2 c^4 + (\vec{\alpha} \cdot \vec{p})\beta mc^3 + \beta(\vec{\alpha} \cdot \vec{p})mc^3. \tag{14.189}$$

Comparing with the relation, $E^2 = p^2c^2 + m^2c^4$, one obtains

$$(\vec{\alpha} \cdot \vec{p})^2 = p^2 \; ; \; \beta^2 = 1 \; ; \; \vec{\alpha}\beta = -\beta\vec{\alpha}. \tag{14.190}$$

Hence, $\vec{\alpha} = i\alpha_x + j\alpha_y + k\alpha_z$ where $\alpha_x, \alpha_y, \alpha_z$ are dimensionless constants and $\vec{p} = ip_x + jp_y + kp_z$.

The relation $(\vec{\alpha}.\vec{p})^2 = p^2$,

$$(\alpha_x p_x + \alpha_y p_y + \alpha_z p_z)^2 = p_x^2 + p_y^2 + p_z^2$$

implies that

$$\alpha_x^2 = \alpha_y^2 = \alpha_z^2 = 1 \tag{14.191}$$

which also implies that

$$\alpha_x \alpha_y = -\alpha_y \alpha_x; \ \alpha_y \alpha_z = -\alpha_z \alpha_y; \ \alpha_z \alpha_x = -\alpha_x \alpha_z. \tag{14.192}$$

$$\vec{\alpha}\beta = -\beta\vec{\alpha}, \tag{14.193}$$

which implies

$$\alpha_x \beta = -\beta\alpha_x; \ \alpha_y \beta = -\beta\alpha_y; \ \alpha_z \beta = -\beta\alpha_z. \tag{14.194}$$

Clearly α's and β are not ordinary constants. They are non-commutative and may be matrices, known as **Dirac matrices**.

From Eq. (14.194), one can easily get by using (14.192) and (14.193) as

$$\beta = -\alpha_i \beta \alpha_i, \tag{14.195}$$

$$\alpha_i = -\beta\alpha_i\alpha. \qquad [\ where \ i = x, y, z\ ] \tag{14.196}$$

These yield

$$Tr\beta = 0 = Tr\alpha,$$

where Tr represents trace of the matrix.

As α_i's and β are traceless matrices, and therefore eigenvalues should be in pairs. As a result they should be even dimensional anti-commutating traceless matrices. However, one can note that α_i and β cannot be 2×2 matrices as we have only three anti-commuting 2×2 Pauli spin matrices.

$$\sigma_x = \begin{bmatrix} 0 & 1 \\ 1 & 0 \end{bmatrix}, \quad \sigma_y = \begin{bmatrix} 0 & -i \\ i & 0 \end{bmatrix}, \quad \sigma_z = \begin{bmatrix} 1 & 0 \\ 1 & -1 \end{bmatrix}.$$

Therefore, the Dirac matrices $\alpha_x, \alpha_y, \alpha_z, \beta$ have the minimum dimension 4.

A particular example of Dirac matrices is

$$\alpha_i = \begin{bmatrix} 0 & \sigma_i \\ \sigma_i & 0 \end{bmatrix}, \quad \beta = \begin{bmatrix} I_2 & 0 \\ 0 & -I_2 \end{bmatrix}.$$

One can write it explicitly as

$$\alpha_x = \begin{bmatrix} 0&0&0&1 \\ 0&0&1&0 \\ 0&1&0&0 \\ 1&0&0&0 \end{bmatrix}, \quad \alpha_y = \begin{bmatrix} 0&0&0&-i \\ 0&0&i&0 \\ 0&-i&0&0 \\ i&0&0&0 \end{bmatrix}, \quad \alpha_z = \begin{bmatrix} 0&0&1&0 \\ 0&0&0&-1 \\ 1&0&0&0 \\ 0&-1&0&0 \end{bmatrix}, \quad \beta = \begin{bmatrix} 1&0&0&0 \\ 0&1&0&0 \\ 0&0&-1&0 \\ 0&0&0&-1 \end{bmatrix}.$$

Now using Eqs. (14.174) and (14.175), Eq. (14.188) yields the time-dependent **Dirac equation** for a free particle as

$$i\hbar\frac{\partial\psi}{\partial t} = (-i\hbar c\vec{\alpha}\cdot\vec{\nabla} + \beta mc^2)\psi. \tag{14.197}$$

Let us introduce,

$$\gamma^0 = \beta \ and \ \vec{\gamma} = \beta\vec{\alpha}. \tag{14.198}$$

Now Eq. (14.197) is multiplied from left by β, then it takes the following form:

$$ic\hbar\gamma^0\frac{\partial\psi}{\partial x^0} + i\hbar c\vec{\gamma}\cdot\vec{\nabla}\psi - mc^2\psi = 0 \tag{14.199}$$

or

$$\left(\gamma^\mu\frac{\partial}{\partial x^\mu} - \frac{mc}{i\hbar}\right)\psi = 0. \quad (x_0 = ct, \ \beta^2 = 1) \tag{14.200}$$

This is the **covariant form of the Dirac equation**.

Dirac equation is relativistically invariant under Lorentz transformation.

Let S' be the moving frame. In S' frame Dirac equation is written as

$$\left(\gamma^\mu\frac{\partial}{\partial x'^\mu} - \frac{mc}{\lambda\hbar}\right)\Psi'(r', t') = 0. \tag{14.201}$$

Let the transformation of the state vector Ψ is given by

$$\Psi'(r', t') = S\Psi(r, t), \tag{14.202}$$

where S is the Lorentz transformation operator, which is a 4×4 matrix.

Under a Lorentz transformation, we know

$$x^\mu \to x'^\mu = L^\mu_\nu x^\nu. \tag{14.203}$$

Here,

$$L^\nu_\mu L^\mu_\sigma = \delta^\gamma_\sigma, \ L^T L = I, \ det(L^\nu_\mu) = \pm 1. \tag{14.204}$$

So for every Lorentz transformation, there exists an inverse transformation such that

$$x^\mu = L^\mu_\nu x'^\nu. \tag{14.205}$$

Using this inverse transformation, we have

$$\frac{\partial}{\partial x^\mu} = L_\mu^\nu \frac{\partial}{\partial x'^\nu}. \tag{14.206}$$

From Eq. (14.201), we have

$$\left(\gamma^\mu S \frac{\partial}{\partial x'^\mu} - \frac{mc}{i\hbar} S\right)\Psi = 0. \tag{14.207}$$

Multiplying both sides from left by S^{-1}, we get

$$\left(S^{-1}\gamma^\mu S \frac{\partial}{\partial x'^\mu} - \frac{mc}{i\hbar}\right)\Psi = 0. \tag{14.208}$$

If this equation is identical to (14.200) then

$$S^{-1}\gamma^\nu S \frac{\partial}{\partial x'^\nu} = \gamma^\mu \frac{\partial}{\partial x^\mu} = \gamma^\mu L_\mu^\nu \frac{\partial}{\partial x'^\nu} \tag{14.209}$$

or

$$S^{-1}\gamma^\nu S = \gamma^\mu L_\mu^\nu. \tag{14.210}$$

Therefore, if one can find a 4×4 matrix S satisfying this equation, the wave functions Ψ and Ψ' satisfy Dirac equation of the same form. Hence, Dirac equation is invariant under Lorentz transformation.

Note that gamma matrices follow the following relations:

$$\gamma_\mu^2 = 1, \quad \mu = 1, 2, 3, 4 \text{ and } \gamma_\mu \gamma_\nu = -\gamma_\nu \gamma_\mu, \quad \nu \neq \mu = 1, 2, 3, 4, \tag{14.211}$$

i.e. gamma matrices are anti-commutative and square of each gamma matrix is unity. Actually,

$$\gamma^0 = \begin{bmatrix} I & 0 \\ 1 & -I \end{bmatrix}, \quad \gamma^i = \begin{bmatrix} 0 & \sigma^i \\ -\sigma^i & 0 \end{bmatrix}. \tag{14.212}$$

It is easy to check γ^0 is Hermitian and γ^i is anti-Hermitian.

The time-independent Dirac equation for a free particle assumes the following form:

$$(c\vec{\alpha} \cdot \vec{p} + \beta mc^2)\psi = E\psi. \tag{14.213}$$

Fig. 14.9 Dirac sea

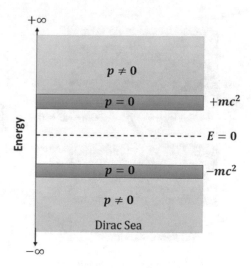

Here, ψ is multi-component wave function. Note that if α_i's and β are $n \times n$ matrices, the Dirac wave function ψ must be a $n \times 1$ matrix, i.e. column vector with n components. In Eq. (14.213), $\vec{p}$ is not an operator rather it is momentum vector.

Dirac equation is physically acceptable relativistic wave equation as it obeys the continuity equation. Taking the Hermitian conjugate of (14.197) and multiplying from right by ψ and subtracting after multiplying (14.197) from its left by $\psi^\dagger$, we get (assuming α, β are Hermitian)

$$i\hbar\frac{\partial(\psi^\dagger\psi)}{\partial t} = -ihc\frac{\partial(\psi^\dagger\alpha\psi)}{\partial x^i}. \tag{14.214}$$

This can be written as a continuity equation

$$\frac{\partial\rho}{\partial t} + \vec{\nabla}\cdot\vec{J} = 0, \tag{14.215}$$

where $\rho = \psi^\dagger\psi$ and $\vec{J} = c\psi^\dagger\alpha\psi$. Note that ρ is positive and thus avoid the trouble faced in the Klein–Gordon equation.

Note 1: The Dirac equation has solutions of negative energy quantum states. To explain this strange result, Dirac proposed a theoretical model of the vacuum as an infinite sea of particles with negative energy so-called **Dirac sea**. Since Dirac's theory also originates from Einstein's relationship, there still exist both negative and positive energy solutions. Even though the negative energy solution is inconsistent with physical interpretations, one cannot ignore as both the solutions constituted a complete set. For stationary particles with positive and negative energy states, have $E = \pm mc^2$ and hence the two states are separated by an energy gap of $2mc^2$. In classical physics, these two states do not mix or interact, however it is not the same

Fig. 14.10 Pair creation and annihilation

in case of quantum physics. In quantum physics, there is some transition probability of particles from positive energy states to negative energy states. If these transitions occur then one might not observe any positive energy particles or, in other words, even atoms might not exist. To void such transitions, Dirac argued that the negative energy states (now known as *Dirac sea* or *Vacuum*, see Fig. 14.9) are completely filled with fermions and the Pauli exclusion principle forbids these transitions. This interpretation cannot apply to Klein–Gordon theory as it described spinless particles (or bosons) where one can occupy the same state, i.e. the phenomena of *Bose–Einstein condensation*. This Dirac's interpretation for negative energy solutions and the idea of Dirac sea lead to the prediction of anti-particles. If a particle in the negative energy state adsorbs sufficient amount of energy from an incident radiation ($\geq 2mc^2$) then the particle can jump to the positive energy state creating a hole in Dirac's sea. Due to the conservation of electric charge during this process, the hole will have an opposite charge to that of the particle jumped to the positive energy state. If the jumped particle is assumed to be an electron (e^-), then the hole will carry positive charge and is the anti-particle of electron now known as *positron* (e^+). The creation of electron–hole pair is termed as '*pair creation*' and the reverse process in the '*pair annihilation*' (see Fig. 14.10). Hence, the Dirac theory originally formulated for single-particle system ended up as many-particle system due to the relativistic effect.

Example 14.1 Show that

$$[\sigma_x, \sigma_y] = 2i\sigma_z \; ; \quad [\sigma_y, \sigma_z] = 2i\sigma_x \; ; \quad [\sigma_x, \sigma_x] = 2i\sigma_y \, .$$

Hint: Use Pauli spin matrices and $[\sigma_x, \sigma_y] = \sigma_x\sigma_y - \sigma_y\sigma_x$.

Example 14.2   Prove that

$$\gamma^a\gamma^b + \gamma^b\gamma^a = 2\eta^{ab}.$$

Hint: Use Eq. 14.198.

Example 14.3   Let us define a matrix

$$\gamma^5 = \gamma^0\gamma^1\gamma^2\gamma^3.$$

Show that

$$\gamma^5\gamma^a + \gamma^a\gamma^5 = 0 \quad : \quad (\gamma^5)^2 = -I.$$

Hint: Use Eq. 14.198.

Example 14.4   Show that

$$\alpha_x = \frac{1}{2}[\alpha_x\alpha_y, \alpha_y] \; ; \quad \alpha_x\alpha_y\alpha_z = \frac{1}{2}[\alpha_x\alpha_y\alpha_z\beta, \beta].$$

Hint: We know,

$$[\alpha_x\alpha_y, \alpha_y] = \alpha_x\alpha_y\alpha_y - \alpha_y\alpha_x\alpha_y.$$

Now, use

$$\alpha_x^2 = \alpha_y^2 = \alpha_z^2 = \beta^2 = 1 \; , \quad \alpha_\mu\alpha_\nu = -\alpha_\nu\alpha_\mu.$$

Chapter 15
Relativistic Mechanics of Continua

15.1 Relativistic Mechanics of Continuous Medium (Continua)

Here we give our attention on the behaviour of the continuous distribution of matter—elastic bodies, liquids and gases. These so-called continuous media possesses energy density, momentum density and stresses. In our previous discussion on relativistic electromagnetic theory, we have noticed that these quantities together are expressed in a four vector form and therefore are Lorentz covariant. So we would expect the same for the continuous medium.

In the mechanics of continuous matter, we will deal with local volume element rather than the individual particle. A volume element contains sufficient number of individual particles that reflect the characteristic of continuous medium.

Let ρ be the density and $\vec{u}$ be the average velocity of matter; then the change of the density in a volume element is given by the non-relativistic continuity equation

$$\frac{\partial \rho}{\partial t} + \nabla \cdot (\rho \vec{u}) = 0. \tag{15.1}$$

This equation implies the influx of matter into the volume element. Let a force acting on a unit volume of matter be denoted by $\vec{g}$. Then one can write the equation of motion using Newton's laws as

$$\rho \frac{d\vec{u}}{dt} = \vec{g}. \tag{15.2}$$

Here the velocity $\vec{u}$ of the unit volume of matter depends on space and time coordinates, i.e. on x_1, x_2, x_3 and t.

Applying chain rule, we have

$$\frac{du_i}{dt} = \frac{\partial u_i}{\partial t} + \sum_{k=1}^{k=3} \frac{\partial u_i}{\partial x_k} \frac{\partial x_k}{\partial t} = \frac{\partial u_i}{\partial t} + u_{i,k} u_k. \tag{15.3}$$

Here we have used summation convention and $u_{i,k} = \frac{\partial u_i}{\partial x_k}$.

By making use of Eq. (15.3), we get from (15.2)

$$\rho \left[\frac{\partial u_i}{\partial t} + u_{i,k} u_k \right] = g_i. \tag{15.4}$$

We can rearrange the left-hand side expression as

$$\rho \left[\frac{\partial u_i}{\partial t} + u_{i,k} u_k \right] = \frac{\partial(\rho u_i)}{\partial t} - u_i \frac{\partial \rho}{\partial t} + (\rho u_i u_k)_{,k} - u_i (\rho u_k)_{,k}$$

$$= \frac{\partial(\rho u_i)}{\partial t} + (\rho u_i u_k)_{,k} - u_i \left[\frac{\partial \rho}{\partial t} + (\rho u_k)_{,k} \right].$$

Using continuity Eq. (15.1), finally we have

$$\rho \left[\frac{\partial u_i}{\partial t} + u_{i,k} u_k \right] = \frac{\partial(\rho u_i)}{\partial t} + (\rho u_i u_k)_{,k}. \tag{15.5}$$

Therefore, Eq. (15.4) yields

$$g_i = \frac{\partial(\rho u_i)}{\partial t} + (\rho u_i u_k)_{,k}. \tag{15.6}$$

There are two types of forces acting on a continuous medium (fluid), namely, external (volume forces) and internal or elastic forces (stresses). The external forces such as the forces due to gravitational field or electric field are acting on the matter in a given infinitesimal volume element proportional to the size of the volume element. Sometimes external forces are called volume or body forces.

Let F_i be the ith component of the external force exerted on the continuous medium (fluid), then

$$F_i = \int_V f_i \, dV, \tag{15.7}$$

where V is the volume of the continuous medium (fluid). Here, f_i is the force per unit volume.

The second type of the forces (internal) is the stress within the material itself, i.e. the forces due to the mutual action of particles in the two adjoining volume elements which are proportional to the area of the surface of separation. These are also known as area forces. The components of the area force, dq_i, depend linearly on the components of the elementary area dA_k.

$$dq_i = t_{ik}dA_k. \tag{15.8}$$

Here dA_k is the element of vector area perpendicular to x_k-direction. Thus dq_i is the force acting in the x_i-direction. The stress t_{ik}, i.e. the force per unit area must be symmetric in its indices

$$t_{ik} = t_{ki}. \tag{15.9}$$

Here the first index i indicates that the direction of the force is along x_i-direction and second index implies the normal to the surface is along x_k. If the components of the stress t_{ik} are not symmetric, then the total angular momentum will change without applying any external (volume forces) f_i at the rate

$$\frac{dM_i}{dt} = \int_V \delta_{ikl} t^{kl} \, dV.$$

[V is the volume of the whole body and δ_{ikl} is the Levi-Civita tensor.]

Now the ith component of the area force exerted on the continuous medium (fluid) inside a closed surface A by the fluid outside is given by

$$H_i = -\int_A t_{ik} \, dA_k = -\int_V t_{ik,k} \, dV \quad \text{(by Gauss law)}, \tag{15.10}$$

where dA_k is an elementary vector surface area and V is the volume enclosed by A. Therefore, area force, i.e. internal force acting on a unit volume of the continuous medium is

$$h_i = -t_{ik,k}. \tag{15.11}$$

Hence, the total force which acts on the matter contained in a volume V is given as

$$G_i = \int_V f_i \, dV - \int_V t_{ik,k} \, dV = \int_V (f_i - t_{ik,k}) \, dV. \tag{15.12}$$

Therefore, the force per unit volume is

$$g_i = f_i - t_{ik,k}. \tag{15.13}$$

From (15.6) and (15.13), we obtain

$$f_i = \frac{\partial(\rho u_i)}{\partial t} + (\rho u_i u_k + t_{ik})_{,k}. \tag{15.14}$$

These three Eqs. (15.12)–(15.14) together with the continuity Eq. (15.1) determine the motion of a continuous medium in non-relativistic mechanics under the influence of internal and external forces.

Thus if one knows the volume or external force f_i such as gravitational force or electromagnetic force and the stresses t_{ik} which depend on the internal deformation, then the continuum system will be completely determined. For example, in case of perfect fluid, $t_{ik} = p\delta_{ik}$ where p is the pressure.

To find the equation of continuity and equation of motion in a continuous medium in special theory of relativity, we use a special coordinate system. In a continuous medium, it is convenient to use a special coordinate system S^0 known as the co-moving coordinate system, with respect to which the matter is at rest at a world point P^0.

At this world point, the formulation of the equations is greatly simplified, because all classical terms containing undifferentiated velocity components vanish. The components of the four velocity U_μ can be expressed as

$$U_i = au_i \quad \text{and} \quad U_4 = aic,$$

where $a = \frac{1}{\sqrt{1-\frac{u^2}{c^2}}}$.

We note that the first derivatives of the spatial part of the velocity U_i are equal to those of u_i, whereas the first derivative of the temporal part of the velocity U_4 vanishes as $U_4 = ic$. In relativistic mechanics, energy flux is c^2 times the momentum density. This follows from mass–energy relation. The non-relativistic momentum density is ρu_i. However, in addition, we have to consider the flux of energy on account of the stress.

Two same and opposite forces $t_{ik}dA_k$ and $-t_{ik}dA_k$ act on either side of the elementary vector surface area dA_k depending on which way the normal to the surface points. The amount of energy gained on one side is equal to that lost on the other. The amount of energy flux vector is $u_i t_{ik}$ and the corresponding momentum density is $\frac{u_i t_{ik}}{c^2}$.

$[E = mc^2; E = t_{ik}, m = \frac{E}{c^2}$, and therefore the momentum density $\rho u_i = \frac{m}{V}u_i = \frac{u_i t_{ik}}{c^2}$ as V = unit volume]

Hence, the total relativistic momentum density of the continuous medium will be

$$P_k = \rho u_k + \frac{u_i t_{ik}}{c^2}. \tag{15.15}$$

Here, we note that P_k vanishes in the special coordinate system but not their derivatives. Therefore, the equation of continuity in a continuous medium in special theory of relativity will be

$$\frac{\partial \rho}{\partial t} + P_{k,k} = 0. \tag{15.16}$$

Substituting the expression for P_k from (15.15), we obtain

$$\frac{\partial \rho}{\partial t} + \rho \acute{u}_{k,k} + \frac{u_{i,k} t_{ik}}{c^2} = 0. \tag{15.17}$$

As we know t is related to x_4 as $x_4 = ict$, therefore above equation can be written as

$$ic\rho_{,4} + \rho u_{k,k} + \frac{u_{i,k} t_{ik}}{c^2} = 0. \tag{15.18}$$

This is the equation of continuity in a continuous medium in special theory of relativity.

Similarly, we can get the equation of motion by adding the contribution from the energy flux as

$$f_i = \frac{\partial \left(\rho u_i + \frac{u_i t_{ik}}{c^2} \right)}{\partial t} + (\rho u_i u_k + t_{ik})_{,k}. \tag{15.19}$$

As before, we replace t by $x_4 = ict$ and obtain *the motion of a continuous medium in relativistic mechanics under the influence of internal and external forces.*

$$f_j = ic\rho u_{j,4} + \frac{i}{c} u_{k,4} t_{kj} + (\rho u_j u_k + t_{jk})_{,k}. \tag{15.20}$$

Appendix A

The investigation of relations which remain valid when we change from coordinate system to any other is the chief aim of tensor calculus. The laws of physics cannot depend on the frame of reference which the physicists choose for the purpose of description. Accordingly it is convenient to utilize the tensor calculus as the mathematical background in which such laws can be formulated.

Transformation of Coordinates:

Let there be two reference systems S and S'. S' system depends on S system, i.e.

$$x^{'i} = \phi^i \left(x^1, x^2, \ldots, x^n \right) \; ; i = 1, 2, \ldots, n. \tag{A.1}$$

Differentiation of (A.1) gives

$$dx^{'i} = \frac{\partial \phi^i}{\partial x^r} dx^r = \frac{\partial x^{'i}}{\partial x^r} dx^r \tag{A.2}$$

or

$$dx^i = \frac{\partial x^i}{\partial x^{'m}} dx^{'m}. \tag{A.3}$$

Vectors in $\left(S' \right)$ system are related with (S) system.

There are two possible ways of transformations:

$$\bar{A}^i = \frac{\partial \bar{x}^i}{\partial x^j} A^j$$

© The Editor(s) (if applicable) and The Author(s), under exclusive license to Springer Nature Singapore Pte Ltd. 2022
F. Rahaman, *The Special Theory of Relativity*, La Matematica per il 3+2 136,
https://doi.org/10.1007/978-981-19-0497-4

it is called **Contravariant vector**.

$$\bar{A}_i = \frac{\partial x^j}{\partial \bar{x}^i} A_j$$

it is called **Covariant vector**.

The expression $A^i B_i$ is an invariant, i.e.

$$\bar{A}^i \bar{B}_i = A^i B_i.$$

$A^i B^j = A^{ij}$ form n^2 quantities.

If the transformation of A^{ij} like

$$\bar{A}^{ij} = \frac{\partial \bar{x}^i}{\partial x^k} \frac{\partial \bar{x}^j}{\partial x^l} A^{kl},$$

then A^{ij} is known as **Contravariant tensor of rank two**.

If

$$\bar{A}_{ij} = \frac{\partial x^k}{\partial \bar{x}^i} \frac{\partial x^l}{\partial \bar{x}^j} A_{kl},$$

then A_{ij} is known as **covariant tensor of rank two**.

Mixed tensor of rank two: A^i_j

Here the transformation,

$$\bar{A}^i_j = \frac{\partial \bar{x}^i}{\partial x^k} \frac{\partial x^l}{\partial \bar{x}^j} A^k_l$$

$$T^\gamma_{\alpha\beta} = A_{\alpha\beta} B^\gamma \quad \rightarrow \quad \textbf{Outer Product}.$$

$$T^\alpha_{\alpha\beta} \quad \rightarrow \quad \textbf{Contraction}.$$

$$A^{ij}_k \, B^k_{mn}$$

or

$$A_k^{ij} \, B_{ij}^m$$

are known as **Inner Product**.

$$T_{\alpha\beta} = T_{\beta\alpha} \quad \longrightarrow \quad \textbf{Symmetric}$$
$$T_{\alpha\beta} = -T_{\beta\alpha} \quad \longrightarrow \quad \textbf{Anti} - \textbf{symmetric}.$$

The Line element:

The distance between two neighbouring points is given by

$$ds^2 = g_{ij} dx^i dx^j. \tag{A.4}$$

Here, g_{ij} are functions of x^i. If $g = |g_{ij}| \neq 0$, then the space is called **Riemannian space**.

g_{ij} is called **fundamental tensor** (covariant tensor of order two).

In Euclidean space:

$$ds^2 = dx^2 + dy^2 + dz^2.$$

In Minkowski flat space–time, the line element

$$ds^2 = dx^{0^2} - dx^{1^2} - dx^{2^2} - dx^{3^2}.$$

In order that the distance ds between two neighbouring points be real, Eq. (A.4) will be amended to

$$ds^2 = e g_{ij} dx^i dx^j \quad [e = \pm 1].$$

Here e is called the indicator and takes the value $+1$ or -1 so that ds^2 is always positive.

The contravariant tensor g^{ij} is defined by

$$g^{ij} = \frac{\Delta_{ij}}{g},$$

here Δ_{ij} is the cofactor of g_{ij}.

Obviously,

$$g_{ab} \, g^{bc} = \delta_a^c.$$

One can raise or lower the indices of any tensor with the help of g^{ab} and g_{ab} as

$$g_{ac} T^{ab} = T^b_c$$
$$T_{ab}\, g^{ac} = T^c_b$$
$$g_{ab} A^b = A_a$$
$$g^{ab}\, A_a = A^b$$
$$g^{ab}\, g^{cd}\, g_{bd} = g^{ac}.$$

Theorem *The expression $\sqrt{-g}\, d^4x$ is an invariant volume element.*

Hints: We know

$$g'_{ab} = \frac{\partial x^c}{\partial x'^a} \frac{\partial x^d}{\partial x'^b}\, g_{cd}$$

$\Rightarrow$

$$\det\left(g'_{ab}\right) = \left|\frac{\partial x}{\partial x'}\right|^2 \det\left(g_{cd}\right)$$

$\Rightarrow$

$$g' = \left|\frac{\partial x}{\partial x'}\right|^2 g. \tag{A.5}$$

Also the volume element d^4x transform into d^4x' as

$$d^4x' = \left|\frac{\partial x'}{\partial x}\right| d^4x. \tag{A.6}$$

From (A.5) and (A.6), we get

$$\sqrt{-g'}\, d^4x' = \sqrt{-g}\, d^4x.$$

Magnitude l of a vector A^μ is defined as

$$l^2 = A^\mu A_\mu = g_{\mu\nu} A^\mu A^\nu = g^{\mu\nu} A_\mu A_\nu.$$

Angle between two vectors A^μ and B^μ:
We know in vector algebra, angle between two vectors $\vec{A}$ and $\vec{B}$ is defined as

$$\cos \theta = \frac{\vec{A} \cdot \vec{B}}{|\vec{A}| |\vec{B}|}.$$

Similarly, the angle between two vectors A^μ and B^μ is defined as

$$\cos \theta = \frac{\text{scalar product of } A^\mu \text{ and } B^\mu}{\text{length of } A^\mu \times \text{length of } B^\mu}$$

$$= \frac{A^\mu B_\mu}{\sqrt{(A^\mu A_\mu)(B^\mu B_\mu)}}$$

$$= \frac{g^{\mu\nu} A_\mu B_\nu}{\sqrt{(g^{\alpha\beta} A_\alpha B_\beta)(g^{\rho\sigma} A_\rho B_\sigma)}}.$$

Levi-Civita Tensor:

$$\epsilon^{abcd} = +1,$$

if a, b, c, d is an even permutation of $0, 1, 2, 3$, i.e. in cyclic order.

$$= -1,$$

if a, b, c, d is odd permutation of $0, 1, 2, 3$, i.e. not in cyclic order.

$$= 0$$

if any two indices are equal.

The components of ϵ_{abcd} are obtained from ϵ^{abcd} by lowering the indices in the usual way, and multiplying it by $(-g)^{-1}$:

$$\epsilon_{abcd} = g_{a\mu} g_{b\nu} g_{c\gamma} g_{d\sigma} (-g)^{-1} \epsilon^{\mu\nu\gamma\sigma}.$$

For example,

$$\epsilon_{0123} = g_{0\mu} g_{1\nu} g_{2\gamma} g_{3\sigma} (-g)^{-1} \epsilon^{\mu\nu\gamma\sigma}$$

$$= (-g)^{-1} det \, g_{\mu\nu}$$

$$= -1.$$

In general,

$$\epsilon_{abcd} = 1,$$

if a, b, c, d is an even permutation of $0, 1, 2, 3$.

$$= -1,$$

if a, b, c, d is odd permutation of $0, 1, 2, 3$.

$$= 0 \quad \text{otherwise.}$$

Generalized Kronecker Delta:
$\delta_{\mu\nu}^{\alpha\beta}$ can be constructed as follows:

$$\delta_{\mu\nu}^{\alpha\beta} = \begin{vmatrix} \delta_\mu^\alpha & \delta_\mu^\beta \\ \delta_\nu^\alpha & \delta_\nu^\beta \end{vmatrix}$$

$$= +1, \quad \alpha \neq \beta, \ \alpha\beta = \mu\nu$$
$$= -1, \quad \alpha \neq \beta, \ \alpha\beta = \nu\mu$$
$$= 0, \quad \text{otherwise.}$$

We can define $\delta_{\mu\nu\xi}^{\alpha\beta\gamma}$ and $\delta_{\mu\nu\xi\sigma}^{\alpha\beta\gamma\delta}$ as follows:

$$\delta_{\mu\nu\xi}^{\alpha\beta\gamma} = \begin{vmatrix} \delta_\mu^\alpha & \delta_\mu^\beta & \delta_\mu^\gamma \\ \delta_\nu^\alpha & \delta_\nu^\beta & \delta_\nu^\gamma \\ \delta_\xi^\alpha & \delta_\xi^\beta & \delta_\xi^\gamma \end{vmatrix}$$

$$= +1, \quad \alpha \neq \beta \neq \gamma, \ \alpha\beta\gamma \text{ is an even permutation of } \mu\nu\xi$$
$$= -1, \quad \alpha \neq \beta \neq \gamma, \ \alpha\beta\gamma \text{ is an odd permutation of } \mu\nu\xi$$
$$= 0, \quad \text{otherwise.}$$

$$\delta_{\mu\nu\xi\sigma}^{\alpha\beta\gamma\delta} = \begin{vmatrix} \delta_\mu^\alpha & \delta_\mu^\beta & \delta_\mu^\gamma & \delta_\mu^\delta \\ \delta_\nu^\alpha & \delta_\nu^\beta & \delta_\nu^\gamma & \delta_\nu^\delta \\ \delta_\xi^\alpha & \delta_\xi^\beta & \delta_\xi^\gamma & \delta_\xi^\delta \\ \delta_\sigma^\alpha & \delta_\sigma^\beta & \delta_\sigma^\gamma & \delta_\sigma^\delta \end{vmatrix}.$$

Likewise the tensor $\delta_{\mu\nu\xi\sigma}^{\alpha\beta\gamma\delta}$ can be expressed in a similar way.

Example 1 Show that

$$\delta^{\alpha\beta}_{\mu\beta} = 3\delta^{\alpha}_{\mu}.$$

Example 2 Show that

$$\delta^{\alpha}_{\alpha} = 4.$$

Example 3 Show that

$$\delta^{\alpha\beta\tau}_{\mu\gamma\tau} = 2\delta^{\alpha\beta}_{\mu\gamma}.$$

Hints:
Here,

$$\delta^{\alpha\beta\tau}_{\mu\gamma\tau} = \begin{vmatrix} \delta^{\alpha}_{\mu} & \delta^{\beta}_{\mu} & \delta^{\tau}_{\mu} \\ \delta^{\alpha}_{\gamma} & \delta^{\beta}_{\gamma} & \delta^{\tau}_{\gamma} \\ \delta^{\alpha}_{\tau} & \delta^{\beta}_{\tau} & \delta^{\tau}_{\tau} \end{vmatrix}.$$

Now, expand along third row and use $\delta^{\tau}_{\tau} = 4$.

Appendix B

Lagrangian and Euler–Lagrange's Equation:

Let Cartesian coordinate x_i can be written in terms of generalized coordinate q_k as

$$x_i = F_i(q_k, t). \tag{B.1}$$

Now, one can get

$$dx_i = \frac{\partial x_i}{\partial q_k} dq_k + \frac{\partial x_i}{\partial t} dt. \tag{B.2}$$

The Cartesian velocity can be found as

$$v_i = \frac{dx_i}{dt} = \frac{\partial x_i}{\partial q_k} \dot{q}_k + \frac{\partial x_i}{\partial t}. \tag{B.3}$$

Therefore, we have the kinetic energy as

$$T = \frac{1}{2} \Sigma m_i v_i^2 = \frac{1}{2} m_i \left(\frac{\partial x_i}{\partial q_k} \dot{q}_k + \frac{\partial x_i}{\partial t} \right) \left(\frac{\partial x_i}{\partial q_k} \dot{q}_k + \frac{\partial x_i}{\partial t} \right). \tag{B.4}$$

Thus, $T = f(q_i, \dot{q}_i, t)$ and hence

$$\frac{\partial T}{\partial \dot{q}_k} = m_i v_i \frac{\partial v_i}{\partial \dot{q}_k} = m_i v_i \frac{\partial x_i}{\partial q_k} \tag{B.5}$$

$$\frac{d}{dt} \left(\frac{\partial T}{\partial \dot{q}_k} \right) = m_i v_i \frac{d}{dt} \left(\frac{\partial x_i}{\partial q_k} \right) + m_i v_i \frac{\partial x_i}{\partial q_k}. \tag{B.6}$$

© The Editor(s) (if applicable) and The Author(s), under exclusive license to Springer Nature Singapore Pte Ltd. 2022
F. Rahaman, *The Special Theory of Relativity*, La Matematica per il 3+2 136,
https://doi.org/10.1007/978-981-19-0497-4

We have the following identity:

$$\frac{d}{dt}\left(\frac{\partial x_i}{\partial q_k}\right) = \frac{\partial^2 x_i}{\partial q_l \partial q_k}\dot{q}_l + \frac{\partial}{\partial t}\left(\frac{\partial x_i}{\partial q_k}\right) = \frac{\partial}{\partial q_k}\left[\frac{\partial x_i}{\partial q_l}\dot{q}_l + \frac{\partial x_i}{\partial t}\right] = \frac{\partial v_i}{\partial q_k}. \quad \text{(B.7)}$$

Putting the result of (B.7) in (B.6), we get

$$\frac{d}{dt}\left(\frac{\partial T}{\partial \dot{q}_k}\right) = m_i v_i \frac{\partial v_i}{\partial q_k} + m_i \dot{v}_i \frac{\partial x_i}{\partial q_k} = \frac{\partial T}{\partial q_k} + m_i \dot{v}_i \frac{\partial x_i}{\partial q_k} \frac{d}{dt}\left(\frac{\partial T}{\partial \dot{q}_k}\right) - \frac{\partial T}{\partial q_k}$$

$$= m_i \dot{v}_i \frac{\partial x_i}{\partial q_k}. \quad \text{(B.8)}$$

If we multiply both sides by virtual changes δq_k, then we have

$$\left[\frac{d}{dt}\left(\frac{\partial T}{\partial \dot{q}_k}\right) - \frac{\partial T}{\partial q_k}\right]\delta q_k = m_i \dot{v}_i \frac{\partial x_i}{\partial q_k}\delta q_k = m_i \ddot{x}_i \delta x_i. \quad \text{(B.9)}$$

Remind D'Alembert's principle

$$(m_i \ddot{x}_i - X_i)\delta x_i = 0. \quad \text{(B.10)}$$

For the conservative force, the potential energy V yields

$$\delta V = X_i \delta x_i = X_i \frac{\partial x_i}{\partial q_k}\delta q_k = Q_k \delta q_k = -\frac{\partial V}{\partial q_k}\delta q_k \quad \text{(B.11)}$$

$Q_k = -\frac{\partial V}{\partial q_k}$ is called generalized force.
Equation (B.9), (B.10) and (B.11) $\Rightarrow$

$$\left[\frac{d}{dt}\left(\frac{\partial T}{\partial \dot{q}_k}\right) - \frac{\partial T}{\partial q_k}\right]\delta q_k = -\frac{\partial V}{\partial q_k}\delta q_k$$

$$\Rightarrow \left[\frac{d}{dt}\left\{\frac{\partial}{\partial \dot{q}_k}(T - V)\right\} - \frac{\partial}{\partial q_k}(T - V)\right]\delta q_k = 0$$

(since V is not a function of $\dot{q}_i$)

$$\frac{d}{dt}\left(\frac{\partial L}{\partial \dot{q}_k}\right) - \frac{\partial L}{\partial q_k} = 0 \tag{B.12}$$

$L = T - V \rightarrow$ Lagrangian and (B.12) **Euler–Lagrangian** equations.
$p_i = \frac{\partial L}{\partial \dot{q}_i} \rightarrow$ generalized momentum.

Note: If the Lagrangian does not involve a certain coordinate that coordinate is called cyclic and the corresponding momentum will be a constant of motion (or a conserved quantity).

Action Integral:

Almost all fundamental equations of physics can be obtained with the help of a variational principle, choosing suitable Lagrangian or action in different cases.

Hamilton's variational principle states that $\delta S = \delta \int_1^2 L dt = 0 \Rightarrow$ Lagrange's equation of motion.

$\delta S = 0 \Rightarrow$ extremization of S $\Leftrightarrow$ equations of motion.

Choose a point particle with mass m whose position is given by $x_i(t)$ at time t. Let it is moving in a time-independent potential $V(x_i)$. The action of this event is given by

$$S([x_i]; t_1, t_2) \equiv \int_{t_1}^{t_2} L dt = \int_{t_1}^{t_2} \left[\frac{1}{2}m\frac{dx_i}{dt}\frac{dx_i}{dt} - V(x_i)\right]. \tag{B.13}$$

It is a functional of the path $x_i(t)$ for $t_1 < t < t_2$ which depends on initial and final times t_1, t_2.

(A functional is an operator whose range lies on the real time R or the complex plane C.)

Thus for a given path $x_i(t)$, we associate a number called the functional (in this case S).

Consider the change of S for a small deformation of path

$$x_i(t) \rightarrow x_i(t) + \delta x_i(t) \tag{B.14}$$

(note that there are no deformation, i.e. no variations at the end points, $\delta x_i(t_1) = \delta x_i(t_2) = 0$)
Then

$$S[x_i + \delta x_i] = \int_{t_1}^{t_2} dt \left[\frac{1}{2}m\frac{d}{dt}(x_i + \delta x_i)\frac{d}{dt}(x_i + \delta x_i) - V(x_i + \delta x_i)\right]. \tag{B.15}$$

Now

$$\frac{d}{dt}(x_i + \delta x_i)\frac{d}{dt}(x_i + \delta x_i) \approx \frac{dx_i}{dt}\frac{dx_i}{dt} + 2\left(\frac{d}{dt}\delta x_i\right)\frac{dx_i}{dt}$$

$$(\text{neglecting } o(\delta x)^2)$$

$$\approx \frac{dx_i}{dt}\frac{dx_i}{dt} - 2\frac{d^2x_i}{dt^2}\delta x_i + 2\frac{d}{dt}\left(\delta x_i\frac{dx_i}{dt}\right)$$

$$V(x_i + \delta x_i) \approx V(x_i) + \delta x_i \partial_i V \qquad \left(\partial_i \equiv \frac{\partial}{\partial x_i}\right).$$

Thus

$$S[x_i + \delta x_i] = S[x_i] + \int_{t_1}^{t_2} dt\delta x_i \left(-\partial_i V - m\frac{d^2x_i}{dt^2}\right) + m\int_{t_1}^{t_2} dt \frac{d}{dt}\left(\delta x_i \frac{dx_i}{dt}\right).$$

$$(\text{B.16})$$

Since at the end points $\delta x_i(t_1) = \delta x_i(t_2) = 0$, the last term vanishes.

Equation (B.16) indicates that S does not change under an arbitrary δx_i if the classical equations of motion are valid.

Thus, we have a clue of correspondence between equations of motion and extremization of action integral S.

Note: Here boundary conditions have to be supplied externally. This example dealt with particle mechanics. It can be generalized to classical field theory as in Maxwell's electrodynamics or Einstein's general relativity.

Lagrangian Formulation:
Newtonian Mechanics:
The action function is given by

$$S[q, \dot{q}] = \int_{t_1}^{t_2} L(q, \dot{q})dt.$$

Here the integration over a specific path $q(t)$. No variations, $\delta q(t)$ of this path at the end points, $\delta q(t_1) = \delta q(t_2) = 0$.

Extremization of the action function $\Rightarrow \delta S = 0$, i.e.

$$0 = \delta S = \int_{t_1}^{t_2} \delta L dt = \int_{t_1}^{t_2} \left(\frac{\partial L}{\partial q} \delta q + \frac{\partial L}{\partial \dot{q}} \delta \dot{q} dt \right)$$

$$= \left[\frac{\partial L}{\partial \dot{q}} \delta q \right]_{t_1}^{t_2} + \int_{t_1}^{t_2} \left(\frac{\partial L}{\partial q} - \frac{d}{dt} \frac{\partial L}{\partial \dot{q}} \right) \delta q \delta t$$

$$\Rightarrow \frac{d}{dt} \frac{\partial L}{\partial \dot{q}} - \frac{\partial L}{\partial q} = 0.$$

This is Euler–Lagrangian equation for a one-dimensional mechanical system.

Field Theory:

Here, we are interested in the dynamical behaviour of a field $q(x^\alpha)$ in curved space–time.

Let W be an arbitrary region in the space–time manifold bounded by a closed hypersurface ∂W. The Lagrangian $L(q, q_{,\alpha})$ is a scalar function of the field and its first derivative.

Thus action function is given by

$$S[q] = \int_W L(q, q_{,\alpha}) \sqrt{-g} d^4 x.$$

Note that the variation of q is arbitrary within W but vanishes on ∂W, $[\delta q]_{\partial W} = 0$. Now,

$$\delta S = \int_W \left[\frac{\partial L}{\partial q} \delta q + \frac{\partial L}{\partial q_{,\alpha}} \delta q_{,\alpha} \right] \sqrt{-g} d^4 x$$

$$= \int_W \left[L' \delta q + (L^\alpha \delta q)_{;\alpha} - L^\alpha_{;\alpha} \delta q \right] \sqrt{-g} d^4 x$$

$(L' \equiv \frac{\partial L}{\partial q}; \quad L^\alpha \equiv \frac{\partial L}{\partial q_{,\alpha}})$.

Using, Gauss divergence theorem, one gets

$$\delta S = \int_W (L' - L^\alpha_{;\alpha}) \delta q \sqrt{-g} d^4 x + \oint_{\partial W} L^\alpha \delta q d\Sigma_\alpha.$$

Since, $[\delta q]_{\partial W} = 0$, finally one can get

$$\delta S = 0 \Rightarrow \nabla_\alpha \frac{\partial L}{\partial q_{,\alpha}} - \frac{\partial L}{\partial q} = 0.$$

This is Euler–Lagrange equation for a single scalar field ϕ (∇ stands for covariant derivative).

Examples of Lagrangian for some fields:

(a)   A scalar field ψ:

This can represent the π^0 meson. The Lagrangian

$$L = \frac{1}{2}\left(g^{\mu\nu}\psi_{,\nu}\psi_{,\mu} + \frac{m^2}{\hbar^2}\psi^2 \right).$$

Euler–Lagrangian equations are

$$g^{\alpha\beta}\psi_{;\alpha\beta} - \frac{m^2}{\hbar^2}\psi = 0$$

$$(L^\alpha = -g^{\alpha\beta}\psi_{,\beta}; \quad L^\alpha_{;\alpha} = -g^{\alpha\beta}\psi_{;\alpha\beta}; \quad L' = -\frac{m^2}{\hbar^2}\psi).$$

This is Klein–Gordon equation. s

(b)   The electromagnetic field:

$$L = \frac{1}{16\pi}F_{ab}F_{cd}g^{ac}g^{bd},$$

where electromagnetic tensor

$$F_{ab} = \partial_a A_b - \partial_b A_a \partial;$$

$A_a \rightarrow$ electromagnetic potential.

Euler–Lagrange equation yields

$$F_{ab;c}g^{bc} = 0.$$

(c) General Relativity:

Einstein's field equations in general relativity can be derived from the principle of Least Action $\delta S = 0$ where

$$S = \int\limits_{W} \sqrt{-g}[L_G + 2kL_F]d^4x.$$

(W be an arbitrary region in the space–time manifold bounded by a closed hypersurface ∂W).

Here we choose $L_G = R$ as the Lagrangian for the gravitational field with $R = g_{\mu\nu}R^{\mu\nu}$. L_F is the Lagrangian for all the other fields (ϕ). k = Einstein's gravitational constant = $\frac{8\pi G}{c^4}$.

The Einstein field equation in general relativity $R_{\alpha\beta} - \frac{1}{2}g_{\alpha\beta}R = kT_{\alpha\beta}$ is recovered by varying $S[g; \phi]$ with respect to $g_{\alpha\beta}$. The variation is subjected to the condition

$$[\delta g_{\alpha\beta}]_{\partial W} = 0.$$

References

1. Bandapadhyaya, N.: Special Theory of Relativity. Academic Publishers, Dordrecht (1998)
2. Banerjee, S., Banerjee, A.: The Special Theory of Relativity. Prentice Hall of India, New Delhi (2002)
3. Barton, G.: Introduction to the Relativity Principle. Wiley, New York (1999)
4. Bergmann, P.G.: Introduction to the Theory of Relativity. Prentice Hall of India, New Delhi (1992)
5. Carmeli, M.: Cosmological Special Relativity. World Scientific, Singapore (2002)
6. Das, A.: The Special Theory of Relativity. Springer, New York (1993)
7. Devanathan, V.: Relativistic Quantum Mechanics and Quantum Field Theory. Narosa Publishing, India (2011)
8. D'Inverno, R.: Introducing Einstein's Relativity. Clarendon Press, Oxford (1992)
9. Einstein, A.: On the electrodynamics of moving bodies (1905). http://www.fourmilab.ch/etexts/einstein/specrel/www/. Accessed 30 1905
10. Einstein, A., Lorentz, H.A., Weyl, H., Minkowski, H.: The Principle of Relativity. Dover Publication, New York (1924)
11. French, A.P.: Special Relativity. W W Norton, New York (1968)
12. Gibilisco, S.: Understanding Einstein's Theories of Relativity. Dover Publication, New York (1983)
13. Goldstein, H.: Classical Mechanics. Narosa, New Delhi (1998)
14. Griffiths, D.J.: Electrodynamics. Wiley Eastern, New Delhi (1978)
15. Jackson, J.D.: Electrodynamics. Wiley Eastern, New Delhi (1978)
16. Kogut, J.B.: Introduction to Relativity. Academic, New York (2000)
17. Landau, L.D., Lifshitz, E.M.: Classical Theory of Fields. Pergamon Press, New York (1975)
18. Lord, E.A.: Tensor Relativity and Cosmology. Tata McGraw-Hill, New Delhi (1976)
19. Misner, C.W., Throne, K., Wheeler, J.: Gravitation. Freeman, San Francisco (1973)
20. Moller, C.: The Theory of Relativity. Oxford University Press, Oxford (1972)
21. Naber, G.L.: The Geometry of Minkowski Spacetime: An Introduction to the Mathematics of the Special Theory of Relativity. Springer, New York (1992)
22. Ney, E.P.: Electromagnetism and Relativity. John Weatherhill, Tokyo (1965)
23. Pathria, R.K.: The Theory of Relativity. Dover Publications, New York (1974)
24. Pauli, W.: Theory of Relativity. Dover Publications, New York (1958)
25. Poisson, E.: A Relativist's Toolkit. Cambridge University Press, Cambridge (2004)
26. Prakash, S.: Relativistic Mechanics. Pragati Prakashan, Meerut (2000)

© The Editor(s) (if applicable) and The Author(s), under exclusive license to Springer Nature Singapore Pte Ltd. 2022
F. Rahaman, *The Special Theory of Relativity*, La Matematica per il 3+2 136, https://doi.org/10.1007/978-981-19-0497-4

27. Puri, S.P.: Special Theory of Relativity. Pearson, India (2013)
28. Rahaman, F.: The General Theory of Relativity, A Mathematical Approach. Cambridge University Press (2021)
29. Resnick, R.: Introduction to Special Relativity. Wiley Eastern Ltd., New Delhi (1985)
30. Rindler, W.: Relativity. Clarendon Press, Oxford (2002)
31. Rosser, W.G.V.: An Introduction to the Theory of Relativity. Butterworths, London (1964)
32. Schwartz, H.M.: Introduction to Special Relativity. McGraw Hill, New York (1968)
33. Sharipov, R.: Classical Electrodynamics and Theory of Relativity. Samizdat Press (2003)
34. Srivastava, M.P.: Special Theory of Relativity. Hindustan Publishing Corporation, New Delhi (1992)
35. Synge, J.L., Griffith, B.A.: Classical Mechanics. McGraw Hill, New York (1949)
36. Taylor, E., Wheeler, J.: Spacetime Physics. Freeman, New York (1992)
37. Taylor, J.G.: Special Relativity. Oxford University, Oxford (1965)
38. Tolman, R.: Relativity. Thermodynamics and Cosmology. Dover, New York (1934)
39. Tourrence, P.: Relativity and Gravitation. Cambridge University Press, Cambridge (1992)
40. Ugaro, V.A.: Special Theory of Relativity. Mir Publishers, Moscow (1979)
41. Weinberg, S.: Gravitation and Cosmology. Wiley, New York (1972)
42. Williams, W.S.C.: Introducing Special Relativity. Taylor and Francis, London (2002)
43. Woodhouse, N.M.J.: Special Relativity. Springer, Berlin (2003)

Index

Printed in the United States
by Baker & Taylor Publisher Services